TUNNELING

THE JERUSALEM SYMPOSIA ON
QUANTUM CHEMISTRY AND BIOCHEMISTRY

Published by the Israel Academy of Sciences and Humanities,
distributed by Academic Press (N.Y.)

Published by the Israel Academy of Sciences and Humanities,
distributed by D. Reidel Publishing Company (Dordrecht, Boston, Lancaster, and
Tokyo)

Published and distributed by D. Reidel Publishing Company (Dordrecht, Boston,
Lancaster, and Tokyo)

VOLUME 19

TUNNELING

PROCEEDINGS OF THE NINETEENTH JERUSALEM SYMPOSIUM ON
QUANTUM CHEMISTRY AND BIOCHEMISTRY HELD IN
JERUSALEM, ISRAEL, MAY 5–8, 1986

Edited by

JOSHUA JORTNER

Department of Chemistry, University of Tel-Aviv, Israel

and

BERNARD PULLMAN

Institut de Biologie Physico-Chimique
(Fondation Edmond de Rothschild), Paris, France

SPRINGER-SCIENCE+BUSINESS MEDIA, B.V.

Library of Congress Cataloging in Publication Data

Jerusalem Symposium on Quantum Chemistry and Biochemistry (19th : 1986)
    Tunneling : proceedings of the Nineteenth Jerusalem Symposium on
Quantum Chemistry and Biochemistry held in Jerusalem, Israel, May 5–8, 1986.

    (Jerusalem symposia on quantum chemistry and biochemistry; v. 19)
    Includes bibliographies and index.
    1. Quantum   biochemistry—Congresses. 2. Tunneling   (Physics)—
Congresses. I. Jortner, Joshua. II. Pullman, Bernard, 1919–   . III.
Title. IV. Series.
QP517.Q34J47   1986   541.2'8   86–17849

ISBN 978-94-010-8611-0   ISBN 978-94-009-4752-8 (eBook)
DOI 10.1007/978-94-009-4752-8

TABLE OF CONTENTS

THE 19TH JERUSALEM SYMPOSIUM ON TUNNELING

May, 1986

PREFACE

The Nineteenth Jerusalem Symposium reflected the high standards of these distinguished scientific meetings, which convene once a year at the Israel Academy of Sciences and Humanities in Jerusalem to discuss a specific topic in the broad area of quantum chemistry and biochemistry. The topic at this year's Jerusalem Symposium was Tunneling, which constitutes a truly interdisciplinary subject of central interest to physicists, chemists and biologists.

The main theme of the Symposium was built around a conceptual framework for the elucidation of the nature of tunneling in chemical and biological physics ; emphasis was placed on the manifestations and implications of tunneling phenomena in broad and diverse areas, such as molecular structure and spectroscopy, chemical dynamics, low-temperature chemical reactivity, surface dynamics, proton transport, electron transfer, primary processes in photosynthesis, tunneling phenomena in biomolecules and tunneling states in proteins. The interdisciplinary nature of these research areas was deliberated by intensive and extensive interactions between scientists from different disciplines and between theory and experiment. This volume provides a record of the invited lectures at the Symposium.

Held under the auspices of the Israel Academy of Sciences and Humanities and the Hebrew University of Jerusalem and the Nineteenth Jerusalem Symposium was sponsored by the Institut de Biologie Physico-Chimique (Fondation Edmond de Rothschild) of Paris. We wish to express our deep thanks to Baron Edmond de Rothschild for his continuous and generous support, which makes him a true partner in this important endeavour. We would also like to express our gratitude to the Administrative Staff of the Israel Academy, and in particular to Mrs. Avigail Hyam, for the efficiency and excellency of the local arrangements.

Joshua Jortner
Bernard Pullman

TUNNELING AND DYNAMIC TUNNELING BY AN ALGEBRAIC APPROACH

R.D. Levine
The Fritz Haber Research Center for Molecular Dynamics
The Hebrew University, Jerusalem 91904, Israel

ABSTRACT. Seemingly tunneling is intimately related to a geometrical approach. That however is not always the case as shown by our first example where tunneling connects regions of phase space which are not separated by a potential yet are disjoint in classical mechanics. This example shows that an algebraic approach can handle dynamic tunneling in a bound state system. Recent work has also considerably firmed the geometric interpretation of the algebraic approach. Hence even such traditional problems as barrier penetration can be discussed. To obtain the tunneling rates we discuss the use of non-unitary representations. Towards the extension of the algebraic approach to unbound states in multidimensional systems, the simpler case of an unbound one dimensional motion is discussed and possible generalizations are indicated.

1. DYNAMIC TUNNELING

For realistic potentials it has been observed that transitions which can take place for quantum dynamics fail to occur in the classical limit. This is true for both intermolecular and intramolecular dynamics. A recent example which has been examined in considerable detail [1-8] is the time evolution of an anharmonic local mode in a symmetric (ABA) triatomic molecule. Consider a zero order description, $|n,m\rangle$, made up of eigenstates of the left and right local modes respectively. In general, all such states are coupled either directly or indirectly through other states. In particular, we can form symmetric and antisymmetric linear combinations

$$|n,m^{\pm}\rangle = (|n,m\rangle \pm |m,n\rangle)/\sqrt{2}, \quad n > m .$$

Say now we excite a particular local mode, $|n,m\rangle$, $n > m$. Then even if the coupling is very weak, the excitation will eventually make it to its degenerate counterpart $|m,n\rangle$. For weak coupling this is readily seen by writing the state of the system at time $t = 0$ (i.e., $|n,m\rangle$) as a coherent, in phase, superposition of $|n,m^{\pm}\rangle$. The symmetric and antisymmetric states are, for weak coupling, nearly good eigenstates and each propagates in time with its own phase factor $\exp(-iE_{\pm}t/\hbar)$. The symmetric and antisymmetric states will be precisely out of phase (and hence the state of the system will be $|m,n\rangle$) by the time $\tau = \pi\hbar/\Delta E$ where $\Delta E = E_{+} - E_{-}$ is the energy difference between the symmetric and

J. Jortner and B. Pullman (eds.), Tunneling, 1–8.
© *1986 by D. Reidel Publishing Company.*

antisymmetric states. If the coupling between the local modes is weak, ΔE will be quite small and indeed this is one of the signatures of a local mode spectrum, Fig. 1. However, even if $\tau = \pi\hbar/\Delta E$ is very long, the energy will always eventually fully exchange between the two modes and will not remain localized in one bond.

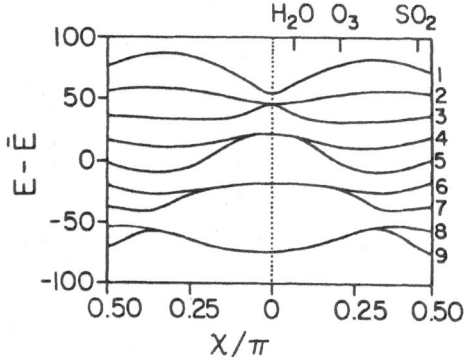

Figure 1. Energy levels computed for an algebraic model Hamiltonian of ABA molecules, [5,6] vs. the coupling constant of the two (anharmonic) local modes. $\chi = 0$ is the pure local mode limit while $\chi = \pm\pi/2$ is the pure normal mode limit. Values of χ for three molecules are indicated. The computation shown is for the multiplet $P \equiv m+n = 8$ and $\bar{E}$ is the mean energy of the multiplet. As $|n-m|$ increases, the degeneracy of $|n,m^+>$ and $|n,m^->$ is being effectively maintained for higher values of χ.

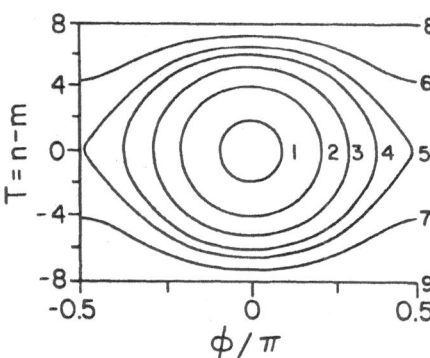

Figure 2. Contours of classical trajectories at the same energies shown in figure 1 for $\chi = 0.2\pi$. For low values of $|T|$, T does change sign as a function of ϕ, the phase difference between the two local oscillators. For higher values of $|T|$, T is nearly conserved and definitely does not change sign. The transition $|n,m> \to |m,n>$ (i.e. $T \to -T$) which is quantum mechanically allowed fails to take place in the classical limit.

The situation is quite different in classical mechanics. Once the energy mismatch (i.e., $n-m$) exceeds a certain threshold value (which depends on $n+m$, the nature of the coupling and the anharmonicity of the local modes), the classical trajectory remains localised, Figure 2. The energy strictly does not transfer to the other side of the molecule. Of course, and as is evident in figure 2, there is a second classical trajectory which is localised in the region of the other bond. These two trajectories have precisely the same energy but cover distinct regions of classical phase space. Only quantum mechanical tunneling can carry the excitation over from one bond to another.

The origin of the classical localization of the excitation energy is readily evident upon examination of the classical Hamiltonian. T, the difference in action of the two bonds is governed by the same Hamiltonian as that for the angular momentum of a hindered rotor. The rotational constant of the 'rotor' is precisely x_e, the anharmonicity of the local mode. Hence nearly harmonic local modes correspond to a rotor of very high inertia and hence, as we shall shortly argue, efficient energy exchange between the two local modes. The height of the barrier for rotation is governed primarily by $(nm)^{1/2} \propto [(n+m)^2-(n-m)^2]^{1/2}$ and hence is small when $T = (n-m)$ is large. Thus, for a large T and particularly so for a large x_e, the motion of the 'rotor' is practically unhindered so that T is nearly a constant of the motion. For a small T, and particularly so for a small x_e, the rotor is hindered and oscillates over a limited angular range. This rapid change of sign of T is precisely the facile energy transfer between the two bonds. In this picture therefore the quantum 'tunneling' is precisely opposite to what we usually mean by tunneling in a one dimensional system. Tunneling here is the 'turning back' of the motion even in the absence of a classical turning point. Such quantal reflection is operative primarily near the extrema in the underlying potential. Hence it serves to confine T to the right or the left of the barrier which is just what is required for energy exchange. Of course, it is possible by a canonical transformation to change into such variables where energy transfer between very asymmetrically excited local modes does correspond to tunneling in the conventional sense through a classical barrier [7,8].

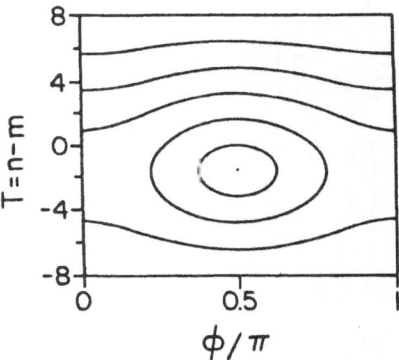

Figure 3. Same as in figure 2 but for an asymmetric linear ABC molecule, [4]. Here too, transitions which correspond to large intramolecular energy transfers are classically forbidden.

The classically restricted energy transfer is by no means unique to symmetric ABA triatomics, Fig. 3. [4], but the analysis is more transparent (i.e., there are fewer independent parameters of the potential) in the symmetric case.

2. THE ALGEBRAIC APPROACH

Interest in realistic potentials suggests that an algebraic approach [9-12] is one way to proceed. As long as attention is confined to the manifold of bound states (intramolecular dynamics), the situation is reasonably in hand and one knows how to proceed. What is now called for are additional examples. Once however the motion is in the continuum, we are still in an exploratory stage. In the following we examine two approaches which have been successfully implemented for one dimensional motion. The first and more general one is the construction of the transfer matrix and hence the scattering matrix [12-14]. This provides not only for the tunneling but the entire transmission amplitude vs. energy. The other is the identification of quasibound states by the use of non-unitary representations [15,16].

 To illustrate the connection between group theory and scattering states consider a one dimensional localised potential. At any given energy E, $E = \hbar^2 k^2/2m$, one can write the asymptotic form of the wavefunction as

$$\psi \rightarrow \begin{array}{ll} A_{in}\exp(ikx) + B_{out}\exp(-ikx) & x \rightarrow -\infty \\ A_{out}\exp(ikx) + B_{in}\exp(-ikx) & x \rightarrow +\infty \end{array} \tag{1}$$

The scattering matrix $\mathbf{S}$ is defined by

$$\begin{pmatrix} B_{out} \\ A_{out} \end{pmatrix} = \mathbf{S} \begin{pmatrix} A_{in} \\ B_{in} \end{pmatrix} . \tag{2}$$

The condition of conservation of flux

$$|A_{in}|^2 + |B_{in}|^2 = |A_{out}|^2 + |B_{out}|^2 \tag{3}$$

implies that $\mathbf{S}$ is an element of the group U(2), the group of 2×2 unitary matrices. On the other hand, the transfer matrix $\mathbf{Q}$, which serves to propagate the wavefunction, is defined by

$$\begin{pmatrix} A_{out} \\ B_{in} \end{pmatrix} = \mathbf{Q} \begin{pmatrix} A_{in} \\ B_{out} \end{pmatrix} . \tag{4}$$

Conservation of flux

$$|A_{in}|^2 - |B_{out}|^2 = |A_{out}|^2 - |B_{in}|^2 \tag{5}$$

plus the condition that $\det\mathbf{Q} = 1$ (which results from time reversal invariance) imply that $\mathbf{Q}$ is an element of the group SU(1,1).

 When the energy is negative, the bound state wavefunction can be normalized only if the 'in' coefficients in (1) vanish. Thus, for example, bound states exist for such energies that $Q_{22}(E) = 0$. Such a condition is also possible for a complex energy with a positive real part and this defines the quasibound states as being 'purely outgoing'.

A very simple example which allows an explicit construction of the transfer matrix as an element of SU(1,1) is the symmetric square well, of depth V_0 extending from $x = -a$ to a. Putting $\tau = -kx$ we can write the Schrödinger equation for ψ'' (where the prime denotes the derivative with respect to τ) as a two component equation for ψ and ψ'

$$i\frac{d}{d\tau}\begin{pmatrix}\psi-i\psi'\\\psi+i\psi'\end{pmatrix} = \mathbf{A}\begin{pmatrix}\psi-i\psi'\\\psi+i\psi'\end{pmatrix}. \tag{6}$$

Here $\mathbf{A}$,

$$\mathbf{A} = \begin{pmatrix}1-a & -a\\a & -(1-a)\end{pmatrix} \tag{7}$$

is a linear combination of two generator matrices of SU(1,1)

$$\mathbf{A} = (1-a)\begin{pmatrix}1 & 0\\0 & -1\end{pmatrix} -ia\begin{pmatrix}0 & -i\\i & 0\end{pmatrix} \tag{8}$$

and $a = -V_0/2E$.

Using (1) we see that outside the region of the potential

$$\frac{1}{2}\begin{pmatrix}\psi-i\psi'\\\psi+i\psi'\end{pmatrix} \rightarrow \begin{pmatrix}A_{out}\exp(ikx)\\B_{in}\exp(-ikx)\end{pmatrix},x \rightarrow +\infty \text{ or } \begin{pmatrix}A_{in}\exp(ikx)\\B_{out}\exp(ikx)\end{pmatrix},x \rightarrow -\infty. \tag{9}$$

Hence integrating (6) will yield what is essentially the transfer matrix.

To integrate (6) we evaluate $\mathbf{Q}$ by

$$id\mathbf{Q}/d\tau = \mathbf{AQ} \tag{10}$$

which shows that, by construction, $\mathbf{Q}$ is an element of SU(1,1). The eigenvalues of $\mathbf{A}$ are $\pm(k'/k)$ where $E+V_0=\hbar k'^2/2m$ or $(1-a)=(1+k'^2/k^2)/2$. To evaluate $\mathbf{Q}=\exp(-i\mathbf{A}\tau)$ we use the eigenvalues of $\mathbf{A}$ to obtain [17]

$$\mathbf{Q} = \mathbf{I}\cos(k'\tau/k) + \mathbf{A}(k/k')\sin(k'\tau/k). \tag{11}$$

Putting $\tau = -kx$ and evaluating the transfer matrix from $-a$ to a we recover the known [18] result.

Starting from a geometrical Hamiltonian, it is possible to determine the corresponding group. Other details for the algebraic treatment of the square well will be found elsewhere [19].

3. QUASIBOUND STATES

The final example is the analytic algebraic determination of the position and width of a tunneling state. The Hamiltonian is a repulsive Coulomb barrier (as in the original α-particle penetration problem [20] which gave rise to considerations of tunneling) and a simple $(\lambda/8)/r^2$ attractive term. λ determines the magnitude of the attraction and for $\lambda > 1$ resonances which can tunnel out through the barrier will turn out to be possible, [21].

This problem is solved in some detail elsewhere [16]. Here we consider only s-wave scattering ($l = 0$) for a particle of unit mass. The substitution $\psi = x^{-3/2}\phi(x)$, $r = x^2$ converts the Schrödinger equation (for r as a radial coordinate), to

$$d^2\phi/dx^2 + ((\lambda - 3/4)/x^2)\phi + 8(Ex^2 - 1)\phi = 0 . \tag{12}$$

To cast the problem in an algebraic form we use the following realization for the generators of SU(1,1) [22]

$$J_1 = d^2/dx^2 + (\alpha/x^2) + x^2/16 \tag{13}$$

$$J_2 = -i(xd/dx + 1/2)/2$$

$$J_3 = d^2/dx^2 + (\alpha/x^2) - x^2/16$$

As is evident almost by inspection, J_1 will have a continuous spectrum. The spectrum of J_3 is discrete [22]

$$J_3|j,m> = m|j,m> \tag{14}$$

$$J^2|j,m> = j(j+1)|j,m> \tag{15}$$

where $J^2 = J_3^2 - J_1^2 - J_2^2$ and the eigenvalue j determines the representation. By explicit evaluation $J^2 = -(\alpha + 3/4)/4$ so that $j = (1/2)(1 \pm (1 - \alpha - 3/4)^{1/2})$.

The differential equation (12) can be written, using the realization (13) in the algebraic form

$$[(64E + 1/2)J_1 - (64E - 1/2)J_3 - 8]\phi = 0 \tag{16}$$

where $\alpha = \lambda - 3/4$. To obtain a discrete spectrum we need to diagonalise J_3. A transformation of the equation (16) under

$$D = \exp(-i\theta J_2) \tag{17}$$

yields

$$D[(64E + 1/2)J_1 - (64E - 1/2)J_3 - 8]D^{-1}D\phi = 0 . \tag{18}$$

But since D is an element of SU(1,1) evaluating the transformation is immediate [12], e.g.

$$DJ_2D^{-1} = J_1\cosh\theta + J_3\sinh\theta \tag{19}$$

leading to

$$\left\{ [(64E + 1/2)\cosh\theta - (64E - 1/2)\sinh\theta]J_1 + [(64E + 1/2)\sinh\theta - (64E - 1/2)\cosh\theta]J_3 - 8 \right\}D\phi = 0 . \tag{20}$$

The choice

$$\tanh\theta = (64E + 1/2)/(64E - 1/2) \tag{21}$$

eliminates the generator J_1 in (20) and brings the equation to the form

$$(J_3 - 1/(-2E)^{1/2})D\phi = 0 . \tag{22}$$

Of course, unless $E < 0$, there is no real θ which is a solution of (21). If, however, one is willing to consider a non-unitary transformation D, cf. (17), then a complex θ can be defined by (21).

Comparing (22) and (14)

$$E = -1/2m^2 . \qquad (23)$$

The range of m, for the spectrum of J_3 which is bounded from below is $-j, -j+1, -j+2,...$ or $m = n-j$ where n is a non negative integer. With $\alpha = \lambda - 3/4$ and for $\lambda > 1$

$$j = -(1/2)(1 \pm i (\lambda-1)^{1/2}) . \qquad (24)$$

The upper sign will be ruled out by the boundary conditions.

The final result is

$$E = -1/2(n+1/2 - i(\lambda-1)^{1/2}/2)^2, \ n = 0,1, . \qquad (25)$$

Putting $E = E_0 - i\Gamma/2$,

$$\Gamma = 4(4n+2)(\lambda-1)^{1/2}/(4n(n+1)+\lambda)^2 . \qquad (26)$$

$$E_0 = (-8(n+1/2)^2 + 2(\lambda-1))/(4n(n+1)+\lambda)^2 . \qquad (27)$$

The highest (and widest) state is for $n = 0$ whose position is $-4/\lambda^2$ below the top of the barrier (which is at $2/\lambda$).

The imaginary part, η of the tilt angle θ is given by

$$tn\,\eta = (\lambda-1)^{1/2}/(2n+1) . \qquad (28)$$

4. MULTIDIMENSIONAL SYSTEMS?

Can one extend the algebraic approach to intermolecular dynamics? As long as one is willing to treat the relative motion classically, such an extension is quite feasible. Indeed, our own work on the algebraic approach originated with studies [23,24] in this direction. The problem is to properly handle the relative motion in a fully quantal approach. The need for doing so is precisely tunneling, both of the familiar 'under the barrier' type but also the reflection or 'over the barrier' variety. One possibility is to use the approach of section 2. The range of the coordinate x for the relative motion is broken into segments. Within each such segment the x-dependence of the potential is approximated by the potential taken at midpoint. (A superior approximation would be to include the next term, linear in x [25] or even the two next terms [26]). The internal degrees of freedom can then be handled algebraically. Propagation between segments is achieved via a transfer matrix. Alternatively, one can treat the propagation along x in a self consistent fashion [27]. Pilot studies are forthcoming [26].

Acknowledgment

I had the pleasure and benefit of collaboration on these and related topics with Y. Alhassid, I. Benjamin, R. Gilmore, F. Iachello, S. Kais, J.L. Kinsey, O.S. van Roosmalen and C.E. Wulfman. This work was supported by the U.S. Air Force Office of Scientific Research Grant AFOSR 81-0030 and by the Volkswagen Stiftung. The Fritz Haber Research Center is supported by the Minerva Gesellschaft für die Forschung, mbH, München, BRD.

References

1. M.S. Child and L. Halonen, Adv. Chem. Phys. **57**, 1 (1984).

2. M.J. Davis and E.J. Heller, J. Chem. Phys. **75**, 246 (1981).

3. J.S. Hutchinson, E.L. Sibert, III and J.T. Hynes, J. Chem. Phys. **81**, 1314 (1984).

4. O.S. van Roosmalen, R.D. Levine and A.E.L. Dieperink, Chem. Phys. Lett. **101**, 512 (1983).

5. O.S. van Roosmalen, I. Benjamin and R.D. Levine, J. Chem. Phys. **81**, 5986 (1984).

6. I. Benjamin, Ph.D. Thesis at the Hebrew University, Jerusalem (1985).

7. M.E. Kellman, J. Chem. Phys. **83**, 3843 (1985).

8. R.D. Levine and J.L. Kinsey, J. Phys. Chem., in press.

9. F. Iachello and R.D. Levine, J. Chem. Phys. **77**, 3046 (1982).

10. O.S. van Roosmalen, F. Iachello, R.D. Levine and A.E.L. Dieperink, J. Chem. Phys. **79**, 2515 (1983).

11. R.D. Levine, Chem. Phys. Lett. **95**, 87 (1983).

12. C.E. Wulfman and R.D. Levine, Chem. Phys. Lett. **97**, 361 (1983).

13. A. Peres, J. Math. Phys. **24**, 1110 (1983).

14. Y. Alhassid, F. Gürsey and F. Iachello, Ann. Phys. (N.Y.) **148**, 346 (1983).

15. Y. Alhassid, F. Iachello and R.D. Levine, Phys. Rev. Lett. **54**, 1746 (1985).

16. I. Benjamin and R.D. Levine, Phys. Rev A**33**, 2833 (1986).

17. R.D. Levine, Mol. Phys. **22**, 497 (1971).

18. E. Merzbacher, *Quantum Mechanics* (Wiley, N.Y., 1960).

19. S. Kais and R.D. Levine, Phys. Rev., in press.

20. G. Gamow, Z. Phys. **51**, 204 (1928), E.U. Condon and R.W. Gurney, Phys. Rev. **33**, 127 (1929).

21. G.D. Doolen, Int'l J. Quant. Chem. **14**, 523 (1978).

22. B.G. Wybourne, *Classical Groups for Physicists* (Wiley, N.Y., 1974).

23. Y. Alhassid and R.D. Levine, J. Chem. Phys. **67**, 4321 (1977); Phys. Rev. A**18**, 89 (1978).

24. R.D. Levine and C.E. Wulfman, Chem. Phys. Lett. **60**, 372 (1979).

25. R.G. Gordon, J. Chem. Phys. **51**, 14 (1969).

26. M. Berman and R.D. Levine, to be published.

27. N.Z. Tishby and R.D. Levine, Phys. Rev. A**30**, 1477 (1984).

UNIFORM QUANTIZATION OF MULTIDIMENSIONAL SYSTEMS

Craig C. MARTENS and Gregory S. EZRA
Department of Chemistry
Cornell University
Ithaca, NY 14853, USA

The need for uniform semiclassical quantization of multidimensional nonseparable systems is briefly reviewed. A numerical method for uniform quantization of states for resonant two degree of freedom coupled oscillator systems, based upon a previously developed Fourier approach to semiclassical mechanics, is described. Preliminary results are presented for the 1:1 resonant Henon-Heiles system.

1. Introduction

The semiclassical mechanics of multidimensional nonlinear systems (e.g., molecules) is a subject that has attracted much recent attention. Questions at issue range from the deep problem of the nature of classical/quantum correspondence in the case that the classical motion is chaotic [1], to the more practical concern of obtaining quantum information, such as bound state energy levels, from knowledge of classical motion in the regular or quasiperiodic regime [2].

In the case that the classical motion is regular, good action-angle variables exist locally in phase space [3], so that motion is quasiperiodic and equivalent to winding on an N-torus [4]. To determine the particular trajectory manifolds associated with quantum states, a single-valuedness condition is imposed on the multidimensional WKB wavefunction [1,2]: this condition implies that the quantizing tori are those for which the N good action variables have certain integer or half-integer values, depending on the shape of the trajectory projection into configuration space [2]. The resulting EBK quantization conditions, a natural generalization of the familiar 1-dimensional Bohr-Sommerfield rule to N dimensions, are due to Einstein [5], who derived them on the basis of the requirement of canonical invariance, and Brillouin [6] and Keller [7], who considered the asymptotic form of the solution to the multidimensional Schrodinger equation. For reviews, see [1,2].

The reliance of the EBK prescription on the existence of good action variables apparently restricts the applicability of semiclassical quantization to regular regions of phase space [3]. Some progress has however recently been made [8-10] on the problem of quantization of chaotic motions using the method of adiabatic switching [8,11] (based upon the procedure for defining quantization motions for nonseparable systems originally suggested by

9

J. Jortner and B. Pullman (eds.), Tunneling, 9–23.

Ehrenfest [12]), where the existence of "vague tori" or long-time correlations in the dynamics provides enough structure in phase space to ensure approximate conservation of quantized actions.

While the EBK procedure has been used with success to determine energy eigenvalues for multidimensional vibrational (see, for example [13] and references therein) and rotation-vibration [14] Hamiltonians, it must be emphasized that the EBK quantization prescription is an entirely <u>primitive</u> method. That is, no account is taken of nonclassical, purely quantum-mechanical phenomena such as tunnelling or above-barrier reflection [15,16]. As discussed below, it is essential for a variety of problems to be able to calculate level splittings etc. due to these nonclassical effects, so that it is important to go beyond the primitive EBK approach to develop methods for <u>uniform</u> quantization of multidimensional systems. (For uniform quantization within the SCF approximation, see [17].) The present paper describes a method for uniform quantization of resonant trajectories in coupled oscillator systems, based upon our Fourier transform approach to semiclassical mechanics developed in [13,18].

We now briefly review several problems for which a uniform approach to quantization is required.

A. Quantization of resonant trajectories

There has been considerable effort devoted to primitive EBK quantization of systems with strong low-order classical resonances ([18] and references therein). For example, using the Fourier approach [13] we have been able to quantize states of 1:1, 2:1 and 3:4 resonant Hamiltonians [18] (see Figure 2a for an example of a resonant 3:4 trajectory). For many resonant states, there is excellent agreement between primitive EBK energies and exact quantum values. However, for several states of the 3:4 resonant system treated in [18], particularly those corresponding to classical motion in the vicinity of the separatrix dividing resonant from nonresonant motion, there is a breakdown of the EBK treatment. In some cases it is found that there is <u>no</u> quasiperiodic trajectory satisfying the primitive EBK quantization condition, whereas in others it is found that there are <u>two</u> tori (one resonant, one nonresonant) corresponding to the same energy level. Note that these difficulties with the EBK procedure occur within the nominally regular region of phase space, and indicate that uniform quantization procedures are required.

B. Local modes and dynamical tunnelling

The work of Lawton and Child [19] has elucidated the classical dynamical significance of local mode behavior as symmetry breaking motion on a symmetric potential surface (cf. also [20]). Symmetry breaking or localization occurs as a result of the existence of dynamical barriers [21]. If localized classical motions are quantized using a primitive method, symmetry related quantizing trajectories lead to strictly degenerate quantum states. Uniform methods are necessary to determine the actual (small) splittings between symmetrized quantum levels, but are complicated by the lack of readily identifiable barriers separating local mode trajectories. A direct caustic to caustic approximate tunnelling calculation has been performed by Lawton et al. [22], while the hindered rotor model of Sibert et al. provides a means for calculation of local mode splittings in simple ABA triatomic Hamiltonians [23]. (The work of Kellman gives a complementary picture of the nature of the local/normal mode transition [24].)

"Dynamical tunnelling" [21] is a general term describing the interaction between quantum states corresponding to classical motions localized in disjoint regions of phase space, not necessarily symmetry related, which gives rise to very small splittings in the energy spectrum and dynamical effects on correspondingly long time scales. The splitting between local mode states is a particular example of a dynamical tunnelling effect.

C. Nonrigid molecule spectra and quantal isomerization

Many features of nonrigid molecule spectra can be understood in terms of dynamical tunnelling: the spectroscopic consequences of nonrigidity are manifest in characteristic splitting of degenerate manifolds corresponding to distinct equilibrium configurations connected via "feasible" internal motions [25-27]. The approach of Dalton [28] directly predicts patterns of splitting of rigid molecules induced by particular nonrigid internal motions, such as the Berry pseudorotation in PF_5 [29], with the overall magnitude of the level splittings being undetermined. Hougen has recently emphasized the need for multidimensional tunnelling integral calculations in the spectroscopy of hydrazine [30]. One-dimensional semiclassical reaction-path methods have been applied by Truhlar and coworkers for calculation of tunnelling splittings in nonrigid molecules such as NH_3 [31] and $(HF)_2$ [32], following work by Miller [33] and coworkers on tunnelling splittings in malonaldehyde [34] and isomerization rates for HNC→HCN [35] and vinylidene→acetylene [36].

D. Rotation-vibration interaction

Semiclassical calculation of energy levels for rigid asymmetric tops is an essential first step towards use of semiclassical methods to calculate highly excited rotation-vibration levels of polyatomic molecules. Although the asymmetric top is at first sight a three degree of freedom system (e.g., three Euler angles and their conjugate momenta), it is known that there are two constants of the motion, j, the magnitude of the rotational angular momentum vector, and m, the projection of the angular momentum onto a space fixed axis. As shown by Augustin and Miller [37], it is possible to find a canonical transformation that shows explicitly the ignorable nature of the angle variables conjugate to j and m, and so reduces the problem to a single degree of freedom. The resulting effective potential has a double minimum plus barrier [38], and accurate calculation of asymmetric top levels close to the top of the barrier requires a uniform treatment ([38], cf. also [39]). It is therefore clear that uniform methods will in general be required for accurate semiclassical quantization of rotation-vibration levels of polyatomics. In their quantization of the rigid bender model Hamiltonian for H_2O, Frederick and McClelland used an effective 1-dimensional treatment of the rotational motion [14]. This approach will not work for strongly coupled resonant quasiperiodic rotation-vibration motions, recently found to be important for centrifugally induced energy transfer between rotation and vibration [40,41].

One way around the necessity for uniformization in the asymmetric top quantization has been suggested by Duchovic and Schatz [42], who propose interpolation of primitive EBK eigenvalues through the separatrix region. The calculations of Harter and Patterson [43] have shown that clustering of rotation-vibration levels often occurs as a result of dynamical symmetry-breaking, and that the associated intracluster level splittings can be so small that uniformization is unnecessary.

E. Radiationless transitions and resonance lifetimes

Heller and Brown [44] have stressed the need for multidimensional tunnelling calculations in the theory of radiationless decay, and have studied effective 1-dimensional tunnelling paths for various quasiperiodic motions on model potential surfaces. EBK methods have been successfully applied to determine energies of compound-state resonances corresponding to bound quasiperiodic motion above a dissociation threshold in several model systems [45-47]. However, a major outstanding problem for semiclassical theory is the determination of resonance lifetimes (widths) using some form of uniform approximation (cf. the recent work of Farrelly [17], who has obtained complex semiclassical eigenvalues for quasibound states within the SCF approximation).

F. Quantization of electronic and vibronic states

In contrast to the many applications noted above of EBK quantization methods to calculate vibrational and (increasingly) vibration rotation levels, considerably less effort has been devoted to electronic or vibronic problems. The work of Strand and Reinhardt [48] and Pajunen [49] (and references therein) on H_2^+ should be noted, as should the treatment of electrons in magnetic fields by Delos, Knudson and Noid [50] and Reinhardt and Farrelly [51]. We have recently applied the self-consistent classical-electron model of Miller and Meyer [52,53] to study the dynamics of the linear Exe Jahn-Teller Hamiltonian [54,55]. In particular, semiclassical energy levels were determined in both the weak [54] and strong [55] coupling limits. While our results suggest that the classical-electron model describes the effects of nonadiabatic interactions in a qualitatively correct fashion, the lack of quantitative agreement with quantum results has been attributed to use of primitive quantization conditions. Miller and White have applied an extended antisymmetrized version of the classical-electron model to obtain perturbative energies for singly excited helium [56]. In this case, uniformization to obtain singlet-triplet splittings was performed by mapping the full problem onto a hindered rotor.

The diverse range of topics covered above reveals the potential importance and utility of multidimensional uniform methods. In the next Section we review our Fourier transform approach to semiclassical mechanics [13,18], while Section III motivates and describes our method for uniform quantization of resonant trajectories, which involves a numerical determination of an effective 1-dimensional resonance Hamiltonian for multidimensional resonant systems. Some preliminary results are presented for the 1:1 resonant Henon-Heiles system.

2. Fourier transform approach to EBK quantization

We now briefly review our Fourier approach to EBK quantization of quasiperiodic trajectories in multidimensional systems. Full details are given elsewhere [13,18] (cf. also [57]).

To formulate a practical method for EBK quantization based on Fourier transforms, two key results are needed. First, we note that in quasiperiodic regions of phase space the cartesian coordinates q and momenta p of an N degree of freedom system can be expanded as Fourier series in the angle variables θ [2,3]:

$$q(\theta,J) = \sum_{k} q_k(J)\, e^{ik\cdot\theta} \qquad (2.1a)$$

$$p(\theta,J) = \sum_{k} p_k(J)\, e^{ik\cdot\theta}. \qquad (2.1b)$$

where $k = (k_1,k_2,\ldots,k_N)$ is a lattice vector with integer components, $\theta = (\theta_1,\theta_2,\ldots,\theta_N)$ is a vector of angle variables, and the Fourier coefficients q_k and p_k satisfy the reality conditions

$$q_k = q_{-k}^{*}, \quad p_k = p_{-k}^{*} \qquad (2.2)$$

The Fourier coefficients $\{q_k\}$ and $\{p_k\}$ are functions of N good action variables J, while the time dependence of the angles θ is given by

$$\theta_\alpha = \omega_\alpha t + \theta_\alpha^{\,0} \qquad \alpha = 1,\ldots N, \qquad (2.3)$$

where ω_i is the ith fundamental frequency

$$\omega_i = \partial H(J)/\partial J_i. \qquad (2.4)$$

Second, the action integral [1-3]

$$J_\alpha = 1/2\pi \int_{0}^{2\pi} d\theta_\alpha \; p\cdot(\partial q/\partial\theta_\alpha) \qquad (2.5)$$

($\theta_{\alpha'}$ constant, $\alpha' \neq \alpha$) is independent of the constant values chosen for the angles $\theta_{\alpha'}$, $\alpha' \neq \alpha$ [4,5], so that it can be written as an integral over all angles as follows:

$$J_\alpha = 1/(2\pi)^{N} \int d\theta_1 d\theta_2 \ldots d\theta_N \; p\cdot(\partial q/\theta_\alpha) \qquad (2.6)$$

Combining equations (2.1) and (2.6) leads to the fundamental expression for the actions J_α in terms of the Fourier coefficients $\{q_k, p_k\}$ [58]:

$$J_\alpha = \sum_{k} ik_\alpha p_k^{*}\cdot q_k \qquad (2.7)$$

In the case that

$$H = \sum_{j} \tfrac{1}{2} p_j{}^2 + V(\mathbf{q}) \qquad\qquad (2.8)$$

e.g., a perturbed harmonic oscillator Hamiltonian, use of Hamilton's equations simplifies the expression for the actions to [58]

$$J_\alpha = \sum_{\alpha'} G_{\alpha\alpha'}\omega_{\alpha'} \qquad\qquad (2.9)$$

with

$$G_{\alpha\alpha'} = \sum_{\mathbf{k}} k_\alpha |q_{\mathbf{k}}|^2 k_{\alpha'}. \qquad\qquad (2.10)$$

This remarkable result, which we have called Percival's equation [58] (cf. also [57,59]), implies that good actions for quasiperiodic motion in nonseparable multidimensional systems can be calculated provided the magnitudes of the coordinate Fourier components can be found. The problem of numerical determination of Fourier coefficients in a stable and accurate fashion using standard Fast Fourier Transform routines has been solved [13].

Assignment of peaks in power spectra of quasiperiodic coordinates and momenta requires the specification of a set of fundamentals $\{\omega_i\}$ such that every frequency in the spectrum can be written in the form

$$\Omega = \mathbf{k} \cdot \boldsymbol{\omega}. \qquad\qquad (2.11)$$

For nonresonant "boxlike" trajectories (cf. Fig 1a), there is a one to one correspondence between the N fundamentals ω_i and the N zeroth-order frequencies of the unperturbed harmonic oscillator system, so that the fundamentals are easily identified (see Fig. 1b,c). For resonant trajectories (cf. Fig 2a), assigning the fundamentals is more complicated. In essence, for 2-dimensional resonant systems (the only case we shall discuss here), two new fundamental frequencies appear in place of the cartesian frequencies. One of these ($\omega_1{}^r$) is the greatest common divisor of the notional resonant cartesian frequencies. Thus, if $n\omega_x{}^o = m\omega_y{}^o$, $\omega_1{}^r \sim \omega_x{}^o/m = \omega_y{}^o/n$. The second fundamental ($\omega_2{}^r$) is a small frequency corresponding to slow precession of the periodic orbit underlying the resonant motion, which gives rise to characteristic spectral patterns (see Figs 2b,c). This choice of fundamental frequencies is the so-called <u>dynamical</u> assignment [18]. A detailed discussion of the classical spectra of strongly resonant trajectories is given in [18].

With the assignment of spectral peaks well understood, at least for 2-dimensional systems, Percival's formula provides a general and efficient way of calculating good actions for both nonresonant [13] and resonant [18] systems. Systematic variation of initial conditions then yields tori satisfying the primitive EBK quantization condition

$$\mathbf{J} = (\mathbf{n} + \boldsymbol{\alpha}/4), \qquad\qquad (2.12)$$

where the components of the Maslov index vector $\boldsymbol{\alpha}$ are 2 for simple librational motions [2], and can be calculated directly for more complicated resonant trajectories [18].

The above procedure has been successfully applied to determine energy levels for nonresonant vibrational Hamiltonians with 2, 3 and 4 degrees of freedom [13], and to quantize states in strongly resonant 1:1, 2:1 and 3:4 systems [18]. As noted in the previous section, primitive EBK eigenvalues for resonant trajectories such as that in Figure 2a are often somewhat less accurate than in the nonresonant case. In fact, there is no quantizing torus corresponding to state number 38 cf the 3:4 Barbanis system treated in [18], whereas there are two primitive quantizing states that correspond to, for example, the 49th state. A uniform quantization treatment is therefore required (see next section).

The Fourier approach to classical and semiclassical mechanics also yields directly the numerical transformation between cartesian coordinates and good action-angle variables [3,18], and is currently being extended to calculate semiclassical wavefunctions for multidimensional systems and highly excited rotation-vibration states of polyatomics [60].

3. Uniform quantization method

To fix ideas, consider the 1:1 resonant Henon-Heiles system

$$H = \tfrac{1}{2}(p_x^2 + p_y^2) + \tfrac{1}{2}(\omega_x^2 x^2 + \omega_y^2 y^2) + \lambda(x^2 y + \eta y^3), \qquad (3.1)$$

where $\omega_x = \omega_x = 1.0$ and $\eta = -1/3$, so that the potential energy surface has C_{3v} symmetry. EBK quantization of the system (1) predicts degeneracy of certain pairs of levels having A_1 and A_2 symmetry [61]. This degeneracy does not persist in the exact (variationally determined) spectrum, however: the A_1 and A_2 states are split, and a uniform method is required to determine the level splittings. Jaffe and Reinhardt [62] have performed a uniform quantization of the 1:1 Henon-Heiles Hamiltonian (1) by using Birkhoff-Gustavson [63] normal form classical perturbation theory to transform the 2 degree of freedom Hamiltonian (3.1) into a function of a set of operators whose Poisson brackets with the zeroth-order Hamiltonian ($\lambda=0$) are zero. For fixed values of one of the good action variables, the resulting expression is equivalent to the Hamiltonian for a 1-dimensional hindered rotor, so that uniform eigenvalues can be obtained using standard 1-dimensional connection formulae [15,16]. The work of Jaffe and Reinhardt makes two significant advances: first, a multidimensional system is reduced to a tractable one degree of freedom problem; second, accurate uniform energies are calculated for all states up to dissociation, including those above the threshold for classical stochasticity. Here, we concentrate attention on the first aspect.

A perturbation theory based quantization of a 3:1 resonant system has recently been given by Uzer, Noid and Marcus [64]. Perturbation theory was used to derive a 1-dimensional resonance Hamiltonian which was then uniformly quantized using a variety of methods. Several approaches to uniform quantization of 1:1 and 2:1 resonant systems have been discussed by Uzer and Marcus [65].

In the studies just cited, use of perturbation theory to reduce the full Hamiltonian to an effective 1-dimensional resonance Hamiltonian is an essential feature. We now apply the Fourier method for computation of actions in resonant systems to determine the 1-dimensional Hamiltonian numerically (cf. the trajectory based procedure in Sec. III of [64]).

Consider again the 1:1 resonant Henon-Heiles Hamiltonian (3.1). A surface of section [3] at energy E = 6.0 (λ = 0.1118) is shown in Fig. 3. The Henon-Heiles system exhibits two main types of regular trajectory: the so-called precessors (Fig. 4a) and librators (Fig. 4b). The existence of the librator trajectories can be understood (in the angular momentum, as opposed to cartesian, viewpoint [18,65]) as the consequence of a resonance that distorts the zeroth order phase space (λ = 0), which consists of two families of precessors with different senses of rotation. From the perspective of standard resonance analysis [3,66], the librators are "trapped" below an effective dynamical barrier, whereas the precessors are able to pass over the barrier and so retain the zeroth order topology.

For precessing trajectories of the 1:1 Henon-Heiles system the dynamical actions J_1 and J_2 are related to the geometrical [18] radial (J_r) and angular (J_ℓ) actions by [18,61]

$$J_1 = J_\ell + 2J_r, \qquad J_2 = J_\ell. \qquad (3.2)$$

J_1 corresponds to the principal quantum number for a degenerate manifold of zeroth-order states [61]. At the center of the resonance associated with the librators are periodic orbits (marked in the surface of section in Fig. 3) for which J_2 is zero. Using the Fourier method, it is possible to determine energy as a function of the dynamical actions J_1 and J_2 in the classical resonance zone:

$$H = H(J_1, J_2). \qquad (3.3)$$

Fixing the action J_1 at a quantizing value

$$J_1^Q = (v + 1) \quad v = 0,1,2 \dots \qquad (3.4)$$

(in general, $J_1^Q = v + \frac{1}{2}(m+n)$ for an m:n resonance), we obtain the energy as a function of J_2 in the vicinity of a periodic orbit with $J_1 = J_1^Q$:

$$E(J_2) \equiv H(J_1 = J_1^Q, J_2). \qquad (3.5)$$

This function is shown in Figure 5 for librator trajectories of the 1:1 Henon Heiles system with J_1 = 6.0. (In fact, we plot E versus β, where $\beta = \pi J_2$ for librators, $2\pi J_2/6$ for precessors, and is continuous through the separatrix.) Note that $E(J_2)$ is linear in the action J_2 for small J_2, a consequence of "harmonic" small-amplitude motion about the stable periodic orbit defining the center of the resonance. Also, the frequency $\omega_2 = \partial E/\partial J_2$ vanishes as E $\rightarrow$ E_s, the energy of the separatrix dividing librational from precessional motion (in general, resonant from nonresonant motion). Close to the separatrix, the actual dynamics exhibits a thin band of chaos, in contrast to the integrable pendulum model of resonance analysis [3,66].

The key point is that the function $E(J_2)$ defines an effective 1-dimensional resonance Hamiltonian

$$h = h(I_2, \phi_2), \qquad (3.6)$$

where

$$J_2 = (1/2\pi) \int_0^{2\pi} I_2 \, d\phi_2 \tag{3.7}$$

and

$$h = \sum_m h_m(I_2) \, e^{im\phi}. \tag{3.8}$$

Our strategy is then to use the numerically determined $E(J_2)$ to obtain the Fourier coefficients in the expansion (3.8), in the spirit of the RKR inversion procedure, and quantize the resulting 1-dimensional Hamiltonian to yield uniform semiclassical eigenvalues (recall that the action J_1 is fixed at a quantizing value).

For the purposes of the exploratory calculations reported here, we shall assume a particular functional form for h, rather than work with the general expression (3.8), and determine the free parameters in the model Hamiltonian from $E(J_2)$. Thus, we take h to be the universal pendulum or hindered rotor Hamiltonian [66] appropriate for an accidental (as opposed to intrinsic [3]) degeneracy:

$$h(I_2,\phi_2) = E_o + AI_2^2 + \tfrac{1}{2}B[1 - \cos(p\phi_2)], \tag{3.9}$$

where

$$E_o \equiv H(J_1 = J_1^Q, J_2 = 0), \tag{3.10}$$

the energy of the stable periodic orbit at the center of the resonance, and the constants A and B can be determined from $E(J_2)$. Once A and B are known, uniform semiclassical eigenvalues for states with $J_1 = J_1^Q$ are obtained from the standard quantization condition for p-fold hindered rotors [15,16]:

$$(1 + e^{2\alpha})^{\tfrac{1}{2}}\cos\beta = \cos(2\pi n/p) \qquad n=0,1,2.... \tag{3.11}$$

where α and β are phase integrals across the barrier and well, respectively (α is defined between complex turning points for energies above the barrier energy B), and can be expressed in terms of complete elliptic integrals [67].

We have applied the above procedure to quantize the manifold of states of the 1:1 Henon-Heiles system with $J_1 = 5.0$ and 6.0 (v = 4 and 5). For $J_1 = 5.0$ (6.0) the parameters $E_o = 4.85732$ (5.790109), A = 0.007215 (0.007207) and B = 0.01860 (0.03250) were obtained by fitting the trajectory data plotted in Figure 5 to the functional form $E(\beta)$ arising from the model Hamiltonian (3.9) with p = 6 [62]. Uniform semiclassical eigenvalues obtained with the quantization condition (3.11) are shown in Table I. It can be seen that our results are in reasonable agreement with the exact quantum energies [61] and the uniform eigenvalues of Jaffe and Reinhardt [62]. In particular, the splitting between the (5,±3) states is obtained to good accuracy.

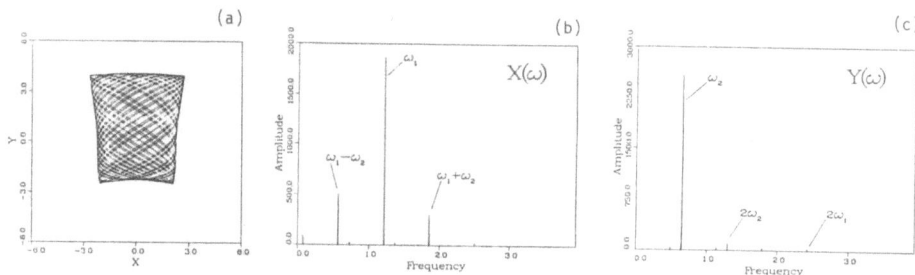

Figure 1   a) Trajectory for the nonresonant system (3.1) with $\omega_x{}^o$ = 1.3, $\omega_y{}^o$ = 0.7, λ = -0.1 and η = 0.1. The energy E = 6.5. b) Fourier transform of the x-coordinate. Peaks are labelled according to the nonresonant frequency assignment. c) Fourier transform of y-coordinate.

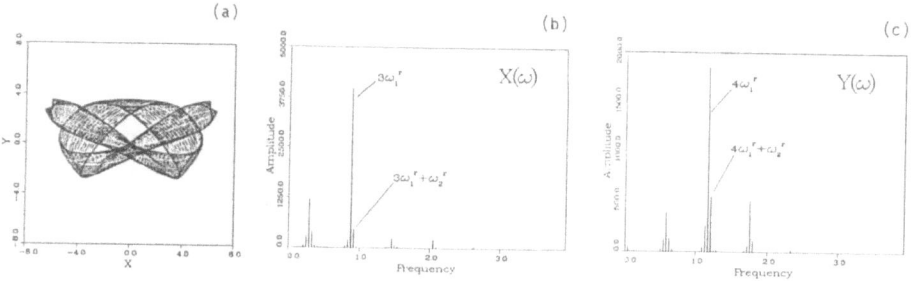

Figure 2   a) Trajectory for the 3:4 resonant system (3.1) with $\omega_x{}^o$ = $(0.9)^{\frac{1}{2}}$, $\omega_y{}^o$ = $(1.6)^{\frac{1}{2}}$, λ = -0.08 and η = 0.0. The energy E = 16.0. b) Fourier transform of the x-coordinate. Peaks are labelled according to the dynamical frequency assignment [18]. c) Fourier transform of the y-coordinate.

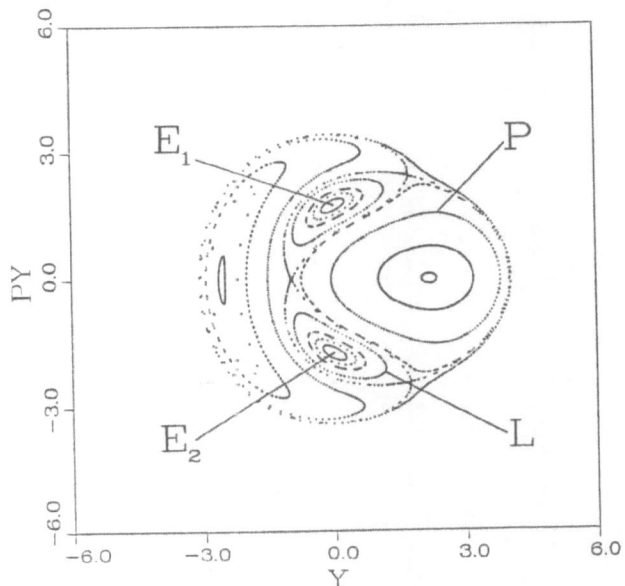

Figure 3   Surface of section for the 1:1 resonant system (3.1) with $\omega_x^o = 1.0$, $\omega_y^o = 1.0$, $\lambda = (0.0125)^{\frac{1}{2}}$ and $\eta = -1/3$. The energy $E = 6.0$. Trajectory P is a precessor (Fig 4a), and trajectory L is a librator (Fig 4b). The periodic orbits E_1 and E_2 are elliptic fixed points at the center of the precessor/librator resonance.

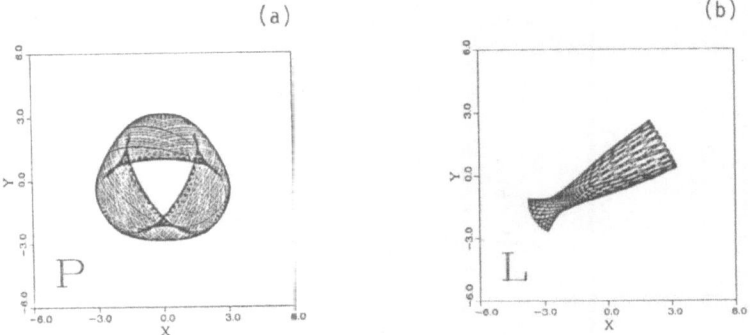

Figure 4   a) Precessing trajectory P marked on surface of section in Fig 3.
b) Librating trajectory L marked on surface of section in Fig 3.

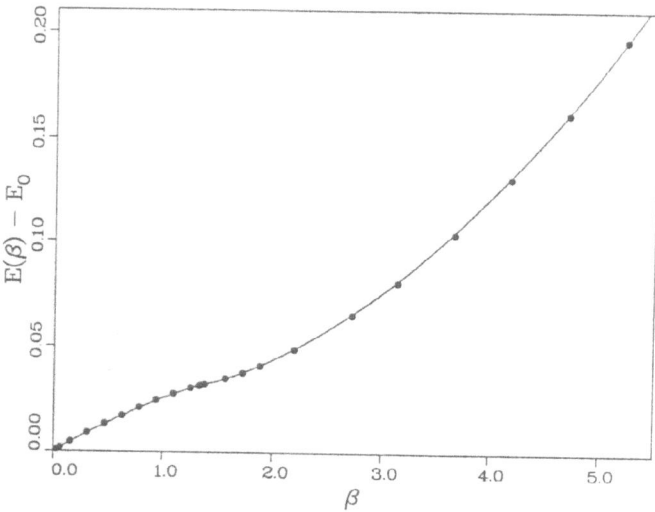

Figure 5   Plot of $E(\beta)$ versus β for the 1:1 resonant Henon-Heiles system, where $J_1 = 5.0$ (β is defined in the text). Points are trajectory data, while the curve is a fit to the 2-parameter pendulum model (3.9).

Table I   Eigenvalues for the 1:1 Henon-Heiles system (3.1).

(n, ℓ)	E_{FFT}^{UQ} (a)	E_{JR}^{UQ} (b)	E_{FFT}^{P} (c)	E_{QM} (d)
(4,0)	4.8730	4.8731	4.8573 (e)	4.8702
(4,±2)	4.8942	4.8942	4.8953	4.8987
(4,±4)	4.9820	4.9821	4.9820	4.9863
(5,±1)	5.8199	5.8202	5.8168 (e)	5.8170
(5,±3)	$\begin{bmatrix} 5.8597 \\ 5.8795 \end{bmatrix}$	$\begin{bmatrix} 5.8608 \\ 5.8778 \end{bmatrix}$	5.8709	$\begin{bmatrix} 5.8670 \\ 5.8815 \end{bmatrix}$
ΔE	0.0198	0.0170		0.0145
(5,±5)	5.9867	5.9871	5.9867	5.9913

a) Uniform semiclassical (this work).
b) Uniform semiclassical [62].
c) Primitive FFT semiclassical [18].
d) Exact quantum [61].
e) Eigenvalue does not exist primitively. Approximate quantization condition
    $J_\ell = \ell/3$ used [61].

4. Conclusion

We have presented a method for uniform semiclassical quantization of resonant trajectories in multidimensional coupled oscillator systems, based upon our Fourier approach to computation of good actions [13,18]. An effective 1-dimensional hindered rotor resonance Hamiltonian was obtained numerically, and quantized using standard connection formulae [15,16]. Good agreement with exact quantum [61] and previous uniform semiclassical [62] eigenvalues was obtained. We are currently applying the method to quantize resonant states of the 3:4 Barbanis system treated in [18].

Acknowledgement

This work was supported by NSF Grant CHE-8410865.

References

[1]   M.V. Berry, in Chaotic Behavior of Deterministic Systems, I. Gerard, R.H.G. Helleman, and R. Stora, Eds. (North Holland, New York, 1983).

[2]   I.C. Percival, Adv. Chem. Phys. 36, 1 (1977).

[3]   A.J. Lichtenberg and M.A. Lieberman, Regular and Stochastic Motion (Springer-Verlag, New York, 1983).

[4]   V.I. Arnold, Mathematical Methods of Classical Mechanics (Springer-Verlag, New York, 1983).

[5]   A.Einstein, Ver. Deut. Phys. Ges. 19, 82 (1917); English translation by C. Jaffe is available as JILA report # 116, U. Colorado, Boulder.

[6]   L. Brillouin, J. Phys. 7, 353 (1926).

[7]   J.B. Keller, Ann. Phys. 4, 180 (1958).

[8]   R.T. Skodje, F. Borondo and W.P. Reinhardt, J. Chem. Phys. 82, 4611 (1985).

[9]   C.W. Patterson, J. Chem. Phys. 83, 4618 (1985).

[10] T.P. Grozdanov, S. Saini and H.S. Taylor, Phys. Rev. A33, 55 (1986); J. Chem. Phys. 84, 3243 (1986).

[11] E.A. Solov'ev, Sov. Phys. JETP 48, 635 (1978); B.R. Johnson, J. Chem. Phys. 83, 1204 (1985); C. Jaffe, preprint.

[12] P. Ehrenfest, Phil. Mag. 33, 500 (1917).

[13] C.C. Martens and G.S. Ezra, J. Chem. Phys. 83, 2290 (1985).

[14] J.H. Frederick and G.M. McClelland, J. Chem Phys. 84, 876 (1986).

[15] W.H. Miller, J. Chem. Phys. 48, 1651 (1968).

[16] M.S. Child, J. Mol. Spec. 53, 280 (1974).

[17] D. Farrelly and A.D. Smith, J. Phys. Chem. 90, 1599 (1986); D. Farrelly, J. Chem. Phys. to appear.

[18] C.C. Martens and G.S. Ezra, to be published.

[19] R.T. Lawton and M.S. Child, Mol. Phys. 37, 1799 (1979); 40, 773 (1980).

[20] C. Jaffe and P. Brumer, J. Chem. Phys. 73, 5646 (1980).

[21] E.J. Heller and M.J. Davis, J. Phys. Chem. 85, 308 (1981).

[22] R.T. Lawton and M.S. Child, Mol. Phys. 44, 709 (1981).

[23] E.L. Sibert III, W.P. Reinhardt and J.T. Hynes, J. Chem. Phys. 77, 3583 (1982); E.L. Sibert III, J.T. Hynes and W.P. Reinhardt, J. Chem. Phys. 77, 3595 (1982).

[24] M.E. Kellman, Chem. Phys. Lett. 113, 489 (1985); J. Chem. Phys. 83, 3843 (1985).

[25] H.C. Longuet-Higgins, Mol. Phys. 6, 445 (1963).

[26] P.R. Bunker, Molecular Symmetry and Spectroscopy (Academic, New York 1979).

[27] G. S. Ezra, Symmetry Properties of Molecules (Springer-Verlag, 1982).

[28] B.J. Dalton, Mol. Phys. 11, 265 (1966).

[29] B.J. Dalton, J. Chem. Phys. 54, 4745 (1971).

[30] J.T. Hougen, J. Mol. Spec. 89, 296 (1981); N. Ohashi and J.T. Hougen, J. Mol. Spec. 112, 384 (1985).

[31] F.B. Brown, S.C. Tucker and D.G. Truhlar, J. Chem. Phys. 83, 4451 (1985).

[32] D.G. Truhlar, 190th ACS National Meeting, Chicago (1985).

[33] W.H. Miller, J. Phys. Chem. 83, 960 (1979).

[34] J. Bicerano, H.F. Schaefer III and W.H. Miller, J. Am. Chem. Soc. 105, 250 (1983).

[35] S.K. Gray, W.H. Miller, Y. Yamaguchi and H.F. Schaefer III, J. Chem. Phys. 73, 2733 (1980).

[36] T. Carrington, Jr., L.M. Hubbard, H.F. Schaefer III and W.H. Miller,

J. Chem. Phys. 80, 4347 (1984).

[37]   S.D. Augustin and W.H. Miller, J. Chem. Phys. 61, 3155 (1974).

[38]   S.M. Colwell, N.C. Handy and W.H. Miller, J. Chem. Phys. 68, 745 (1978).

[39]   P. Pajunen, J. Chem. Phys. 83, 2363 (1985).

[40]   J.H. Frederick, G.M. McClelland and P. Brumer, J. Chem. Phys. 83, 190 (1985); J.H. Frederick and G.M. McClelland, J. Chem. Phys. 84, 4347 (1986).

[41]   G.S. Ezra, Chem. Phys. Lett., to appear.

[42]   R. Duchovic and G.C. Schatz, J. Chem. Phys. 84, 2239 (1986).

[43]   W. Harter and C.W. Patterson, J. Chem. Phys. 80, 4241 (1984).

[44]   E.J. Heller and R.C. Brown, J. Chem. Phys. 79, 3336 (1983).

[45]   R.J. Wolf and W.L.Hase, J. Chem. Phys. 72, 316 (1980); 73, 3779 (1980).

[46]   D.W. Noid and M.L. Koszykowski, Chem. Phys. Lett. 73, 114 (1980).

[47]   R.M. Hedges, R.T. Skodje, F. Borondo and W.P. Reinhardt, in Resonances, D.G. Truhlar, Ed. (ACS, Washington, D.C., 1984).

[48]   M.P. Strand and W.P. Reinhardt, J. Chem. Phys. 70, 3812 (1979).

[49]   P. Pajunen, Mol. Phys. 43, 753 (1981).

[50]   J.B. Delos, S.K. Knudson and D.W. Noid, Phys. Rev. Lett. 50, 579 (1983); Phys. Rev. A28, 7 (1983); D.W. Noid, S.K. Knudson and J.B. Delos, Chem. Phys. Lett. 10J, 367 (1983); S.K. Knudson and D.W. Noid, Chem. Phys. 89, 353 (1984).

[51]   W.P. Reinhardt and D. Farrelly, J. Phys. (Colloque) 43, C2-29 (1982).

[52]   Meyer, H.-D., and W.H. Miller, J. Chem. Phys. 70, 3214 (1979); 71, 2156 (1979); 72, 2272 (1980).

[53]   R.L. Whetten, G.S. Ezra and E.R. Grant, Ann. Rev. Phys. Chem. 36, 277 (1985).

[54]   J.W. Zwanziger, R.L. Whetten, G.S. Ezra and E.R. Grant, Chem. Phys. Lett. 120, 106 (1985).

[55]   J.W. Zwanziger, E.R. Grant and G.S. Ezra, J. Chem. Phys., to appear.

[56]   W.H. Miller and K. White, J. Chem. Phys. 84, 5059 (1986).

[57]   C.W. Eaker, G.C. Schatz, N. De Leon and E.J. Heller, J. Chem. Phys. 81, 5913 (1984); C.W. Eaker and G.C. Schatz, J. Chem. Phys. 81, 2394 (1984).

[58]   I.C. Percival, J. Phys. A7, 794 (1974).

[59]   J. Binney and D. Spergel, Astrophys. J. 252, 308 (1982); Mon. Not. Roy. Astr. Soc. 206, 159 (1984).

[60]   C.C. Martens and G.S. Ezra, work in progress.

[61]   D.W. Noid and R.A. Marcus, J. Chem. Phys. 67, 559 (1977).

[62]   C. Jaffe and W.P. Reinhardt, J. Chem. Phys. 77, 5191 (1982).

[63]   G.D. Birkhoff, Dynamical Systems (A.M.S. Colloqium Publications, New York, 1927); F.G. Gustavson, Astron. J. 71, 670 (1966).

[66]   T. Uzer, D.W. Noid and R.A. Marcus, J. Chem. Phys. 79, 4412 (1983).

[65]   T. Uzer and R.A. Marcus, J. Chem. Phys. 81, 5103 (1984).

[66]   B.V. Chirikov, Phys. Rep. 52, 263 (1979).

[67]   P.F. Byrd and M.D. Friedman, Handbook of Elliptic Integrals for Engineers and Scientists (Springer-Verlag, Berlin, 1954)

MEAN-FIELD APPROACH TO VIBRATIONAL ENERGY LEVELS AND TUNNELING DYNAMICS IN POLYATOMIC SYSTEMS

[a]R.B. Gerber, [b]M.A. Ratner and Z. Bacic
[a]Department of Physical Chemistry and
The Fritz Haber Research Center for Molecular Dynamics
The Hebrew University of Jerusalem, Jerusalem 91904, Israel
[b]Department of Chemistry
Northwestern University
Evanston, Illinois 60201, USA

ABSTRACT. Tunneling processes in the vibrational motions of polyatomic molecules, which are affected by the coupling between different modes, are studied by a mean field approximation. The method used is based on a self-consistent field (SCF) approximation in which framework each mode experiences an effective potential produced by the averaged motions of the other modes. Also, a variational principle is used to select coordinates that are optimal for the SCF approach. The method is applied to the calculation of the bending-stretching energy levels of HCN, and of the related HCN -> HNC tunneling isomerization rates. An interesting feature of the results is strong decrease of the tunneling rate with increasing excitation of the HC stretching mode. Coupling between modes in HCN is essentially static, with no energy transfer occurring on the timescale of tunneling. A generalized version of the above approximation, the time-dependent self-consistent field (TDSCF) approximation is proposed for cases where energy transfer affects the tunneling process. This method is illustrated for a model problem. it is concluded that the mean-field methods are powerful quantitative tools for calculations of tunneling in systems of several interacting modes.

1. INTRODUCTION

Tunneling isomerization of molecules is interesting from a fundamental point of view as an inherently quantum-mechanical unimolecular reaction process. Much of the available theoretical knowledge on unimolecular reaction dynamics was derived·from classical trajectory studies, which are obviously inapplicable to tunneling Rigorous quantum-mechanical calculations of such reaction processes are, however, extremely difficult and feasible at present only for some very simple systems. There is thus strong motivation for pursuing new algorithms and approximation methods to quantitatively study tunneling

25

J. Jortner and B. Pullman (eds.), Tunneling, 25–38.

dynamics in systems of several coupled degrees of freedom. One interesting approach, proposed by Gray *et al.* [1] employed the reaction path Hamiltonian to calculate the microcanonical rate constant below (as well as above) the classical reaction threshold. This method was illustrated in a calculation for the HCN -> HNC isomerization process. A path-Hamiltonian framework was used also in calculations on state-specific tunneling isomerization, in the case of the vinylidene/acetylene rearrangement [2]. Waite [3] used a mixed classical/quantum-mechanical treatment to investigate mode-specific behaviour in the tunneling regime for the HCN -> HNC reaction. The validity of this approximation remains to be tested. Several authors have used in recent years the self-consistent field (SCF) approximation for coupled vibrational modes [4] to study tunneling dynamics in models of polyatomic systems [5-7]. In the present paper, we propose application of the SCF approach to tunneling in systems of several degrees of freedom, that contains two important improvements over previous treatments: First, the results of the SCF approximation depend on the coordinates used. Here as variational principle is employed to determine the coordinates which are optimal for the SCF. These coordinates are, in fact, state-dependent. The variational principle can be applied to any parametrized family of coordinate systems and physical considerations must be involved in choosing an appropriate set of "trial coordinates". The quantitative consequences of using optimal coordinates turn out to be extremely important, as will be shown later. Another improvement proposed here involves the treatment of energy transfer between modes in the context of a tunneling process. Previous SCF studies of tunneling were based on the assumption that energy transfer between the modes can be neglected on the timescale considered, and thus only "static" interaction between the modes was included. It is proposed here that the time-dependent self-consistent field (TDSCF) approximation [4] can be used when energy migration between the modes can affect the tunneling process. The TDSCF is thus presented as an approximation for treating tunneling when the latter may have the same timescale as intramolecular energy redistribution. The structure of the paper is as follows: The "optimal coordinates" improvement of SCF is discussed in Sec. II, and applications to the calculation of bending-stretching energy levels and of state-specific tunneling isomerization rates for HCN -> HNC are given in Sec. 3. In particular, we emphasize the question of the effect of C-H stretching excitation on the tunneling rate. In Sec. IV we discuss the TDSCF approximation, and present results for a model system. Concluding remarks are brought in Sec. V.

2. OPTIMAL-COORDINATES, SELF-CONSISTENT FIELD METHOD

The SCF approximation makes the assumption that the wavefunction of the system can be factorized into a product of single-mode wavefunctions [4]. This leads to a dynamical picture in which each mode is governed by a formally separate Hamiltonian pertaining to it. The potential acting on each mode is the result of its interactions with all the other modes averaged, however, over the motions of the latter. The equations for the various modes must be solved simultaneously and consistently. It should be stressed that the effective potential acting on any mode depends on the states of the other modes. The SCF approach is thus a step beyond simplistic separation of the modes. While the SCF method for electrons in atoms and molecules dates back to the early stages of quantum theory, the corresponding approach for vibrational modes in polyatomic systems was developed only in recent years, in both quantum-mechanical [9] and semiclassical [10] versions. The results of the approximation were found to depend considerably on the choice of the coordinates to which the SCF separability was applied, e.g. [11]. From a physical point of view this is obvious: correlation effects between the motions of the various modes are likely to depend, e.g. on the geometric nature of the modes. Intuitive guidelines can be proposed for the selection of "good" coordinates for the SCF approximation [4]: (i) Separation of frequency magnitudes between the various modes is an advantage. (ii) It is essential to choose coordinates that prevent unphysical crowding of atoms together (with the resulting large correlation due to the hard-core repulsions between the atoms). Individual atom coordinates are not useful as SCF modes for that reason. (iii) Collective modes are desirable, as their motions are spread over the molecule, giving reduced mutual instantaneous correlations between any two modes, and better mutual screening. Different collective modes may diverge greatly in their behaviour in this respect.

We proceed now to consider the SCF approximation for the HCN molecule. The results on this system are all taken from a recent study by Bacic *et al.* that will be published in more detail elsewhere [8]. Our interest will be focused on the bending mode (pertinent to the isomerization process) and on its coupling to the CH stretching vibration. Inspection of the potential energy surface of this system [12] suggests that the CN stretching vibration plays no role in the isomerization process, or in the spectroscopic transitions studied here. The CN separation will thus be treated as fixed at equilibrium value. With this simplification the vibrational Hamiltonian of HCN has the following form in polar coordinates:

$$ H = \frac{\hbar^2}{2\mu_1} t_R + \frac{\hbar^2}{2\mu_1 R^2} t_\theta + \frac{\hbar^2}{2I_{CN}} t_\theta + V(R,\theta) \tag{1} $$

here R is the distance between H and the center-of-mass of CN, θ is the polar angle between $\vec{R}$ and the $C-N$ axis, μ_1 - the reduced mass of H and CN, I_{CN} - the moment of inertia of CN, $V(R,\theta)$ - the potential surface. T_R, t_θ are defined by:

$$ t_R = -\frac{1}{R^2} \frac{\partial}{\partial R} (R^2 \frac{\partial}{\partial R}); t_\theta = -\frac{1}{\sin\theta} \frac{\partial}{\partial \theta} (\sin\theta \frac{\partial}{\partial \theta}). \tag{2} $$

We consider now on an intuitive basis the choice of coordinate systems that should be most suitable for SCF calculations of the bending-stretching system in HCN. It seem reasonable to assume that a good choice of coordinates is likely to be one which conveniently describes motion along the minimum energy path. This appears obvious when attempting to calculate the HCN -> HNC reaction rate, and should apply also to calculations of the bending energy levels since the reaction occurs by bending motion. Moreover, in the quantum-mechanical regime bending levels and tunneling rates are inter-related. Therefore, *a choice of coordinate that works well in an approximate calculation of the HCN→HNC tunneling rate, should also be suitable for computing the energy levels.* Another reason for the relevance of motion along the minimum energy path in choosing coordinates for SCF calculations for the system and states investigated here is related to the structure of the potential energy function [12]: As one moves away from the minimum energy path, the potential energy function of HCN increases rather steeply [12]. In general, as the total potential energy increases, so does also the correlation between modes (which constitute the error in the SCF approximation). Exceptions for this may occur for potential energy surfaces of certain structure, but the surfaces available for HCN do not suggest such a behaviour. Consider now the minimum energy path for HCN → HNC, shown in Fig. 1, which was computed from the potential surface of Murrell *et al.* [12].

The shape is approximately that of an ellipse which suggests the use of spheroidal (ellipsoidal) coordinates. In these coordinates, the H atom can be located in terms of distances r_1, r_2 from two points r_1, r_2, on the C-N axis, the coordinates used then are:

Minimum Energy Path for HCN ⇌ HNC

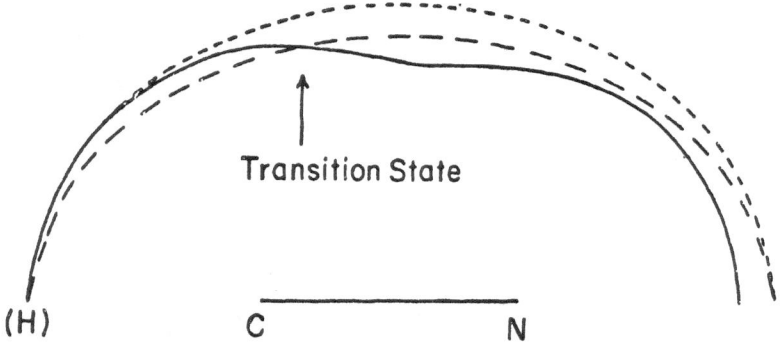

Figure 1.   The Minimum Energy Path for HCN.
  ____ Exact
  ____ Ellipse approximation with a = 1.2273 Å .
  ____ Ellipse approximation with a = 1.063 Å .
a - half interfocal distance

$\xi=(r_1+r_2)/2a$, $1\leq\xi<\infty$; $\eta=(r_1-r_2)/2a, -1\leq\eta\leq1$. *We stress that we cannot identify the two reference points for the coordinate system necessarily with the equilibrium positions of C and N, and 2a is not necessarily the CN equilibrium distance, but a parameter characterizing the coordinate system.* For simplicity we now make the approximation $I_{CN} \to \infty$ in Eq. (1). (Corrections due to the term so ignored can be introduced perturbatively). The approximate Hamiltonian H' so obtained has the following form in spheroidal coordinates

$$H' = -\frac{\hbar^2}{2\mu_1 a^2}\left\{\frac{l}{\xi^2-\eta^2}\frac{\partial}{\partial\xi}\left[(\xi^2-1)\frac{\partial}{\partial\xi}\right]+\frac{1}{\xi^2-\eta^2}\frac{\partial}{\partial\eta}\left[(1-\eta^2)\frac{\partial}{\partial\eta}\right]\right\}+V(\xi,\eta). \quad (3)$$

The quantum SCF approximation [9] consists of making the factorization assumption:

$$\psi_{mn}(\xi,\eta) = \chi_m(\xi)\phi_n(\eta) \quad (4)$$

for the eigenfunctions of the Hamiltonian. This leads to two equations, for $\chi_m(\xi)$ and $\phi_n(\eta)$, respectively that must be solved consistently. We use here the semiclassical SCF approximation (SC-SCF) [10] which for the present case leads to the following equations (see Bacic *et al.* [8] for details):

$$\int_{\xi_a}^{\xi_b} p_\xi(\xi)d\xi=(m+\frac{1}{2})\bar{n}\pi \ ; \ \int_{\eta_b}^{1} p_\eta(\eta)d\eta = \frac{1}{2}(n+\frac{1}{2})\pi\hbar \quad (5)$$

where $\xi_a(b)$, η_b are classical turning points and the symmetry of p_η with regard to the η domain was used in writing the second equation of (5).

The classical momenta p_ξ, p_η of the ξ and η modes respectively are given by:

$$p_\xi(\xi)=(d_n(\xi))^{\frac{1}{2}}\left\{2\mu_1 a^2[\epsilon'_m(n)-(\bar{V}_n(\xi)+\frac{\hbar^2}{2\mu_1 a^2}\bar{t}_\eta(\xi))]\right\}^{\frac{1}{2}} \quad (6\text{-}a)$$

$$p_\eta(\eta)=(d_m(\eta))^{-\frac{1}{2}}\left\{2\mu_1 a^2[\epsilon'_n(m)-(\bar{V}_m(\eta)+\frac{\hbar^2}{2\mu_1 a^2}\bar{t}_\xi(\eta))]\right\}^{\frac{1}{2}} . \quad (6\text{-}b)$$

Eqs. (5) for the SCF energy levels $\epsilon'_m(n)$, $\epsilon_n(m)$ must be solved self-consistently with the evaluation of the following classically averaged quantities:

$$\bar{V}_n(\xi)=\frac{2}{C_\eta}\int_{\eta_b}^{1}\frac{V(\xi,\eta)}{p_\eta(\eta)}d\eta \ ; \quad (7)$$

$$\bar{V}_m(\eta)=\frac{1}{C_\xi}\int_{\xi_a}^{\xi_b}\frac{V(\xi,\eta)}{p_\xi(\eta)}d\xi$$

$$\bar{t}_\eta(\xi) = \frac{2}{\hbar^2 C_\eta}\int_{\eta_b}^{1}d\eta\,\frac{1-\eta^2}{\xi^2-\eta^2}p_\eta(\eta) : \quad (8)$$

$$\bar{t}_\xi(\eta)=\frac{C_\xi}{\hbar^2}\int_{\xi_a}^{\xi_b}d\xi\,\frac{\xi^2-1}{\xi^2-\eta^2}p_\xi(\xi)$$

$$d_m(\xi) = \int_{\eta_b}^{1} \frac{\xi^2 - 1}{(\xi^2 - \eta^2) p_\eta(\eta)} \, d\eta \; ; \tag{9}$$

$$d_m(\eta) = \int_{\xi_a}^{\xi_b} \frac{(1 - \eta^2) d\xi}{(\xi^2 - \eta^2) p_\xi(\xi)}$$

$$C_\eta = 2 \int_{\eta_b}^{1} \frac{d\eta}{p_\eta(\eta)} \; ; \; C_\xi = \int_{\xi_a}^{\xi_b} \frac{d\xi}{p_\xi(\xi)} \tag{10}$$

It can be shown that the eigenvalues E'_{mn} of the Hamiltonian H' of (3) are within the SCF approximation given by the SCF energies:

$$E'_{mn} = \epsilon'_m(n) = \epsilon'_n(m) . \tag{11}$$

The energies obtained by solving the SCF equations (5) depend on the parameter a, the focal half-distance parameter of the prolate spheroidal coordinate system (or rather half the distance between the two "centers" from which the H motion is measured. As stressed earlier, this is really a parameter, for there is need to identify the "centers" mentioned with the C and N atoms exactly. This immediately suggests the idea of the Optimal Coordinates SCF (OC-SCF) approximation: Choose the parameter of the coordinate system variationally, so as to determine the coordinate system that is best for SCF. In the present case:

$$\left(\frac{\partial E'_{mn}}{\partial a} \right) = \left(\frac{\partial \epsilon'_m(n)}{\partial a} \right) = 0 . \tag{12}$$

In the OC-SCF approach the SCF equations for the energies, (5), are solved simultaneously with the variational equations (12) for the optimal coordinates. The case of HCN with the one-parameter family of spheroidal coordinates is, of course, merely one illustration of a general approach. The possibility of using coordinate transformations to improve SCF results was recognized previously but only in the limited case of linear mappings of Cartesian coordinates [13-15]. The optimal coordinates found depend in general on the state (m,n): In the present case, for instance, different bending states probe different portions of the minimum energy path, and as seen in Fig. 1, different ellipses should best be used in fitting different regions of the m.e.p.

So far, the discussion was limited to energy level calculations. The SC-SCF scheme presented involves a primitive semiclassical approximation (see Eq. (5)), that does not include tunneling effects. However, given the fact that the SCF equations are one-dimensional, extensions that cover also tunneling can readily be given [5-7], e.g. by using a uniform semiclassical approximation [7]. The simplest approach uses a familiar approximate semiclassical expression for the tunneling rate $k_{m,n}$ [8]:

$$k_{m,n} = 2\omega_{m,n} \cdot P_{m,n} \; ; \; P_{m,n} = e^{-2\delta_{m,n}} \tag{13}$$

$$\delta_{m,n} = \int_{\eta_3}^{\eta_2} d\eta |d_m(\eta)|^{-\frac{1}{2}} \times \tag{14}$$

$$\left\{ 2\mu_1 a^2 [(\bar{V}_m(\eta) + \frac{\hbar^2}{2\mu_1 a^2} \bar{t}_\xi(\eta)) - \epsilon_n(m)] \right\}^{\frac{1}{2}}$$

where $\omega_{m,n}$ is the vibrational frequency of the bending mode for the HCN isomer, $P_{m,n}$ is the tunneling probability each time the bending mode reaches the isomerization barrier, η_2 and η_3 are the final and initial bending configurations on both sides of the barrier at energy $E_{m,n}$. The various quantities that appear in (14) were defined earlier. The factor 2 that appears in Eq. (13) for $k_{m,n}$ is due to the fact that there are *two* equivalent barriers through which the system can tunnel in each vibrational period. Eq. (13) is valid when $P_{m,n}$, the tunneling probability factor is small, $P_{m,n} \ll 1$.

3. CALCULATIONS FOR HCN

The calculations reported here were carried out for the potential energy surface of Murrell *et al.* [12]. SC-SCF energy levels were obtained for polar coordinates, and also for different spheroidal coordinate systems each corresponding to a different value of the parameter a. The SC-SCF results for the "optimal" coordinates, which satisfy Eq. (12) were also obtained. These correspond to solving Eqs. (5) simultaneously with (12). It should be stressed that SCF and even OC-SCF calculations are numerically very simple compared e.g. with full semiclassical quantization of the same system or with exact quantum-mechanical calculations.

3.1 SCF results for different coordinates

The results presented here are for the $I_{CN} \to \infty$ limit of the HCN Hamiltonian (see Eq. (3)). Fig. 2 shows the SC-SCF results obtained for the states $(m,n) = (0,2)$ and $(0,4)$ for a function of the spheroidal coordinate parameter. Shown also are the energies obtained in a bare-mode approximation using the same coordinate. (The bare-mode approximation involves a simplistic separation of the coordinates, without any SCF correction for the coupling). The energy obtained in SCF calculations using polar coordinates are also indicated. It is seen here that spheroidal SCF, and in particular the *optimal* spheroidal SCF (for which $\partial E_{m,n}/\partial a = 0$) offers considerable improvement over the polar SCF. The improvement mounts for the higher excitation states. The advantage gained by optimizing the coordinates (compared with SCF in arbitrary spheroidal coordinates) does also increase with increased excitation. Fig. 3 shows the spherical SCF results for the $(m,n) = (0,16)$, $(0,18)$. The polar results are rather poor in comparison with those of the spheroidal case in that they are out of scale in Fig. 3. As shown by the results of Fig. 3, the improvement gained by optimizing the coordinates is far greater for highly excited states. We thus conclude that OC-SCF can produce much better results than SCF in non-optimized coordinates.

3.2 SCF vs. bare-mode results

As demonstrated in the results of Figures 2 and 3, choosing the right coordinates for mode separation requires the framework of the SCF approximation: The simple straightforward approximate separation of modes, to which we refer as the "bare-mode" treatment gives much poorer results than SCF for the same modes. Unlike the base-mode results, the OC-SCF energies are in very good agreement with the exact values for the few levels where the latter are available [8]. The errors involved in OC-SCF energies do not seem to exceed 3%, and are usually much smaller than the latter value (the relative error for the $(m,n) = (0,2)$ excited bending states is of the order of 0.15%!). Many current calculations of tunneling rates in molecules are performed using a bare-mode type approximation. Given the relation between bending energies and tunneling rates, the above findings suggest that the SCF should prove an excellent alternative for such tunneling calculations, offering far

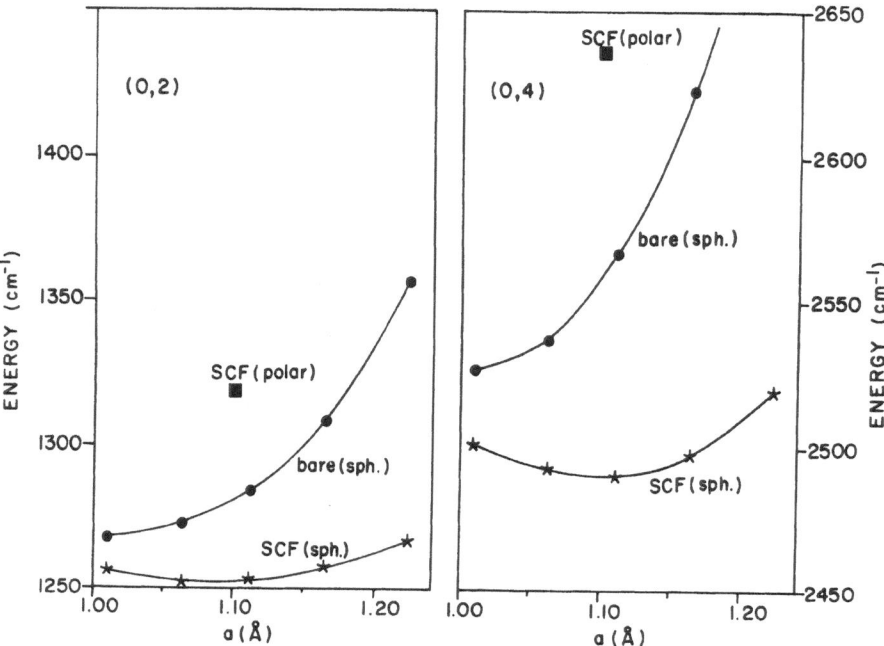

Figure 2. Optimal coordinate behaviour of the SCF energies of the (0,2) and (0,4) states. The spherical SCF results are plotted vs. the parameter that characterizes the coordinate systems. The base-mode results vs. a are also shown, and the polar SCF energy is given for comparison.

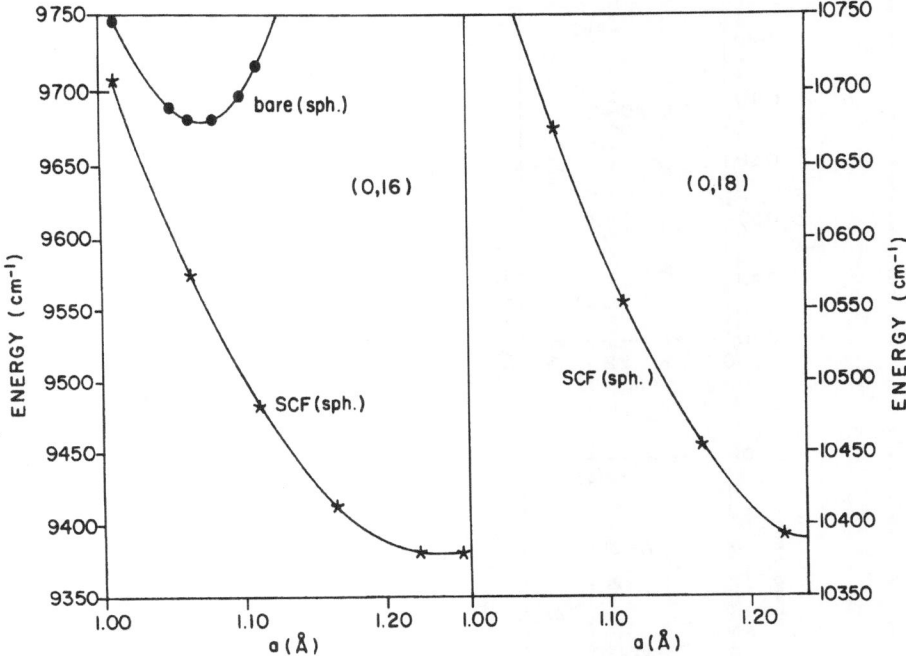

Figure 3. Optimal-coordinate behaviour of the SCF energies of the (0,16) and (0,18) states. The spheroidal SCF results are plotted vs. the coordinate parameter a. For (0,16) the bare-mode results are shown as well.

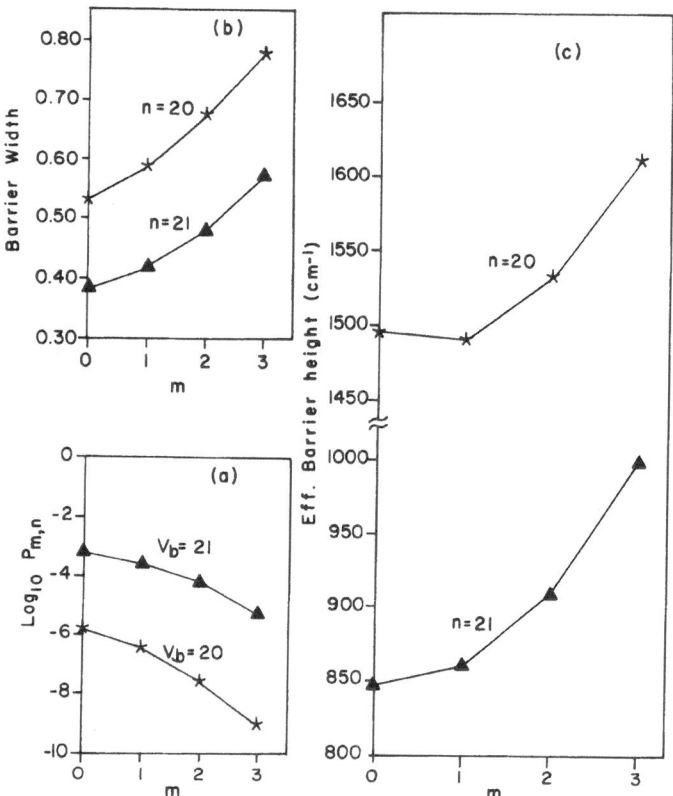

Figure 4. Effect of stretching excitation on tunneling
(a) Tunneling probability factor $P_{m,n}$ vs. stretch level;
(b) Width of effective barrier for tunneling vs. m;
(c) Height of effective barrier for tunneling vs. m.

greater reliability and accuracy yet still being computationally simple.

3.3 SCF state-specific tunneling rates

The SCF approach outlined in Sec. II includes the effect of "static" interactions between the modes. It can therefore be used to predict the effect of coupling between vibrational modes on tunneling rates, provided no effect of "dynamical interaction" (energy transfer) is involved. In the case of HCN in the tunneling regime, energy transfer between the stretching and bending modes can be improved because of the absence of resonances or near-resonances between degrees of freedom involved. Fig. 4a shows OC-SCF results for the isomerization tunneling probabilities of the $n = 20$ and $n = 21$ levels of HCN, and their variation with the stretching state. *The results show a strong state specificity and the tunneling probability factor is seen to fall strongly with increasing initial excitation of the stretch mode.* Very roughly, the tunneling rate decreases by almost an order of magnitude per each additional quantum of stretching excitation. The interpretation seems straightforward: The higher the stretching excitation, the larger is the average C-H separation. As a result, the system moves further away with increasing m from the minimum energy path (which in fact requires a slight shortening of the C-H distance in the first part of the isomerization process). This effect enters in the SCF treatment via the effective potential for the bending-type motion η. This potential $V_m(\eta)$ is defined in Eq. (7). As shown in parts (b) and (c) of Fig. 4, both the *height* and *width* of the effective barrier for tunneling contained in $V_m(\eta)$ increase rapidly as m increases.

4. TDSCF APPROXIMATION FOR TUNNELING RATES

It is obviously desirable to extend the SCF treatment of tunneling also to systems where energy transfer between the modes occurs on a timescale comparable with the tunneling process and can effect the latter. This can be done by applying the time-dependent self-consistent field (TDSCF) approximation to tunneling in polyatomic systems. As the SCF approximation, so the TDSCF also dates back to the earliest stages of quantum theory. Its applications to vibrational dynamics in molecules are, however, very recent [4]. We mention in particular the application of the method, with suitable adaptation, to unimolecular dissociation of Van der Waals complexes [16]. We outline here the TDSCF approximation for a system of two coupled modes, x_1 and x_2. The time-dependent Schrödinger equation for the system:

$$i\hbar \frac{\partial \Psi(x_1,x_2,t)}{\partial t} = H \Psi(x_1,x_2,t) \tag{15}$$

is treated approximately by assuming the factorization:

$$\Psi(x_1,x_2,t) = \phi_1(x_1,t)\phi_2(x_2,t) . \tag{16}$$

The Hamiltonian is written as

$$H = T_1 + T_2 + V(x_1,x_2) \tag{17}$$

where T_i is the kinetic energy operator in the mode x_i. Substituting (16) into (15), one obtains after some algebra:

$$i\hbar \frac{\partial \hat{\phi}_i(x_i,t)}{\partial t} = [T_i + \bar{V}_i(x_i,t)]\hat{\phi}_i(x_i,t) \tag{18}$$

with

$$\bar{V}_i(x_i,t) = <\hat{\phi}_j(x_j,t)|V(x_i,x_j)|\hat{\phi}_j(x_j,t)> \tag{19}$$

where $i \neq j$, $i,j = 1,2$ and the integral in (19) is over x_j. The $\hat{\phi}_i(x_1,t)$ of Eq. (18) differ from the $\phi_i(x_i,t)$ by time-dependent phase factors that are of no importance here. In the TDSCF each mode is governed by a Hamiltonian that depends explicitly on that mode alone. However, the effective potential acting on any mode depends on the other modes, thus the equations of motion of all the modes must be solved simultaneously, so as to be mutually consistent. Energy transfer between modes within the TDSCF is possible due to the time-dependence of the potentials that affect each mode. The remaining question is whether the amount of energy transfer predicted by the TDSCF is correct for the processes considered. Tests for several classically-allowed unimolecular processes [4,9], gave an affirmative answer to this question.

We consider here the application of TDSCF to reorientation by tunneling of molecules trapped in a solid cage. The potential function of a molecule occupying a site in a host crystal (e.g. a rare-gas solid matrix) has typically several minima, the number of which depends on the symmetry of the site, the nature of the molecular interaction with the host, etc. At low temperatures a molecule occupying one of the minima in the librational potential field cannot alter its orientation to occupy another minimum, except by tunneling. (Barriers for rotation of molecules such as Cl_2 or NH_3 in a rare gas matrix are typically of the order of ~100cm^{-1}. The classical probability of surmounting such a barrier at a low cryogenic temperature. e.g. $\leq$ 20K is negligible). There is considerable experimental evidence to the effect that molecular reorientation in solids by tunneling does indeed occur [17]. Strong temperature dependence was found in several examples for the reorientation rate, raising the question of a possible role for phonon-to-libration energy transfer on the tunneling process.

It is with the above motivation that we consider here a very simple model for energy transfer from an oscillator ("phonon mode") to a molecule in a libration well. Systems involving a double-well coupled to a harmonic oscillator were investigated also as models for many other processes [18] so the problem investigated here is of broad interest. The model Hamiltonian was

$$H = T_0 + V_0(1+\alpha u)(1-\cos 2\theta) + H_{osc}(u) \tag{20}$$

where θ is the rotation (libration) angle; T_0 - the free rotor Hamiltonian; u - the displacement of the oscillator from equilibrium, H_{osc} - the harmonic oscillator Hamiltonian; V_0 and α were estimated from the study of Girardet et al. [19] so as to correspond to the case of NH_3 in Kr (the pertinent NH_3 motion is proper rotation, or spinning around the C_3 axis).

Both exact quantum-mechanical calculations (using a stationary basis set) and TDSCF calculations for the tunneling rate in the coupled double-well/oscillator system were carried out. We report here results for the tunneling rates when the oscillator is initially in its first excited level $n=1$, and the librator is localized in one of the wells with ground-state energy. Table 1 shows the tunneling rate for three different cases in the oscillator frequency ω_{osc}. The relevant parameter is $r_\omega = \hbar\omega_{osc}/\Delta E$, where ΔE is the energy gap between the lowest (split) level in the well and the first excited state. r_ω measures the proximity to resonant transfer between the oscillator and the librator. The tunneling time is given in terms of $\tau_r = \tau_{n=1}/\tau_{n=0}$, where $\tau_{\omega=0}$ is the (exact quantum-mechanical) tunneling time when the oscillator is initially at its ground state.

Table 1: TSCF and Exact Quantum Tunneling Rates for Coupled Librator/Oscillator Systems

Frequency parameter $\hbar\omega_{osc}/\Delta E$	Relative tunneling time (TDSCF) τ_r	Relative tunneling time (Exact) τ_r
0.333	0.880	0.866
0.666	0.675	0.642
1.000	0.074	0.069

Not surprisingly, resonance energy transfer from the oscillator to the librator in the first well greatly enhances the tunneling rate into the second well. The interesting fact is that the TDSCF reproduces the enhancement effect to good accuracy. It seems reasonable to suggest that TDSCF should prove an adequate tool for tunneling calculations in coupled-mode systems when energy transfer between different degrees of freedom significantly affects the process.

5. CONCLUDING REMARKS

Self-consistent field methods were considered as tools for quantitative tunneling calculations in systems that involve several coupled degrees of freedom. In cases where energy migration between modes can be ignored on the tunneling timescale, the main new contribution has been to propose the use of coordinates in which SCF errors are minimized. This was illustrated in the case of HCN, where "optimal coordinates" were found by first making a physically-motivated guess, and then refining the choice by a rigorous variational principle. Optimization of the coordinates was shown to produce an enormous quantitative improvement in the SCF method for the example studied. The strategy adopted here is, in principle, general although clearly physical considerations may suggest optimal coordinates of a type very different from those used here for other systems.

Energy transfer between modes can affect tunneling processes, and for such cases the use of the TDSCF approximation was advocated. The method reproduced in a test case the correct enhancement effect of resonance energy transfer on the tunneling rate. The conditions of the validity of the TDSCF for tunneling in coupled mode systems are, however, not yet known and should be explored by further examples and perhaps by more general arguments. We believe that this study further indicates the considerable power and promise held by mean-field methods as tools for studying many aspects of intramolecular dynamics, including energy level structure, tunneling rates, energy redistribution rates and other related topics.

Acknowledgement: The Fritz Haber Research Center is supported by the Minerva Gesellschaft für die Forschung, mbH, München, BRD.

REFERENCES

1. S.K. Gray, W.H. Miller, Y. Yamaguchi and H.F. Schaefer, J. Chem. Phys. **73**, 2733 (1980).

2. T. Carrington, Jr., L.M. Hubbard, H.F. Schaefer and W.H. Miller, J. Chem. Phys. **80**, 4347 (1984).

3. B.A. Waite, J. Phys. Chem. **88**, 5076 (1984).

4. A recent review is: M.A. Ratner and R.B. Gerber, J. Phys. Chem. **96**, 20 (1986).

5. K.M. Christoffer and J.M. Bowman, J. Chem. Phys. **76**, 5370 (1982).

6. D. Farrelly, R.M. Hedges and W.P. Reinhardt, Chem. Phys. Lett. **96**, 599 (1983).

7. D. Farrelly, J. Chem. Phys. (in press).

8. Z. Bacic, R.B. Gerber and M.A. Ratner, J. Phys. Chem. (in press).

9. J.M. Bowman, J. Chem. Phys. **68**, 608 (1978); G.D. Carney, L.I. Spandel and C.W. Kern, Adv. Chem. Phys. **37**, 305 (1978).

10. R.B. Gerber and M.A. Ratner, Chem. Phys. Lett. **68**, 195 (1979).

11. R.M. Roth, R.B. Gerber and M.A. Ratner, J. Phys. Chem. **87**, 2376 (1983).

12. J.N. Murrell, S. Carter and L.O. Halonen, J. Mol. Spectros. **93**, 307 (1982).

13. T.C. Thompson and D.G. Truhlar, J. Chem. Phys. **77**, 3031 (1982).

14. R. Lefebvre, Int. J. Quant. Chem. **23**, 543 (1983).

15. N. Moiseyev, Chem. Phys. Lett. **98**, 233 (1983).

16. R.B. Gerber, V. Buch and M.A. Ratner, J. Chem. Phys. **77**, 3022 (1982).

17. V. Narayanamurti and R.O. Pohl, Revs. Mod. Phys. **42**, 201 (1970).

18. K.M. Christoffer and J.M. Bowman, J. Chem. Phys. **74**, 5057 (1981).

19. C. Girardet, L. Abouaf-Marguin, B. Gauthier-Roy and D. Maillard, Chem. Phys. **89**, 415 (1984).

CONFORMATIONAL CHANGES ON EXCITATION OF MnO_4^- TYPE d^0 TETROXO ANIONS IN CRYSTALS

W. Barendswaard, R.T. Weber and J.H. van der Waals
Center for the Study of the Excited States of Molecules
Huygens Laboratory University of Leiden,
P.O. Box 9504, 2300 RA Leiden
The Netherlands

ABSTRACT. Electron paramagnetic resonance experiments have been carried out on luminescent crystals of $CaMoO_4$, YVO_4 and $Ba_3(VO_4)_2$ at liquid helium temperature. The results prove that the luminescence originates from a spin triplet of the tetroxo ion, in accordance with a suggestion by Blasse. The zero-field splitting tensor in the spin hamiltonian does not reflect the symmetry of the crystal field but points to a distortion of the anion through a static Jahn-Teller effect. It is suggested that a similar Jahn-Teller instability should be taken into account in the assignment of the first spin-allowed transition in the absorption spectra of the d^0 ions, and MnO_4^- in particular.

1. NATURE OF THE PROBLEM

The identification of the lower excited states of the tetrahedral d^0 oxy-anions, VO_4^{3-}, CrO_4^{2-}, MnO_4^-, and the analogous $4d^0$ compounds, represents a classical but only partly solved problem in ligand field theory. For an isolated ion of tetrahedral symmetry (T_d) the theoretical predictions of Ballhausen and Liehr [1] are generally accepted. In their model the first excitation is the $t_1 \rightarrow 2e$ charge transfer, in which the t_1 mo is made up exclusively of $2p_\pi$ ao's on the oxygen ligands, whereas in the 2e mo two of the σ-type ao's of the metal participate ($d_{x^2-y^2}$ and d_{z^2} in the commonly used axis system [1]). From this excitation four electronic states result, $^1T_2$, $^1T_1$ and $^3T_2$, $^3T_1$, with energies thought to lie in the order $^1T_2 > {}^1T_1 > {}^3T_2 \approx {}^3T_1$ [2], see fig. 1. More sophisticated Hartree-Fock type calculations [3] lead to the same first excited configuration, but they also show that the simple charge-transfer picture is an oversimplification because of a readjustment of the doubly occupied mo's. In solution spectra the $^1T_1 \leftarrow {}^1A_1$ and $^1T_2 \leftarrow {}^1A_1$ transitions are identified with the first and second absorption bands; of these the first is the weaker one because it is electric-dipole forbidden and must acquire its intensity through vibronic coupling and environmental perturbation from the electric-dipole-allowed $^1T_2 \leftarrow {}^1A_1$ transition(s).

As regards the $^3T_1$ and $^3T_2$ spin triplets, in which we are primarily

J. Jortner and B. Pullman (eds.), Tunneling, 39–47.
© *1986 by D. Reidel Publishing Company.*

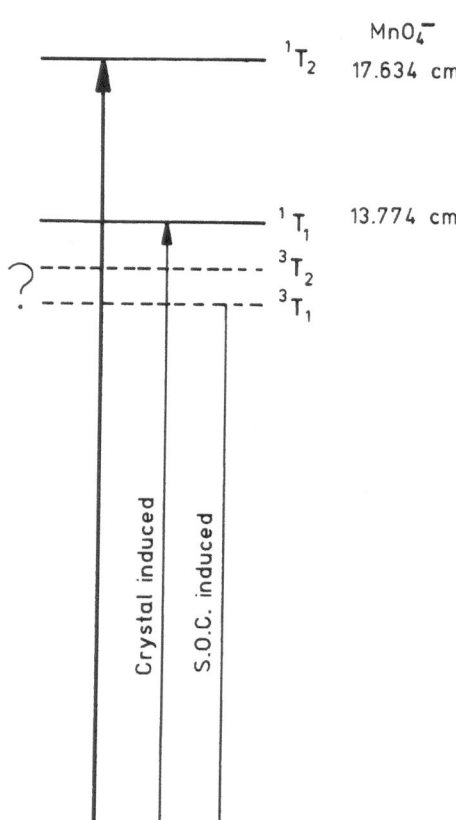

$$\text{MnO}_4^-$$

$^1T_2$ 17.634 cm^{-1}

$^1T_1$ 13.774 cm^{-1}

$^3T_2$

$^3T_1$

Crystal induced

S.O.C. induced

$^1A_1$

Figure 1. Electronic states arising from the excitation $t_1 \rightarrow 2e$ in a d^0 tetroxo anion in T_d symmetry. The energies of the excited singlet states refer to MnO_4^- in $\text{LiClO}_4 \cdot 3\text{H}_2\text{O}$ [7]. The positions of the triplet states are not known and the separation between the first excited singlet and triplet states may be quite small (for $\text{Cr}_2\text{O}_7^{2-}$ in $\text{K}_2\text{Cr}_2\text{O}_7$ this separation is 425 cm^{-1} [16]).

interested, the situation remained unclear. Teltow, in a remarkable piece of work [4], had already shown that CrO_4^{2-} and MnO_4^- when incorporated in crystalline hosts like K_2SO_4 or KClO_4 have absorption spectra with a wealth of vibronic features, which may be grouped in band systems corresponding to the solution spectra. However, efforts to identify the spin-forbidden $^3T_2 \leftarrow {}^1A_1$ and (or) $^3T_1 \leftarrow {}^1A_1$ transitions in such spectra have remained fruitless thus far [5].

Even for the long wavelength band system thought to correspond to the $^1T_1 \leftarrow {}^1A_1$ spin-allowed transition no detailed assignment has been given. Although vibrational modes of the tetrahedral ion are recognized, there still is no agreement about the size of the crystal field splitting of the $^1T_1$ multiplet in various hosts. For the widely studied system MnO_4^- in $\text{LiClO}_4 \cdot 3\text{H}_2\text{O}$, for instance, in which the Mn site symmetry is C_{3v}, a splitting of 10 cm^{-1} between the $^1A_2$ and ^1E components of the $^1T_1(T_d)$ state has been deduced by Ballhausen and Trabjerg [6], whereas Johnson first proposed a value of 520 cm^{-1} [7] and later concluced that "the position of the $^1A_2$ state is still uncertain" [8]. Further we are convinced that too little thought has been given to the possibility of a sizable, static Jahn-Teller effect (e.g. in the ^1E component of the system just discussed), which would invalidate several of the arguments put forward in previous discussions.

VO_4^{3-} and the $4d^0$ and $5d^0$ ions differ in two important aspects from CrO_4^{2-} and MnO_4^-: i) their absorption spectra, even in host crystals at 4.2 K, consist of broad featureless bands starting in the uv; ii) they are, in general, strongly luminescent in crystals at low temperature. The emission spectra likewise are completely featureless with very little intensity in the 0-0 region. This situation, together with a large Stokes shift (0.5 - 1 eV), is indicative of strong coupling between the guest and the host lattice.

Although the luminescence of vanadates, molybdates and tungstates has been studied for some forty years [9,10], it was only relatively recently that Ronde and Blasse [11] put forward a general explanation by assigning it to the spin-forbidden $^3T_1$ and (or) $^3T_2 \rightarrow {}^1A_1$ transitions within the tetroxo anions. According to this assignment the emitting state is paramagnetic (S=1) and we saw it as a challenge to try and identify the elusive spin triplets of the d^0 tetroxo anions by EPR with optical detection.

The $^3T_2$ and $^3T_1$ spin triplets must obtain radiative character from the $^1T_2 \rightarrow {}^1A_1$ electric-dipole allowed transitions through spin-orbit coupling (SOC), and we enquired into the principal coupling path. In a first approximation, applicable to the $3d^0$ ions in particular, the dominant coupling should be that between the multiplets of fig. 1 arising from the common $t_1 \rightarrow 2e$ excitation. When analyzing this SOC [12] one discovers that it occurs exclusively on the oxygen atoms. This at first sight somewhat surprising conclusion derives from the circumstance that the significant terms in the expansion of the SOC matrix elements in integrals over mo's are of the one-electron type. Since the excited configuration has one "hole" in the t_1 set of mo's and one electron in the 2e mo's, the orbital integrals that might give a contribution are of two types only: $< t_1'|\underline{\Omega}|t_1''>$ and $<e'|\underline{\Omega}|e''>$. The primes and double primes indicate that the specific mo's appearing in bra and ket should not be identical, and $\underline{\Omega}$ stands for the orbital part of the SOC operator (for its specific form see e.g. eqn. (5) of [12] where further references to the literature on SOC in polyatomic molecules are given). All what matters here is that the components of Ω transform like the components of angular momentum. The three mo's of the set t_1 (expressed in real form) likewise transform as the components of angular momentum, and the mo's e' and e" as the spherical harmonics P_2^0 and P_2^2. According to group theory the only integrals that need not vanish then are of the type

$$< t_{1,u} |\Omega_v| t_{1,w}> = \tfrac{1}{2}i\zeta_{ox} \quad (u,v,w) = (x,y,z \text{ cyclic}) \quad (1)$$

The result, in which ζ_{ox} stands for the SOC constant for a 2p electron in oxygen, expresses the fact that the t_1 mo's, by symmetry, are restricted to the ligands; for an O^- ion $\zeta_{ox} \sim 100$ cm^{-1}.

According to the above analysis the SOC mechanism that causes the VO_4^{3-} ion to be luminescent with a radiative lifetime of the order of a millisecond is remarkably similar to that responsible for the phosphorescence of ketones like acetone [13] and benzophenone [14]. Using techniques which proved successful for calculating the phosphorescence lifetime of heteroaromatic molecules, it has in fact been possible to give a satisfactory quantitative estimate of the lifetime of the $^3T_1$ state of VO_4^{3-} on the basis of the SOC integral (1) [12]. When SOC of the $^3T_1$ ($t_1 \rightarrow 2e$) state with the $^1T_2$ states arising from some of the more highly excited configurations is also taken into account the central metal no longer is excluded from contributing to the coupling. Whereas for the 3d metals this contribution is thought to be relatively small, it should become appreciable for the 4d metals as in MoO_4^{2-}. For the 5d ions the present simple picture should lose its validity altogether.

2. EXPERIMENTS CARRIED OUT IN LEIDEN.

Our first investigation concerned the $Cr_2O_7^{2-}$ ion in $K_2Cr_2O_7$. This ion, which consists of two chromate tetrahedra sharing a common oxygen, in a sense represents an exception: it becomes luminescent at low temperature and, nevertheless, exhibits highly structured absorption and emission spectra [15]. With their experiments Van der Poel et al. [15,17] established the S = 1 character of the emitting state and, moreover, showed the excitation to be localized on one chromate moiety of the composite ion; the results therefore appear of considerable relevance for the present problem. Optical experiments in which the sample was simultaneously irradiated with microwaves resonant with one of the transitions between two sublevels of the spin triplet further enabled them to give assignments for the emission and absorption spectra [16].

From these assignments it follows that in $Cr_2O_7^{2-}$ the metastable spin triplet from which the luminescence originates lies a mere 425 cm^{-1} below the first excited singlet state. For the tetrahedral ions one thus may have to reckon with a situation in which the energy separation between the $^3T_1$ and $^3T_2$ multiplets (due to the Coulomb term in the Hamiltonian) is of the order of a few hundred cm^{-1} only. If so, experimental reality for a tetroxo ion in a crystal may be far from the idealized picture of fig. 1 valid for a tetrahedral ion with neglect of SOC. The 18 spin-orbital components of the $^3T_1$, $^3T_2$ multiplets then should be prone to heavy scrambling by three additional, non-commuting interactions: the crystal field and Jahn-Teller effects, and SOC. Consequently, EPR experiments on the metastable triplet state of the tetroxo ions, even if successful, should be far more difficult to interpret in terms of the electronic structure of the emitting state than in the familiar case of phosphorescent organic molecules.

To date successful EPR experiments with optical detection on the tetroxo ions proper have been completed for MoO_4^{2-} in $CaMoO_4$ [18], while such experiments are in progress for VO_4^{3-} in YVO_4 and $Ba_3(VO_4)_2$ crystals [19]. In all three systems strong EPR signals are observed at liquid helium temperature, which prove the S = 1 character of the luminescent state. But, initially to our surprise, the observed zero-field splittings do not reflect the site symmetry of the ion expected

from crystallographic data. The EPR results show that in each of the
three systems a lowering of symmetry occurs on excitation. Apparently,
the orbital degeneracy is lifted and we think the Jahn-Teller
instability to be the major cause.

The $CaMoO_4$ and YVO_4 crystals both possess a tetragonal structure in
which the tetrahedral ion is slightly compressed along the
crystallographic c axis. The S_4 axis of the tetrahedron that coincides
with the c axis, however, is preserved and all four anions in the unit
cell are magnetically equivalent. In $CaMoO_4$ the structure is as
indicated in fig. 2a with S_4 symmetry at the Mo site. In YVO_4 it is
somewhat higher: the angle ϕ here is zero and the ac and bc planes now
are mirror planes of the anion with D_{2d} symmetry at the V site.

If on incorporation of the tetrahedral ion into either crystal the
fate of the $^3T_1$ (or, conceivably $^3T_2$) state were solely determined by
the crystal field and SOC, one would expect to observe a single $S = 1$
species in the EPR experiments, with one of the principal axes of its
zero-field splitting tensor ("spin axes") pointing along c. In reality
the EPR experiments reveal *four* distinct species with none of their spin
axes along c. Whereas in a given crystal the size of the zero-field
splitting is the same for all four species, the directions of the spin
axes are different though interrelated by a fourfold rotation about c.

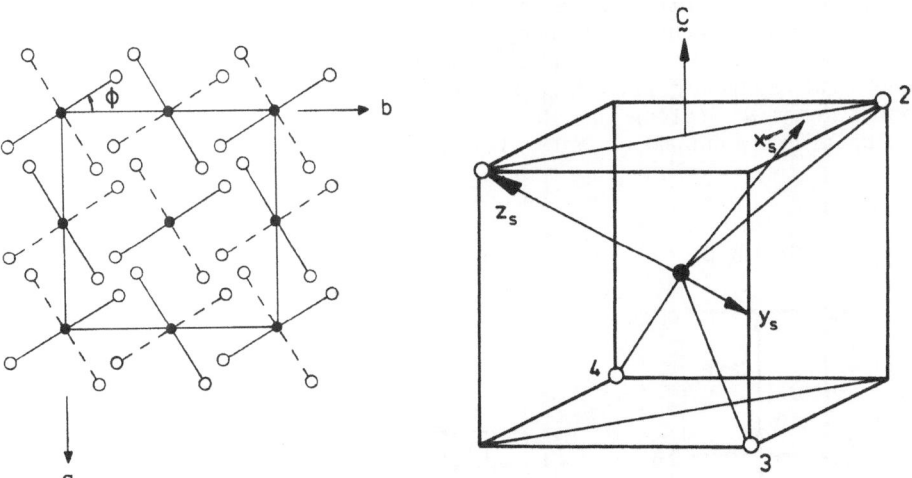

Figure 2. Structure of the $CaMoO_4$ and YVO_4 crystals.
a) projection along c of the $CaMoO_4$ (Scheelite) crystal) where ϕ =
31.3°; in the related structure of YVO_4 (zircon) ϕ = 0. The solid lines
are MoO bonds pointing upwards, the broken lines represent MoO bonds
pointing down. The Ca and Mo atoms alternate along the c direction.
b) view of the anion. The four oxygens are symmetry related via the
operation S_4 parallel to c. The directions of the spin axes of VO_4^{3-} in
YVO_4 are as indicated; the planes O_1-V-O_2 and O_3-V-O_4 here coincide with
the ac and bc crystallographic planes.

In YVO$_4$ two of the spin axes for each species lie in either of the σ_d planes (here coincident with the ac and bc crystal planes) while the third is normal to this plane. We think it particularly significant that one of the in-plane spin axes (z_s) points along the bond connecting the central metal with one of its oxygen ligands. Since there are four such ligands, four species thus arise, see fig. 2b. Apparently the D$_{2d}$ symmetry of the ion is destroyed on excitation. While one of the oxygen ligands becomes distinct fron the other three, a σ_d mirror plane seems to be preserved. A natural, but at this stage still tentative, explanation for these observations is provided by a (static) Jahn-Teller effect. Suppose the Jahn-Teller instability of the T$_d$ ion on excitation were such that it prefers to assume a pyramidal conformation of C$_{3v}$ symmetry, then the only symmetry element that remains for the excited species in the crystal is a single mirror plan (σ_d). If, in a simple picture, one further thinks of the Jahn-Teller effect in terms of a localization of the "hole" in the t$_1$ mo on the base (or apex) of the pyramid, then the orientation of one of the spin axes along a V-O bond becomes plausible as well.

In the Scheelite structure of CaMoO$_4$ the Mo site symmetry is lowered to S$_4$ since the σ_d planes of the tetrahedral ions no longer coincide with the ac and bc crystallographic planes. Nevertheless, the x-ray data show that in their ground state the MoO$_4{}^{2-}$ ions in the crystal are indistinguishable from D$_{2d}$ symmetry. The EPR results for the photoexcited triplet state of the MoO$_4{}^{2-}$ ion in this crystal have been discussed in extenso elsewhere [18] and reveal a situation very similar to that just sketched for VO$_4{}^{3-}$ in YVO$_4$. But, because of the lowering of the site symmetry from D$_{2d}$ to S$_4$, the spin axes z_s and y_s no longer point *exactly* along a Me-O bond and the normal of a σ_d plane, respectively, but do so only within an angle of about 10°.

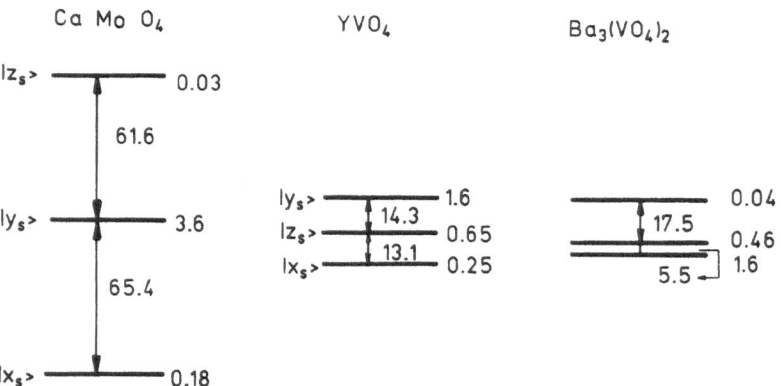

Figure 3.   Zero-field splitting (GHz) and lifetimes (ms) of the luminescent triplet states. For the two vanadates the order of the levels may have to be reversed.

The zero-field splittings and lifetimes of the individual spin sublevels that follow from the EPR experiments have been assembled in fig. 3. The splittings are quite large compared to those normally encountered in phosphorescent organic molecules. As noted, the mean lifetime of the order of 1 ms observed for the VO_4^{3-} ion [19] is in surprising agreement with an estimate of the radiative lifetime obtained from an analysis of SOC between the $^3T_1$ and $^1T_2$ multiplets in fig. 1 [12]. Also, the finding that one of the sublevel lifetimes tends to be much longer than the other two is in accord with the result of this analysis when applied to the Jahn-Teller distorted ion [18].

To test the validity of the explanation just given, we have looked for a system in which the symmetry of the crystal field acting on the tetroxo ion is altogether different. We were fortunate enough to enlist the interest of Dr. L. Brixner of the Dupont Experimental Station, who grew some $Ba_3(VO_4)_2$ crystals for us. These crystals have a hexagonal structure in which the crystal c axis coincides with one of the trigonal axes of the VO_4^{3-} tetrahedron, which now contains two types of oxygen, O(1) at 1.70 Å from the V atom and 3 O(2) at 1.71 Å, see fig. 4 [20].

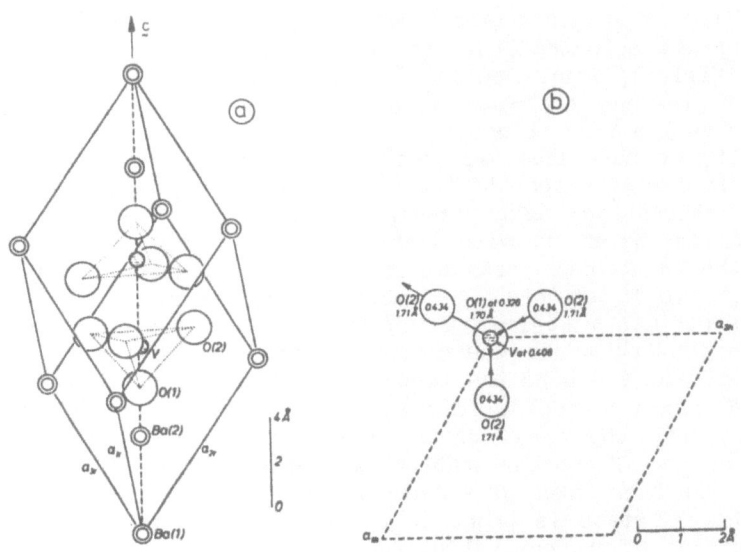

Figure 4. Structure of $Ba_3(VO_4)_2$ according to Süsse and Buerger [20].
a) view of the unit cell with the two VO_4^{3-} anions interrelated by inversion.
b) projection of VO_4^{3-} ion along c. The arrows represent coinceivable displacements on excitation.

EPR experiments on the luminescent state of $Ba_3(VO_4)_2$ crystals at low temperature are still in progress, and a complete picture cannot yet be given. A major result already stands out: three sets of EPR signals appear which have to be attributed to three deformed VO_4^{3-} species that are interrelated by a threefold rotation about the c axis. Apparently the Jahn-Teller effect now operates in such a way that one of the $O(2)$ atoms becomes inequivalent to its two partners. If the crystal field were dominant in lifting the degeneracy, one of the spin axes of each species would be parallel to c. Again, this is not the case, and the axis orientations reflect the lowering of symmetry produced by a static Jahn-Teller effect.

3. CONCLUDING REMARKS

Our EPR experiments prove Blasse's contention [10,11] that the low-temperature luminescence of vanadate and molybdate crystals originates in a metastable triplet of the tetroxo ion. As we have seen, the zero-field splitting tensor of this triplet state – which represents a "fingerprint" of the symmetry of the excited species – shows that this species has a symmetry lower than that of the ground state ion in the crystal and we attribute this to a static Jahn-Teller effect. A crucial test of this explanation would be to demonstrate a transition from a static to a dynamic Jahn-Teller effect on raising the temperature. Whereas in the static limit the zero-field splitting reflects the symmetry of the deformed ion, the dynamic limit should be characterized by a zero-field splitting corresponding to the site symmetry. We are at present setting up EPR experiments over a wide temperature range to search for such a transition.

Finally we note that our results on the spin triplet may have an important implication for the interpretation of the long wavelength part of the absorption band assigned to the $^1T_1 \leftarrow \,^1A_1$ spin-allowed transition. In one of the widely studied systems, MnO_4^- in $LiClO_4 \cdot 3H_2O$ [4-6,8], the Mn atom occupies a site with C_{3v} symmetry, much like the V atom in $Ba_3(VO_4)_2$ of fig. 4. In Stark experiments [8] and uniaxial pressure experiments [6], in which the electric field or pressure was applied perpendicular to the c axis, a field (or pressure) dependent splitting of the 0-0 band has been observed. So far it has been assumed tha the $^1T_1$ state is split into a $^1A_2$ and ^1E state by the C_{3v} crystal field, and that the first order Stark effect represents the further splitting of the ^1E orbital doublet. We think this explanation, in which no account has been taken of a Jahn-Teller instability, needs revision.

If the $^1T_1$ state is prone to a similar Jahn-Teller instability as the $^3T_1$ multiplet, the orbital degeneracy of the $^1T_1$ state of the ion in the crystal will be completely removed. Absorption in the 0-0 band will then lead to three excited species carrying equal but differently oriented dipole moments. When an electric field is applied these differently oriented dipoles will give different positive as well as negative Stark shifts, the sum of which may well result in the poorly resolved Stark splitting observed experimentally [8]. In terms of the picture of fig. 4b this process may be visualized as follows. Suppose the tetroxo ion is deformed on excitation according to the arrows in the figure, for instance by a localization of the hole in the t_1 shell on the inward moving oxygen atoms. The excited conformation of the anion will then carry a net dipole moment having a component in the (001) crystallographic plane in which the electric field is applied. For each $VO_4{}^{3-}$ ion three such conformations will exist with dipole moments interrelated by a threefold rotation about $\underline{c}$. Whether this interpretation is correct might perhaps be detected via small anisotropies in the Stark pattern when the $\underline{E}$ field is rotated in the (001) plane.

This work was supported by the Netherlands Foundation for Chemical Research (SON) with financial aid from the Netherlands Organization for the Advancement of Pure Research (ZWO).

REFERENCES

[1]   C.J. Ballhausen and A.D. Liehr, J. Mol. Spectr. 2 (1958) 342.
[2]   C.J. Ballhausen, Theoret. Chim. Acta 1 (1963) 285.
[3]   T. Ziegler, A. Rauk and E.J. Baerends, Chem. Phys. 16 (1976), 209.
[4]   J. Teltow, Z. Physik. Chem. B 40 (1938) 397; B 43 (1939) 198.
[5]   S.L. Holt and C.J. Ballhausen, Theoret. Chim. Acta 7 (1967) 313.
[6]   C.J. Ballhausen and I. Trabjerg, Mol. Phys. 24 (1972) 689.
[7]   L.W. Johnson and S.P. McGlynn, J. Chem. Phys. 55 (1971) 2985; L.W. Johnson, E. Hughes, Jr. and S.P. McGlynn, J. Chem. Phys. 55 (1971) 4476.
[8]   L.W. Johnson, J. Chem. Phys. 79 (1983) 1096.
[9]   F.A. Kröger, Some Aspects of the Luminescence of Solids, Elsevier, Amsterdam 1948.
[10] G. Blasse, Struct. Bonding 42 (1980) 1.
[11] H. Ronde and G. Blasse, J. Inorg. Nucl. Chem. 40 (1978) 215.
[12] J.H. van der Waals, Int. Rev. Phys. Chem. 5 (1986) 219.
[13] H.F. Hameka and L.J. Oosterhoff, Mol. Phys. 1 (1958) 358.
[14] J. Sidman, Chem. Rev. 58 (1958) 689.
[15] A. Freiberg and L.A. Rebane, J. Luminescence 18/19 (1979) 702.
[16] W.A.J.A. van der Poel, M. Noort, J. Herbich C.J.M. Coremans and J.H. van der Waals, Chem. Phys. Lett. 103 (1984) 245.
[17] W.A.J.A. van der Poel, J. Herbich and J.H. van der Waals, Chem. Phys. Lett. 103 (1984) 253.
[18] W. Barendswaard and J.H. van der Waals, Mol. Phys. to be published.
[19] W. Barendswaard, J. van Tol and J.H. van der Waals, Chem. Phys. Lett. 121 (1985) 361.
[20] P. Süsse and M.J. Buerger, Z. Kristallogr. 31 (1970) 161.

THE JAHN-TELLER EFFECT IN ISOLATED MOLECULES : FROM THE DYNAMICAL EFFECTS OCCURING FOR SMALL BARRIERS TO TYPICAL TUNNELING IN DEEPLY DISTORTED CONFIGURATIONS.

Claudina COSSART-MAGOS
Laboratoire de Photophysique Moléculaire du CNRS
Bâtiment 213 - Université de Paris-Sud
91405 - ORSAY CEDEX.

ABSTRACT. The commonly used model for the Jahn-Teller effect, in the case when only double degeneracies are present, is briefly described in order to introduce the dimensionless coupling parameters to which the adiabatic potential linear and quadratic barrier heights E^{JT} and Δ, respectively, are related. According to the relative magnitudes of E^{JT} and Δ and the deperturbed quantum energy $h\nu$ of the unique vibrational mode present in the model, three distinct situations are defined : E^{JT} and Δ less than $h\nu$; $E^{JT} > h\nu$, $\Delta < h\nu$; E^{JT} and Δ greater than $h\nu$. Examples of gas phase studies of systems belonging to each case are presented.

1. INTRODUCTION

Although the Jahn-Teller (JT) effect might be considered as a tunneling effect only for the lowest energy levels of a strong coupling case, all degrees of coupling will be considered here, even those for which every energy level is above the barrier top. After all, one of the most striking properties of the classic square barrier in quantum mechanics, namely its perfect transparency for widths equal to an integer number of half wavelengths of the incident wave, occurs above the barrier top [1].

In the JT effect [2], the potential barriers are created by the system searching for a more stable nuclear configuration than that of highest symmetry which corresponds to a degeneracy of the electronic wave function.

Although conceived for isolated molecules, the JT theorem has mostly been applied in condensed phases where crystal and ligand field effects play an important role. Only recently [3-8] have experiments of high resolution gas phase spectroscopy been carried out which really test the theory in a detailed way. The studies of the JT effect in H_3 [5] and Cu_3 [7] are exceptional in the sense that there exist *ab initio* calculations and gas phase experimental results simultaneously. Usually experimental data are only interpreted in terms of a simple model first defined by Moffit and

49

J. Jortner and B. Pullman (eds.), Tunneling, 49–64.
© *1986 by D. Reidel Publishing Company.*

Liehr [9] but better known through the review paper of Longuet-Higgins
[10]. In section 2, the principal characteristics of that model are
briefly described. Only double electronic and vibrational degeneracies
(Exe JT effect) will be considered. Higher orders of degeneracy are
treated in the classic book of Englman [11] and in the recent book of
Bersuker [12]. The particular case where the vibrational mode inducing
the JT effect is non-degenerate (systems with a fourfold principal axis of
symmetry) was treated in detail by Hougen [13].

The model will be applied to calculate potential energy surfaces and
vibronic levels for two different degrees of vibronic coupling strength :
i) weak coupling, when all the vibronic levels of the system are found
above the top of the central barrier ; ii) stronger coupling, with some
levels below the top, which in many contexts are the most interesting. In
section 3, two examples of the dynamical effects occuring in case i) are
selected in the emission spectra of the sym-trifluorobenzene ion $\widetilde{B}^2A_2" \rightarrow$
$\widetilde{X}^2E"$ transition [3,4]. In section 4, case ii) is first considered for an
example of hindered internal rotation of the JT deformation found in the
recent study of the s-triazine $3s^1E'$ Rydberg state [8], and then for the
case of true tunneling observed in the gas phase, in the ground state
$\widetilde{X}^2E'$ of Cu_3 [7].

2. REVIEW OF Exe JAHN-TELLER EFFECT CALCULATION MODELS

2.1. Potential energy surfaces

It is assumed that the system possesses only one doubly degenerate
vibrational mode inducing the Jahn-Teller effect, with components $Q_{x,y}$.
In complex form, and using polar coordinates ρ and φ, one defines
successively :

$$Q_\pm = Q_x \pm iQ_y = \rho e^{\pm i\varphi} \qquad\qquad (1)$$

A "crude adiabatic" – (or "diabatic") – harmonic basis is used :

$$\{ \psi_\pm (q,Q_0) \cdot \eta_{v,l}(\rho,\varphi) \cdot \prod_i \chi_{v_i}(Q_i) \} \qquad\qquad (2)$$

where $\psi_\pm$ are the complex fonctions built from the degenerate solutions
$\psi_e(q, Q_0)$ and $\psi_{e'}(q, Q_0)$ of the electronic wave equation for Q_0, a
configuration corresponding to $\rho = 0$ in the expression (1) above.

$$\psi_\pm = \psi_e (q, Q_0) \pm i \psi_{e'} (q, Q_0) \qquad\qquad (3)$$

The χ_{v_i} are harmonic oscillator wavefunctions, one for each normal mode
Q_i, and $\eta_{v,l}$ is the isotropic two-dimension harmonic oscillator
wavefunction corresponding to the mode inducing the JT effect.
The vibrational angular momentum quantum number l is related to v : l = v,
v-2, ... 1 or 0.

In the hamiltonian operator :

$$H = H_e + T_n = T_e + V + T_n \tag{4}$$

V, the coulombic interaction term, is developed in a Taylor series in the neighborhood of Q_0, up to second order in Q_+ and Q_-. The form of the non-vanishing terms in the matrix elements of H_e in the basis formed by the functions ψ_+ and ψ_- can be deduced from the symmetry properties of these functions and of the coordinate components $Q_\pm$. According to the detailed analysis of Child and Longuet-Higgins [14], one obtains :

$$\begin{pmatrix} E_0 + \dfrac{1}{2} f \rho^2 & k \rho e^{-i\varphi} + g \rho^2 e^{2i\varphi} \\ k \rho e^{i\varphi} + g \rho^2 e^{-2i\varphi} & E_0 + \dfrac{1}{2} f \rho^2 \end{pmatrix} \tag{5}$$

where E_0 is the electronic energy for $Q = Q_0$ ($\rho = 0$) and

$$f = 2 < \psi_\pm | (\frac{\partial^2 V}{\partial Q_+ \, Q_-})_0 | \psi_\pm > \tag{6}$$

$$k = < \psi_+ | (\frac{\partial V}{\partial Q_-})_0 | \psi_- > = < \psi_- | (\frac{\partial V}{\partial Q_+})_0 | \psi_+ > \tag{7}$$

$$g = \frac{1}{2} < \psi_+ | (\frac{\partial^2 V}{\partial Q_+^2})_0 | \psi_- > = \frac{1}{2} < \psi_- | (\frac{\partial^2 V}{\partial Q_-^2})_0 | \psi_+ > \tag{8}$$

The potential surfaces are the eigenvalues of the matrix (5) above :

$$\left. \begin{array}{l} U_1 (\rho, \varphi) \\ U_2 (\rho, \varphi) \end{array} \right\} = E_0 + \frac{1}{2} f\rho^2 \pm \sqrt{k^2\rho^2 + g^2\rho^4 + 2kg\rho^3 \cos 3\varphi} \tag{9}$$

To second order in ρ, one obtains :

$$\left. \begin{array}{l} U_1 (\rho, \varphi) \\ U_2 (\rho, \varphi) \end{array} \right\} = E_0 + \frac{1}{2} f\rho^2 \pm (k\rho + g\rho^2 \cos 3\varphi) \tag{10}$$

or :

$$
\left.\begin{aligned}
U_1\,(\rho,\ \Psi) \\
U_2\,(\rho,\ \Psi)
\end{aligned}\right\} = E_0 + \frac{1}{2}\ (f \pm 2g\cos 3\Psi)\ (\rho \pm \frac{k}{f \pm 2g\cos 3\Psi})^2
$$

$$
- \frac{k^2}{2(f \pm 2g\cos 3\ \Psi)} \qquad (11)
$$

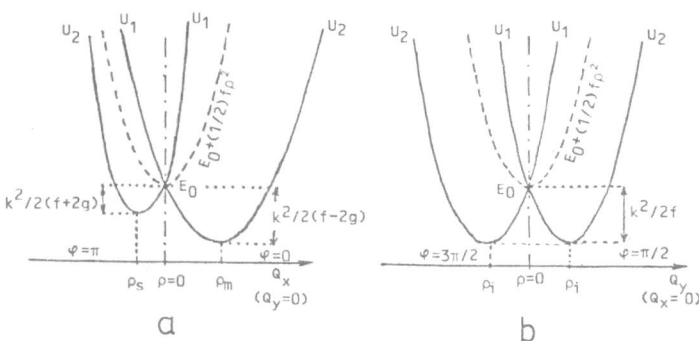

Figure 1. Linear plus quadratic JT effect : adiabatic potential surfaces calculated to second order in ρ – intersection with : a) the Q_x plane and b) the Q_y plane.

Expression (11) is useful for analyzing the form of the potential surfaces U. Intersection of U_1 or U_2 by any vertical half-plane (defined by a particular value of Ψ) is part of a parabola which, relative to the zero-order curve $E_0 + f\rho^2/2$, corresponds to a modified force constant $f \pm 2g\cos 3\Psi$ and displaced minimum positions. In Fig. 1a. are shown the intersections of the two surfaces with the plane containing the Q_x axis. k and g are supposed positive. There are three equivalent minima at : $\Psi = 0$ (shown in Fig. 1a.), $2\pi/3$ and $4\pi/3$, with ρ equal to :

$$
\rho_m = \frac{k}{f - 2g} \qquad (12)
$$

and three saddle-points for $\Psi = \pi$ (shown in Fig. 1a.), $5\pi/3$ and $7\pi/3$ ($\pi/3$), with ρ equal to :

$$
\rho_s = \frac{k}{f + 2g} \qquad (13)
$$

In the planes bissecting the angles formed by the above planes and corresponding to $\Psi = \pi/6$, $3\pi/6$ ($\pi/2$), $5\pi/6$, $7\pi/6$, $9\pi/6$ ($3\pi/2$) and $11\pi/6$, U_1 and U_2 are as shown in Fig. 1b for the case of the $Q_x = 0$ plane ($\Psi=\pi/2$ and $3\pi/2$). One has then two identical parabolas corresponding to equal force constants f and having minima at ρ equal to :

$$\rho_i = k/f \qquad (14)$$

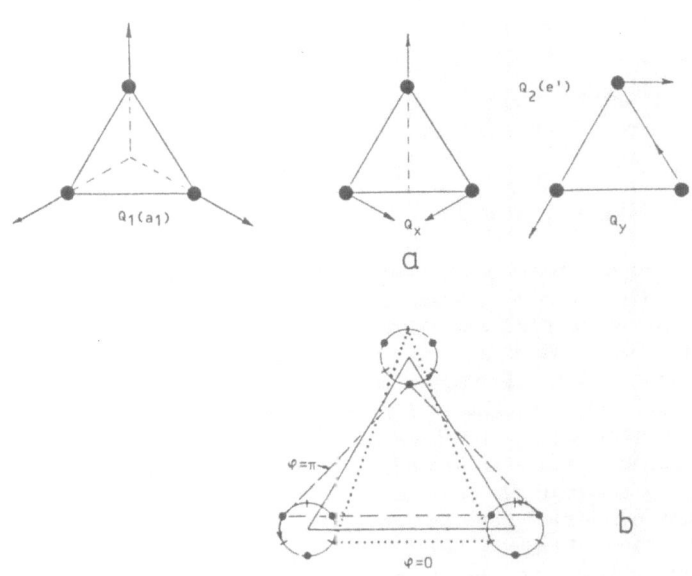

Figure 2. a) Normal modes of vibration of the X_3 molecule in the $\mathcal{D}_{3h}$ point group ; b) Lower symmetry geometries corresponding to the minimum (φ=0) and to the saddle point (φ=π) of the lower energy surface in Fig. 1a., with $\rho_m = \rho_s$ here, for ease of illustration.

In Fig. 2a. the forms of the normal modes of vibration of an $X_3(\mathcal{D}_{3h})$ molecule are recalled, and in Fig. 2b. the geometries of the molecule at the minimum and saddle-point of Fig. 1a. are represented, respectively, an acute and an obtuse isosceles triangle. In the figure, $\rho_m = \rho_s$ for ease of illustration.

If g is supposed zero, the JT effect is only <u>linear</u>, and U_1 and U_2 have perfect cylindrical symmetry ; their intersection by any vertical plane is as represented in Fig. 1b. U_2 forms the so-called "mexican hat". The dimensionless linear coupling parameter currently used is D defined as :

$$D = \frac{k^2}{2f\,h\nu} \qquad (15)$$

In cm^{-1} $h\nu$, often called ω, is the zero-order frequency of the JT mode corresponding to the force constant <u>f</u> defined above. The height of the uniform potential barrier of U_2 in the linear case or its average value for linear plus quadratic coupling, is called the <u>Jahn-Teller energy</u> E^{JT}, and is related to D and ω :

$$E^{JT} = D\omega \qquad (16)$$

As dimensionless quadratic coupling parameter, one may use [15]

$$q = g/2f \qquad (17)$$

In the case of small quadratic coupling [15], the height Δ of the potential barriers in the valley of the lower surface U_2 has a simple relation with q and E^{JT} :

$$\Delta = 2gk^2 \ / \ (f^2 - 4g^2) \simeq 8q \ E^{JT} \qquad (18)$$

A parameter $K = 2q$ has also been used [16], as well as 4q which is denoted g by Whetten et al [8] and b by Thompson et al [17]. The parameter β used by Englman [11], Bersuker [12] and O'Brien [18] is not dimensionless. It is defined as the coefficient of $\cos 3\varphi$ in the potential energy surfaces, which may contain terms of higher order in ρ than those considered in equation (10) and (11). In the study of Cu_3 which will be presented with some detail in section 4, Smalley and co-workers [7] include in the potential surfaces a term in $\rho\cos\varphi$, which does not appear in equation (10). One more parameter is thus introduced in the model, but later in their paper, the number of parameters is reduced by one by making $\rho_m = \rho_s$. Thompson et al. [17] obtain a better fit of the data of ref. 7 using an expansion of the adiabatic energy equivalent to (10).

2.2 The vibronic levels

To obtain the vibronic matrix of the hamiltonian (4) developed in a Taylor series as indicated above, and using the basis functions defined in (2), one has to use the well-known expressions of the vibrational integrals of $Q_\pm$ [9] :

$$< \eta^{\mp}_{v,l} \ |Q_\pm| \ \eta^{\pm}_{v+1, l\mp1} > = (h\nu/2f)^{1/2} \ (v \mp l + 2)^{1/2}$$

$$< \eta^{\mp}_{v,l} \ |Q_\pm| \ \eta^{\pm}_{v-1, l\mp1} > = (h\nu/2f)^{1/2} \ (v \pm l)^{1/2} \qquad (19)$$

where the $\pm$ signs on each η function are only used to recall to which electronic wave function it is associated. One also needs the matrix elements of $Q_\pm^2$, which may be obtained by matrix multiplication from those of $Q_\pm$.

$$< \eta^{\pm}_{v,l} \ |Q_\pm^2| \ \eta^{\mp}_{v+2, \ l\mp2} > = (h\nu/2f) \ [(v \mp l +2) \ (v \mp l+4)]^{1/2}$$

$$< \eta^{\pm}_{v,l} \ |Q_\pm^2| \ \eta^{\mp}_{v, l\mp2} > = (h\nu/f) \ [(v \mp l+2) \ (v +l)]^{1/2}$$

$$< \eta^{\pm}_{v,1} |Q^2_{\pm}| \eta^{\mp}_{v-2,\, 1\mp2} > = (h\nu/2f)\, [(v \pm 1)(v \pm 1-2)]^{1/2} \qquad (20)$$

From (19), it is concluded that a quantum number j defined as $j = 1 \pm 1/2$ for $\psi_{\pm}$ electronic components respectively, is a good quantum number for the pure linear JT effect because the zero-order levels "mixed" have the same value of j. But quadratic coupling, through the matrix elements defined in (20), "mixes" all levels having the same value of 2j mod 3.

2j mod 3 = 1 means $j = +1/2, -5/2, +7/2, \ldots$, whereas for 2j mod 3 = 2, $j = -1/2, +5/2, -7/2, \ldots$ – two identical vibronic matrices are obtained corresponding to intrinsically degenerate E levels. Finally, 2j mod 3 = 0 for $j = \pm3/2, \pm9/2, \ldots$ – the vibronic matrix obtained has eigenvectors which are linear combinations of j values of opposite sign with well-defined A_1 or A_2 symmetry ; accidentally degenerate for linear coupling, these levels are split by the quadratic JT effect.

In Fig. 3, the potential surfaces and vibronic levels of a Exe system are represented in zero-order and for two cases of linear JT effect : i) $E^{JT} <$ hν and ii) $E^{JT} >$ hν, with q = 0 in both cases. The vibronic levels were

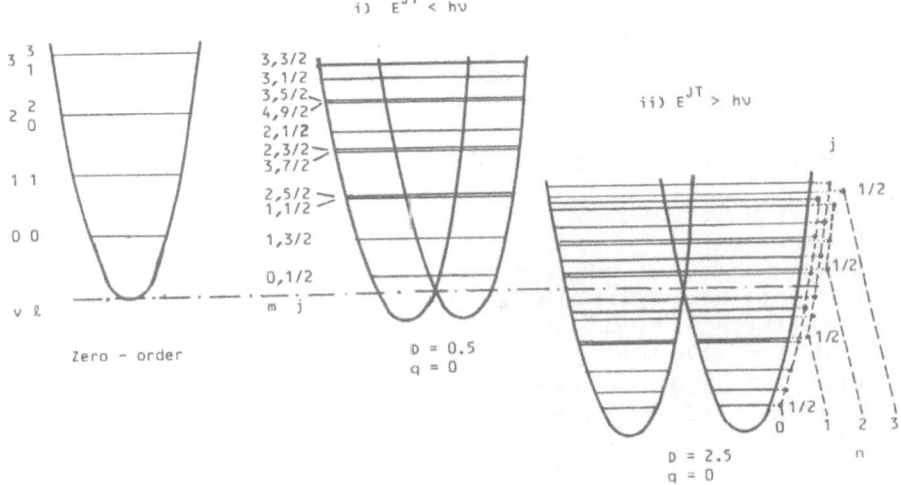

Figure 3. Linear Exe JT effect : vibronic levels and cross-sections of the adiabatic potential surfaces for zero-order and for two different degrees of coupling. Parameters E^{JT}, hν, D and q, and numbers v, 1, j, m and n are defined in section 2.

obtained by diagonalization of vibronic matrices established for each j value according to the formulae (19) with a maximum value of v, equal to 50. In both cases j, as demonstrated above, is a good quantum number, and the only one ; however, in the figure, for case i) j is preceded by an

integer m giving the order of the level within the particular j stack and
beginning at the minimum value of v present in the corresponding sub-set
of basis functions. For case ii) a different integer n is given which
corresponds to the radial oscillator-free rotor model to be described in
section 4. Examples of cases i) and ii) will be given in sections 3 and 4,
respectively.

3. $E^{JT} <$ hν : THE CASE OF THE 1,3,5-$C_6F_3H_3^+$ $\tilde{X}^2E''$ STATE

Analysis of the gas phase discharge emission and laser induced
fluorescence excitation spectra of the $\tilde{B}^2A_2''$ - $\tilde{X}^2E''$ transition of
1,3,5-$C_6F_3H_3^+$ and D_3^+ in the visible has shown [16,19] that several e'
modes induce JT effect in the E ground state. Mode 6, corresponding to CCC
deformation has the highest JT D constant : $D_6 = 0.8$ with $\omega_6 = 475$ cm^{-1}
[16]. Another ring mode ν_8, corresponding to CC stretch, is also active :
$D_8 = 0.38$; $\omega_8 = 1620$ cm^{-1} [16]. It should be noted that in the multimode
analysis of ref. 16 only one observed level directly related to excitation
of ν_8 could be considered. Indeed, the D_8 constant was mainly determined
by the resulting displacement of the first j = 3/2 level of mode 6 [16,19].
The intermode effects on the first j = 1/2 and j = 5/2 levels are found to
be equivalent to a lowering of the D_6 constant. In fact, these levels
closely approach those obtained in a single-mode calculation with D = 0.5
(given in Fig. 3-i.) at least sufficiently for our present purposes, viz.
to show the more salient manifestations of the linear and quadratic JT
effect in the sym-trifluorobenzene ion spectra.

3.1. Irregularity in vibrational progression intervals due to linear JT effect

In Fig. 4 are shown low-resolution discharge emission spectra [20] of two
fluorobenzene ions : a) sym-$C_6F_3H_3^+$ $\tilde{B}^2A''$ → $\tilde{X}^2E''$ transition and b)
1,3,4,5-$C_6F_4H_2^+$ $\tilde{B}^2B_2$ → $\tilde{X}^2B_2$ and $\tilde{B}^2B_2$ → $\tilde{A}^2A_2$ transitions. On one hand,
substitution effects split the degenerate $^2E''$ ground state of the first
ion in two states $\tilde{X}$ $^2B_2$ and $\tilde{A}^2A_2$ in the ion of lower symmetry ($\mathscr{C}_{2v}$ with
x axis in the molecular plane). On the other hand, only the more
symmetrical ion may present JT effect. This is apparent in the spectra, if
one compares the regular frequency intervals (427, 429, 431 cm^{-1} [20]) in
the 6a (a_1) mode progression of the tetrafluorobenzene ion, with those of
mode 6 (e') j = 1/2 progression in the sym-trifluorobenzene ion, where a
decrease of 130 cm^{-1} is observed from the first to the second frequency
spacing which respectively equal to 550 and 420 cm^{-1} [20].

In the case of $E^{JT} <$ hν (D < 1) the band progression $N_{m,1/2}^{0,0}$ for a
single mode N inducing linear JT effect (v', l' values given as an
exponent and m, j in index) is always characterized by frequency
differences $O_0^0 - N_{1,1/2}^{0,0} > N_{1,1/2}^{0,0} - N_{2,1/2}^{0,0}$. The ratio $R_{1/2} = (O_0^0 - N_{2,1/2}^{0,0})$ /
$(O_0^0 - N_{1,1/2}^{0,0})$ was studied as a function of D in ref. 21. $R_{1/2} < 2.0$ for D <
1.0, with a minimum value 1.72 for D = 0.20 ; $R_{1/2} = 2.0$ for D = 1.0 and
$R_{1/2} > 2.0$ for D > 1.0, with a maximum value 2.06 for D = 1.78 [21].

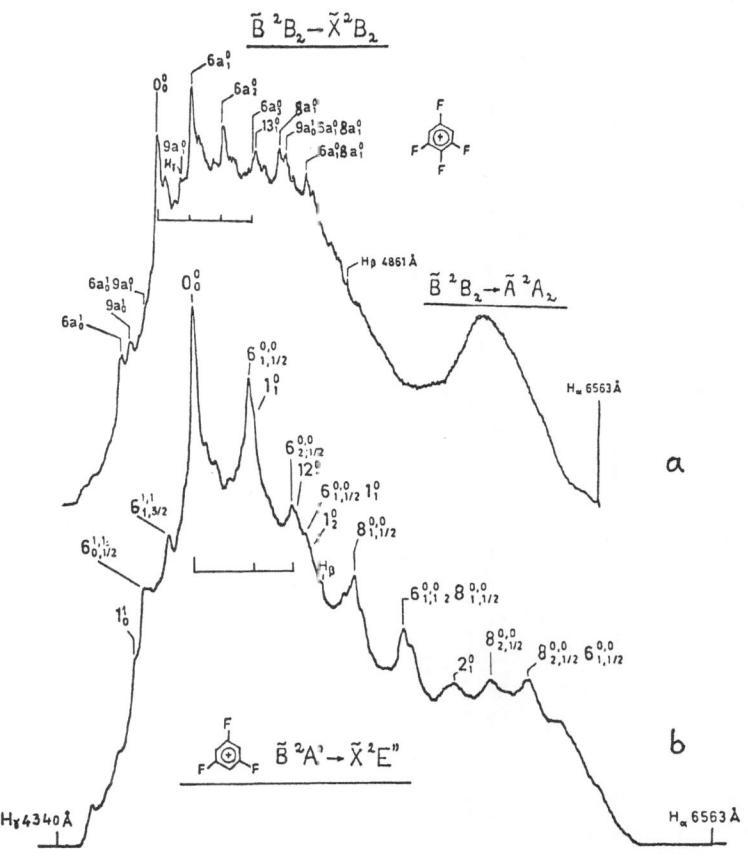

Figure 4. Gas-phase discharge emission spectra of two fluorobenzene ions:
a) 1,3,4,5-$C_6F_4H_2^+$; **b)** 1,3,5-$C_6F_3H_3^+$ (after C. Cossart-Magos *et al.*
[20]). In **b** an <u>irregular ground state progression</u> is observed as a
consequence of <u>linear JT effect</u>.

3.2. Anti-crossing of levels 1,1/2 and 2,5/2 due to quadratic JT effect : comparison of gas phase and Ne matrix emission spectra

The second $j = 1/2$ level $(1, 1/2)$ and the first $j = 5/2$ (or 2, 5/2 level, as
$v = 2$ is the lowest v value contributing to $j = 5/2$ levels) are quasi-
degenerate in Fig. 3i. At zero-quadratic coupling, these two levels cross
freely at $D = 0.59$ as shown by the straight lines of Fig. 5. Only the
$N_{1,1/2}^{0,0}$ transition is allowed with 46 % of the 0_0^0 transition intensity
(single-mode calculation). When quadratic coupling is switched on, the
two E' vibronic levels repel each other, and $N_{1,1/2}^{0,0}$ shares its intensity
with $N_{2,5/2}^{0,0}$. The level positions calculated for $q = 0.006$ (slightly
greater than the trifluorobenzene ion q_6 value, 0.004 [16,19] and so

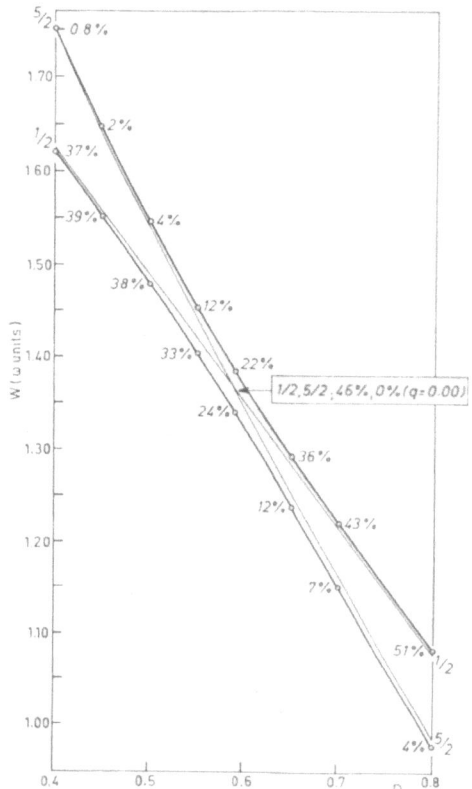

Figure 5. Crossing of the second energy eigenvalue of $j = 1/2$ and the first of $j = 5/2$ in the linear coupling approximation ($q = 0$) and avoided crossing when the quadratic coupling parameter $q \neq 0$. In the figure, $q = 0.006$. For this value of q, percentage intensities of $N_{1,1/2}^{0,0}$ and $N_{2,5/2}^{0,0}$ transitions relative to the 0_0^0 transition intensity are given, for some D values, and, for $q = 0$, at the crossing point (after Klapstein *et al* [3]).

chosen for ease of illustration) are given by the two non-crossing curves of Fig. 5. Percentage intensities (relative to the 0_0^0 transition intensity) are given for the same q value and for some selected D values. Such an evolution of the relative intensity of two bands, $6_{1,1/2}^{0,0}$ and $6_{2,5/2}^{0,0}$, as a function of the frequency difference between them, is actually observed in going from the trifluorobenzene ion gas phase spectra to the spectra in solid Ne matrix. Fig. 6 shows the relevant sections of the gas phase emission spectra of $1,3,5-C_6F_3H_3^+$ and D_3^+, supercooled in an Ar jet expansion [3], and of the same ions in a solid Ne matrix [22]. It is clear that we are at D_6 values greater than that of the crossing point of Fig. 5 and that D_6 <u>decreases in the matrix</u> where the two bands $6_{1,1/2}^{0,0}$ and $6_{2,5/2}^{0,0}$ are for both isotopic species, nearer to each other than in the gas phase, and where $6_{2,5/2}^{0,0}$ has gained more intensity. It can be deduced [3] that this outstanding effect corresponds to a change in D_6 not greater than ≈ 0.02. Other data indicate that q_6 may also decrease in the matrix but the effects are of the order of the uncertainty of the measured frequency differences [3].

4. $E^{JT} > h\nu$: FROM FREE AND HINDERED ROTATION OF THE DEFORMATION TO TUNNELING BETWEEN EQUIVALENT DISTORTED CONFIGURATIONS

Low linear and quadratic JT barriers cause the most irregular level displacements, splittings and avoided crossings, as was shown in the preceding section. On the contrary, when $E^{JT} > h\nu$, if the quadratic

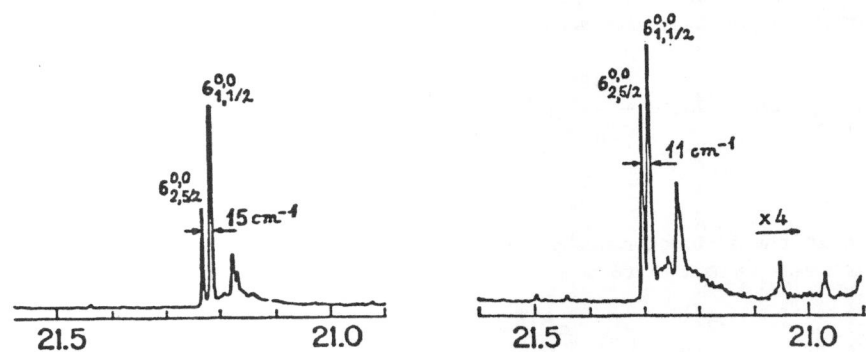

a. sym-TRIFLUOROBENZENE⁺ EMISSION IN NE MATRIX

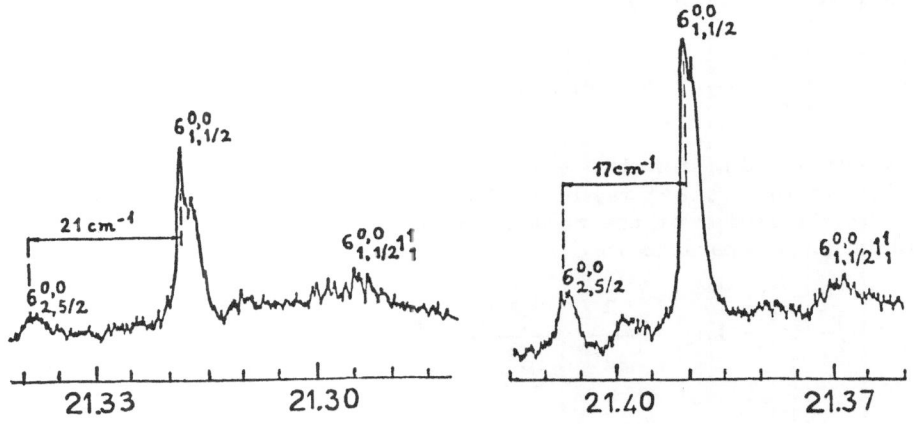

b. sym-TRIFLUOROBENZENE⁺ EMISSION IN AR JET

Figure 6. Quadratic JT effect in the $\tilde{B}^2A_2" \rightarrow \tilde{X}^2E"$ emission spectra of sym-trifluorobenzene–h_3 and –d_3 cations – Avoided crossing of $6_{1,1/2}$ and $6_{2,5/2}$ levels and intensity sharing between the $6^{0,0}_{1,1/2}$ and $6^{0,0}_{2,5/2}$ bands : a) in a Ne matrix (after Bondybey et al. [22]) and b) in an Ar supersonic jet (after Klapstein et al. [3]). The frequency unit for all spectra is $10^3 cm^{-1}$ but different scales were used in a and b.

coupling remains negligibly small, the pattern of low-lying levels shows a new regularity. Indeed, as first noted by Longuet-Higgins et al. [23], as D increases the energy level pattern approaches that of a harmonic radial oscillator plus a free rotor whose characteristics are determined as follows.

For zero quadratic coupling it is possible to make an approximate separation of the variables ρ and Ψ in the nuclear equation of motion, for motion on the lower sheet of the adiabatic potential U_2 (given by expression (10) and decoupled from U_1). With the nuclear kinetic energy expressed in the cylindrical coordinates ρ and Ψ, that equation becomes :

$$[E_0 + \frac{1}{2} f\rho^2 - k\rho - \frac{\hbar^2}{2\mu} (\frac{\partial^2}{\partial \rho^2} + \frac{1}{\rho} \frac{\partial}{\partial \rho} + \frac{1}{\rho^2} \frac{\partial^2}{\partial \Psi^2})] \chi (\rho,\Psi) =$$

$$E \chi (\rho,\Psi) \qquad (21)$$

where μ is the reduced mass of the JT mode. A possible solution χ of this equation may take the form :

$$\chi_{nj} (\rho,\Psi) = \rho^{-1/2} \tilde{\chi}_n (\rho) \frac{e^{ij\Psi}}{\sqrt{2\pi}} \qquad (22)$$

The factors $\rho^{-1/2}$ and $1/\sqrt{2\pi}$ are introduced in order to replace the normalization relation :

$$\int_0^{2\pi} \int_0^\infty \chi^*(\rho,\Psi) \chi(\rho,\Psi) \rho \, d\rho \, d\Psi = 1 \quad by \quad \int_0^\infty \tilde{\chi}^*(\rho) \tilde{\chi}(\rho) \, d\rho = 1$$

In equation (21), for the energy region below the barrier top, the coefficient of $\partial^2/\partial \Psi^2$ may be held constant by fixing $\rho = \rho_i = k/f$, its value in the bottom of the trough (cf. expressions (12) and (13)). Then equation (21) separates into :

$$\left[E_0 + \frac{1}{2} f\rho^2 - k\rho - \frac{\hbar^2}{2\mu} (\frac{d^2}{d\rho^2} + \frac{1}{4\rho^2}) \right] \tilde{\chi}_n (\rho) = E_n \tilde{\chi}_n (\rho) \quad (23)$$

$$- \frac{\hbar^2}{2\mu\rho_i^2} \frac{d^2}{d\Psi^2} e^{ij\Psi} = E_j e^{ij\Psi} \qquad (24)$$

with $E = E_n + E_j$

Neglecting the centrifugal energy $-\hbar^2/8\mu\rho^2$, equation (23) becomes that of a <u>displaced one-dimension harmonic oscillator in ρ (radial oscillator)</u> with energy levels E_n :

$$E_n = (n + \frac{1}{2}) \, h\nu - \frac{k^2}{2f} \qquad (25)$$

and equation (24) gives :

$$E_j = \frac{\hbar^2 \, j^2}{2\mu \, \rho_i^2} \qquad (26)$$

which is the energy of a free rotor of rotational constant

$$\frac{\hbar^2}{2\mu\rho_i^2} = \frac{\hbar^2 \, f^2}{2\mu \, k^2} = \frac{1}{4D} \, h\nu$$

In $h\nu$ units :

$$E = n + \frac{1}{2} - D + \frac{j^2}{4D} \qquad (27)$$

Inspection of the lower energy levels in Fig. 3ii) shows that this simple picture is reasonably valid. However, when D is large, it becomes unrealistic to continue ignoring quadratic coupling ; the higher the quadratic barrier height Δ the more the motion along the warped trough differs from free rotation. As indicated above, a major effect of quadratic coupling is the splitting into A_1 and A_2 levels of each j level corresponding to 2j mod 3 = 0. For the first j = 3/2 level, the A_1 and A_2 level separation is calculated by first-order perturbation theory as $8qh\nu \simeq \Delta/D$ (cf. eq. (18)). Two cases have to be considered.

4.1. $\Delta \ll h\nu$: the system goes through hindered internal rotation of the JT distortion.

An example of this situation is found in the 3s $^1E'$ Rydberg state of s-triazine recently studied [8] in the gas phase by two-photon absorption with ionization detection (two-photon resonant three-photon ionization). The spectrum is interpreted on the basis of a single active mode, the ring-deformation mode ν_6 (e') inducing a JT effect characterized by parameter values D = 2.29 and q = 0.0115 [8]. $\omega = 661$ cm^{-1} [8], so that the barrier due to the linear effect is $E^{JT} = D\omega = 1514$ cm^{-1} and the pseudo-rotation barrier $\Delta \simeq 8qE^{JT} = 139$ cm^{-1}. The zero point level is well above the saddle-points, but several levels, including some of the observed ones, are below the top of the central barrier.

4.2. $\Delta \gg h\nu$: the system undergoes true tunneling from one distorted configuration of minimal energy to another

This phenomenum is analogous to the inversion of a non-rigid molecule such as ammonia. The tunneling splitting or tunneling energy, δ [12] or 3Γ

[11], is the separation between the lowest vibronic level, of E symmetry, as seen above, and the next level, A_2 or A_1, according to the sign of the quadratic coupling parameter, positive or negative respectively. By diagonalization of the 3 x 3 matrix corresponding to first-order perturbation of the three lowest levels, one for each equivalent distorted configuration of minimal energy, one obtains [12] :

$$\delta = \frac{3U}{(1 + 2S)(1 - S)} \tag{28}$$

where $U = H_{12}$ is the interaction energy matrix element between the zero-order functions of any two minima and S their overlap. Expressions for δ are given in Bersuker's book [12] for several limiting cases.

Experimental studies of the JT effect in the gas phase are in general not very abundant and even less so if one is looking for systems presenting large linear and quadratic JT barriers. Octahedral complexes of Cu(II), of transition metal ions or rare earth ions are good examples of such systems but have been studied exclusively in the solid state (cf. data collected at the end of Englman's book [11]). In the solid state, random strain intervenes and tunneling may be prevented if external distorting forces are strong enough to make the minima in the adiabatic potential nonequivalent [12].

Ab initio calculations of the electronic properties of the alkali trimers Li_3 [24,25], Na_3 [26,27] and K_3 [25] give high values for E^{JT} and Δ, both greater than $h\nu$ except Δ in Li_3 [17]. In this context, Li_3 is expected to behave as a symmetric rotor but not Na_3 or K_3 which are stabilized as obtuse isosceles triangles [25]. The gas phase absorption spectrum of Na_3 produced in a supersonic molecular beam was obtained by two-photon ionization together with mass-spectrometric detection [27]. This spectrum has not been analyzed and the ESR spectrum of the same species in an Ar matrix [28] was found to be consistent with both linear and obtuse isosceles triangle geometries.

The study of the copper trimer is more advanced. Recent *ab initio* calculations [29] predict D_{3h} symmetry with the $^2E'$ ground state having a linear JT barrier of the order of 1000 cm^{-1}, equivalent to a D value of 11.99 according to the analysis of Thompson *et al.* [17]. The gas phase absorption spectrum of Cu_3 produced in a supersonic molecular beam was recorded by the usual resonant two-photon (two-color) ionization and a new depletion technique [7]. The latter is a very elegant way to study predissociated states. An intense tunable laser is used to saturate the transition of molecules into the excited state. After sufficient time (100 ns) ground state (instead of excited state) molecules are ionized by a second laser : the excited state predissociated levels appear with a negative sign in the ionization signal [7]. The observed spectra, lying in the 5430 - 5225 Å region, were assigned to a $^2E'' \leftarrow {}^2E'$ transition of Cu_3 with a very weak JT effect in the excited state [7,17] and a much stronger one in the ground state. Only one hot band was assigned in the spectrum involving a ground state JT level (other than the E' zero-vibration level). From its frequency a tunneling splitting of 12 ± 7 cm^{-1} was obtained which may correspond to the slightly bent JT distorted state of $^2B_1$ symmetry deduced from the E.S.R. spectrum [30] of the same species trapped in adamantane at 77 K.

V. CONCLUSION

It has been shown in the present paper that it is now possible to find examples of gas phase studies of JT systems going continuously from the low linear and quadratic potential barriers to the deep deformations where tunneling is active. However the best examples of the latter case are still found in solid state studies of systems such as coordination complexes, impurity centers in crystals, etc.. Our conclusion is that the study of the strong JT effect, which is a type of tunneling, is practically unexplored in the gas phase.

REFERENCES

1. L.I. Schiff, "Quantum Mechanics" (Mc Graw Hill, 2nd ed., 1955), p. 95.
2. H.A. Jahn and E. Teller, Proc. R. Soc. London, Ser. A **161**, 220 (1937).
3. D. Klapstein, S. Leutwyler, J.P. Maier, C. Cossart-Magos, D. Cossart and S. Leach, Mol. Phys. **51**, 413 (1984) and references therein.
4. T.A. Miller, and V.E. Bondybey, Appl. Spectrosc. Rev. **18**, 105 (1982) ; T.A. Miller, Ann. Rev. Phys. Chem. **33**, 257 (1982).
5. G. Herzberg, H. Lew, J.J. Sloan and J.K.G. Watson, Can. J. Phys. **59**, 428 (1981).
6. S.R. Long, J.T. Meek and J.P. Reilly, J. Chem. Phys. **79**, 3206 (1983).
7. M.D. Morse, J.B. Hopkins, P.R.R. Langridge-Smith and R.E. Smalley, J. Chem. Phys. **79**, 5316 (1983).
8. R.L. Whetten, K.S. Haber and E.R. Grant, J. Chem. Phys. **84**, 1270 (1986).
9. W. Moffitt and A.D. Liehr, Phys. Rev. **106**, 1195 (1956).
10. H.C. Longuet-Higgins, "Advances in Spectroscopy", vol. II, ed. H.W. Thompson (Interscience, 1961) p. 429.
11. R. Englman, "The Jahn-Teller Effect in Molecules and Crystals" (Wiley-Interscience, 1972).
12. I.B. Bersuker, "The Jahn-Teller Effect and Vibronic Interactions in Modern Chemistry" (Plenum Press, 1984).
13. J.T. Hougen, J. Mol. Spectr. **13**, 149 (1964).
14. M.S. Child and H.C. Longuet-Higgins, Phil. Trans. Roy. Soc. London, Ser. A **254**, 259 (1961).
15. C. Cossart-Magos and S. Leach, Chem. Phys. **48**, 349 (1980).
16. T. Sears, T.A. Miller and V.E. Bondybey, J. Chem. Phys. **72**, 6070 (1980).
17. T.C. Thompson, D.G. Thruhlar and C.A. Mead, J. Chem. Phys. **82**, 2392 (1985).
18. M.C.M. O'Brien, Proc. Roy. Soc. London, Ser. A **281**, 323 (1964).
19. C. Cossart-Magos, D. Cossart, S. Leach, J.P. Maier and L. Misev, J. Chem. Phys. **78**, 3673 (1983).

20. C. Cossart-Magos, D. Cossart and S. Leach, Mol. Phys. **37**, 793 (1979).
21. S. Leach, D. Cossart and C. Cossart-Magos, Chem. Phys. **41**, 345 (1979).
22. V.E. Bondybey, T.A. Miller and J.H. English, J. Chem. Phys. **71**, 1088 (1979).
23. H.C. Longuet-Higgins, U. Opik, M.H.L. Pryce and R.A. Sack, Proc. Roy. Soc. London, Ser. A **244**, 1 (1958).
24. W.H. Gerber and E. Schumacher, J. Chem. Phys. **69**, 1692 (1978) ; G.H. Gerber, Ph. D. Thesis, University of Bern, 1980.
25. J.L. Martins, R. Car and J. Buttet, J. Chem. Phys. **78**, 5646 (1983).
26. W. Schulze, H.U. Becker and H. Abe, Chem. Phys. **35**, 177 (1978).
27. A. Herrmann, M. Hofmann, S. Leutwyler, E. Schumacher and L. Wöste, Chem. Phys. Lett. **62**, 216 (1979).
28. D.M. Lindsay, D.R. Herschbach and A.L. Kwiram, Mol. Phys. **32**, 1199 (1976).
29. E. Miyoshi, H. Tatewaki and T. Nakamura, J. Chem. Phys. **78**, 815 (1983).
30. J.A. Howard, K.F. Preston, R. Sutcliffe and B. Mile, J. Phys. Chem. **87**, 536 (1983).

SIMPLIFIED PATH INTEGRAL TREATMENT OF TUNNELLING

A. Ranfagni, D. Mugnai
Istituto di Ricerca sulle Onde Elettromagnetiche del C.N.R.
Firenze, 50127
Italy

and

R. Englman
Soreq Nuclear Research Center
Yavne 70600
Israel

ABSTRACT. Utilizing a simple procedure previously developed by A. Ranfagni and D. Mugnai for the evaluation of quantum fluctuations in path integrations, we derive the decay rate for a metastable state in a strongly asymmetric potential and the tunnelling splitting in a (nearly) symmetric potential. The two limits are used to interpolate to intermediate cases of moderately asymmetric double-well potentials (without dissipation). Miller's nearly periodic orbit description transformed to the imaginary time domain suggests a unification scheme for the various situations.

1. INTRODUCTION

Path-integral methods are suitable for describing tunnelling processes, which are known to play a central role in many physical systems. A crucial point in this kind of analysis is the evaluation of quantum fluctuations which enter the expression of the probability-amplitude (kernel) in going between initial and final points, or states. [1-3] This represents a rather complicated problem even in simple tunnelling processes: the complexity of the usual procedure (instanton approach) increases in passing from symmetrical to asymmetrical potentials. [4-8] In spite of this, several works have been devoted to the study of decay-rates in asymmetrical cases, both at zero [7] and finite temperatures, also including dissipative terms. [9-14] It seems however, that a connection between symmetrical and asymmetrical cases, or in other words between resonant and non-resonant problems is still lacking, even for isolated systems at zero temperature.

In this work we attempt to establish a connection on the basis of a simple procedure for evaluating quantum fluctuations, which has been

J. Jortner and B. Pullman (eds.), Tunneling, 65–80.

successfully tested in the symmetric case. [15] After a brief description of the procedure we look at the opposite extreme: the strongly asymmetric case and obtain an expression for the kernel as well as the decay-rate for a metastable level. Then we consider double-well potentials of moderate asymmetry and we compare the results with symmetrical and asymmetrical cases.

2. KERNEL

2.1. General considerations

The wave function at the coordinates of arrival (x_b, t_b) is expressed in terms of the kernel K as

$$\Psi(x_b, t_b) = \int K(b,a)\ \Psi(x_a, t_a)\ dx_a$$

The kernel can be put in the form

$$K(b,a) = \frac{1}{\sqrt{2\pi h}}\ \left| \left(\frac{\delta^2 S}{\delta x^2}\right)_{\bar{x}(t)} \right|^{1/2} \exp\left[-\ |S_0(b,a)|/\not h\ \right] \quad (1)$$

where S_0 is the classical action in going from the initial (a) to the final (b) point. The pre-exponential factor represents quantum fluctuations; it can be calculated according to a standard variational procedure, as follows.

$$\left(\frac{\delta^2 S}{\delta x^2}\right)_{\bar{x}(t)} = \left(\frac{\delta^2 S}{\delta(aT)^2}\right)_{aT=0} \frac{(aT)^2}{y^2}$$

$$= \frac{(aT)^2}{y^2}\int_{-T}^{T}\ \left[\frac{\partial^2}{\partial(aT)^2}\ \mathcal{L}_E(\bar{x}+\Delta x,\ \dot{\bar{x}}+\Delta\dot{x},t)\right]_{aT=0} dt \quad (2)$$

where y and aT represent respectively the spatial and temporal displacements between a generic and the classical path [15]. The integration is carried out by considering only quasi-translational variations, which are thought to be the most important [4]. With such variations one can plausibly conjecture that

$$(\Delta x)^2 = (\dot{\bar{x}})^2\ (aT)^2$$

$$(\Delta\dot{x})^2 = (\ddot{\bar{x}})^2\ (aT)^2$$

2.2. Strongly asymmetric potential

We represent the relevant part of the potential function, shown by a heavy continuous line in Fig.1, by

$$V(x) = \varepsilon x^2 - \alpha x^4 \qquad (3)$$

Here $\varepsilon = M\omega^2/2$, ω is the angular frequency near the bottom of the initial well (x = 0), M is the mass and α is a positive parameter determining the position of the barrier top $x_b \equiv \frac{1}{2}x_o = \sqrt{\varepsilon\alpha}$ and of the non-zero root $x_B = \sqrt{\varepsilon/\alpha}$. By considering the euclidean lagrangian

$$\mathcal{L}_E = E_{kin} + V,$$

we find that the classical path for the motion in the reversed potential U = -V in going from x = 0 to x = x_B and coming back to x = 0 is given by (see Fig. 2)[3]

$$\bar{x}(t) = x_B \operatorname{sech}\left[\omega(t - t_o)\right] \qquad (4)$$

A classical trajectory of this kind is called instanton bounce: t_o is the time of the bounce at x = x_B. The classical action for the bounce turns out to be [3]

$$S_B = \frac{(2\varepsilon)^{3/2}}{3\alpha} \qquad (5)$$

The pre-exponential factor in Eq. (1) is evaluated by the saddle-point method. In the present case, however, we have to proceed cautiously because of the presence of another saddle point (corresponding to the classical solution $\bar{x}(t) \equiv 0$) near the saddle point that we are considering (corresponding to the bounce, Eq. (4)). The right application of the saddle point method [16] implies that the steepest-descent path is in the complex plane, so that we have in the result a factor $(1 \pm i)/2$.[3,6] The real part causes a shift of the energy ground state $\hbar\omega/2$, while the imaginary part is related to the decay rate.
Then Eq. (2) reduces to

$$\left(\frac{\delta^2 S}{\delta x^2}\right)_{\bar{x}(t)} = -12\,\omega^2\,\frac{(aT)^2}{y^2}\,S_B\,e^{-2\omega T} + O(e^{-4\omega T}) \qquad (6)$$

where S_B is given by Eq. (5). Note that Eq. (6) has been obtained by integrating over a large but finite time interval 2T. For T → ∞ Eq. (6) goes to zero, as it must, since in an infinite time-interval the variation of the action for a translational motion is zero. By substituting Eq. (6) into Eq. (1) we have for the kernel of the single bounce the following expression

$$K(0,T; 0,-T) \;=\; \frac{aT}{y} \, \omega \left(\frac{6 \; S_B}{\pi} \right)^{1/2} e^{-\omega T} \, e^{-S_B / \not h} .\qquad(7)$$

This expression will allow us to derive the decay rate of the metastable level, as we shall see later.

2.3. Slightly asymmetric potential

Let us consider now, by the same procedure, the case of an asymmetric double-well potential, like that represented by dashed line in Fig.1, which can be written as

$$V(x) \;=\; 2\varepsilon x^2 (1 - \frac{x}{x_o})^2 \;-\; \sigma \, \frac{x}{x_o} \qquad(8)$$

For slight asymmetry ($\sigma \ll V_o = \varepsilon^2 /4\alpha$, V_o being the barrier height at $x_o/2$) the classical path, in going from $x = x_B$ and coming back, can be expressed approximately as [15]

$$\bar{x}(t) \;=\; \frac{x_o}{2} \left[1 \pm \tanh \frac{\omega}{\sqrt{2}} \, (t \pm t_b) \right] \qquad(9)$$

where the sign + holds for $t < 0$ and the sign − for $t > 0$, and

$$t_b \;=\; \frac{1}{2 \, \sqrt{2}\omega} \; \ln (16 \, \frac{V_o}{\sigma}) \qquad(10)$$

We thus have a classical path (the dashed line in Fig.2) shaped like an instanton bounce, consisting of a kink (for $t < 0$) for a quasi symmetric potential and of an analogous antikink (for $t > 0$). The quantum fluctuations are evaluated from Eq. (2) with the potential given by Eq. (8). After integration we find

$$\left(\frac{\delta^2 S}{\delta x^2} \right)_{\bar{x}(t)} \;=\; - \, 48 \; S_o \; \omega^2 \frac{(aT)^2}{y^2} \left(\frac{\sigma}{16 V_o} + \frac{16 V_o}{\sigma} e^{-2\sqrt{2}\omega T} \right) \qquad(11)$$

where S_o is the classical action for a single symmetric kink. We note that the result of Eq. (11) is just twice that obtained for a single quasi-symmetric kink, provided one replaces (in the last term of Eq.(34)

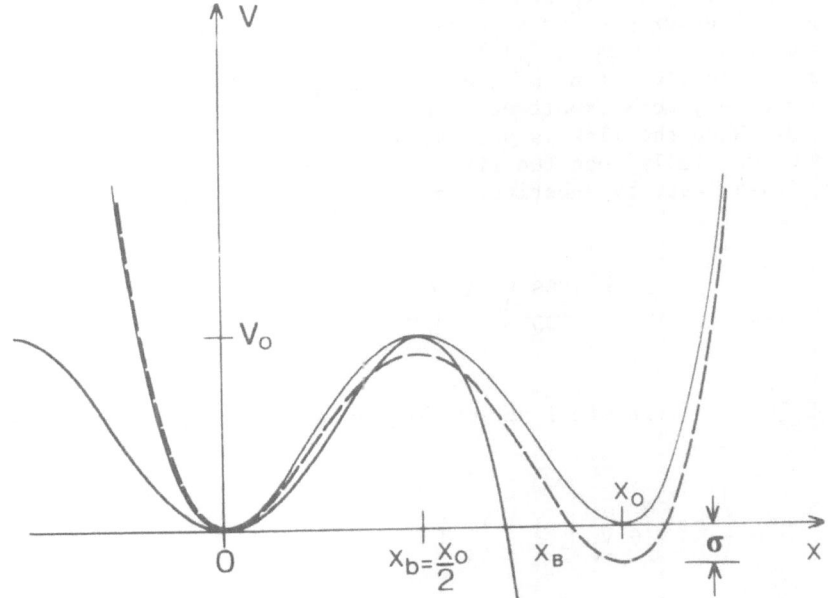

Fig. 1   Potential functions for tunnelling processes starting from the initial
          well at $x = 0$. Complete asymmetrical case, Eq. (1): heavy continuous
          line. Asymmetrical double-well potential, Eq. (8): dashed line. Perfect
          symmetrical case, Eq. (8) with $\sigma = 0$: light continuous line.

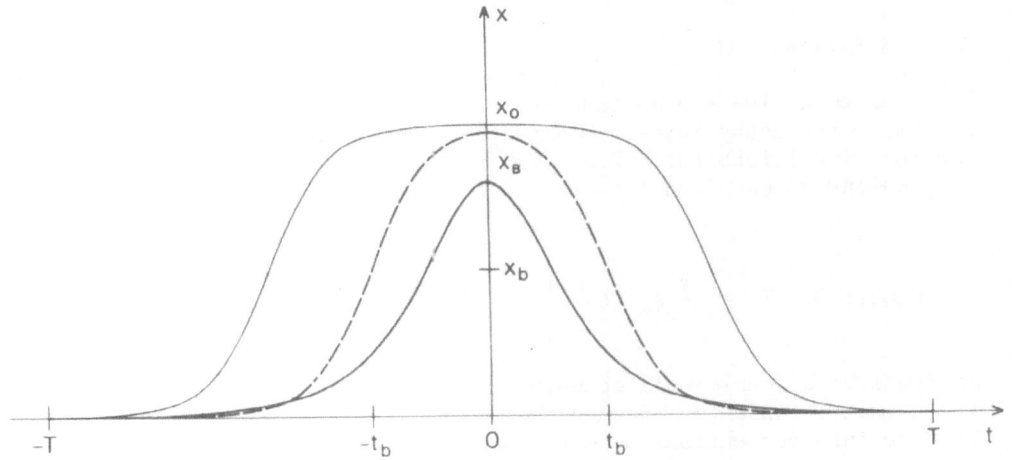

Fig. 2   Classical trajectories of tunnelling events starting from the initial well
          at $-T$ and ending in the smae well at $+T$. Instanton bounce, Eq. (2): heavy
          continuous line. Quasi-bounce as a sum of an asymmetric kink and antikink,
          Eq. (8): dashed line. Symmetric kink and antikink: light continuous line.

in Ref, 15) T by $T - t_b$ and uses the value of t_b in Eq. (10); hence this result encourages us to proceed. However, we have to consider that the first term in Eq. (11) is actually spurious since it derives from the non-perfect link of the two kinks, at $t = 0$, where there is a small cusp and, more important, a discontinuity in the velocity $\dot{x}(t)$ at $t = 0$. When the link is smooth, and the velocity correctly vanishes at $t = 0$, we really lose the first term in Eq. (11). Taking into account this fact, by substituting into Eq. (1), we get for the kernel

$$K(0,T;0,-T) = \frac{aT}{y} \omega \left(\frac{6S_o}{\pi \hbar} \right)^{1/2} \left(\frac{16 \ V_o}{\sigma} \right)^{1/2} e^{-\omega T} e^{-2S_{as}^{(o)}/\hbar} \quad (12)$$

where $S_{as}^{(o)}$ is the classical action for the quasi-symmetric kink given by [15]

$$S_{as}^{(o)} = S_o \left\{ 1 - \frac{3}{16} \frac{\sigma}{V_o} \left[1 + \ln 4 \left(\frac{V_o}{\sigma} \right)^{1/2} \right] \right\}.$$

We note that Eq. (12) resembles Eq. (7) (since $2 S_{as} \sim 2 S_o \sim S_B$) but it is not possible to regain (7) from (12) for $\sigma \to 0$ since the latter holds only for moderate asymmetries, that is for $\sigma \ll V_o$. On the other hand, Eq. (12) cannot be used in the perfect symmetric case ($\sigma = 0$), due to the divergence of the factor $(16 \ V_o/\sigma)^{1/2}$. In truth, this factor has as an upper limit the value $\exp(\sqrt{2} \ \omega T)$, corresponding to $t_b = T/2$, for which the kink reverts to the symmetric case.

3. THE FEYNMAN SUM

In order to derive expressions for observable quantities, such as energy-shifts and decay-rates, we have to compare the kernels as given by the functional integrals, Eqs. (7) and (12), with the corresponding expressions as obtained from the Feynman sum which, for imaginary time, is [1]

$$K(0,T; 0, -T) = \sum_n \psi_n(0) \ \psi_n^*(0) \ e^{-2E_n T/\hbar} \quad (13)$$

We consider a double-well potential the symmetric case ($\sigma = 0$) and the completely asymmetric one ($\sigma \to \infty$). We take into account only the contributions from the two lowest eigenstates which can be written as

$$\psi_a = (\sin \phi)\psi_1 + (\cos \phi)\psi_2$$

$$\psi_b = (\cos \phi)\psi_1 - (\sin \phi)\psi_2 \qquad (14)$$

where ψ_1 and ψ_2 are ground-state functions in the two separate wells. Using a truncated matrix formulation [7] we have the energy matrix in the form

$$
\begin{array}{c}
\psi_1 : \\
\psi_2 :
\end{array}
\left(
\begin{array}{cc}
\dfrac{\sigma}{2} & \gamma \\
\gamma & -\dfrac{\sigma}{2}
\end{array}
\right)
$$

in which the energy zero is placed at $E_o = (\hbar\omega - \sigma)/2$. Here $\tan 2\phi = 2\gamma/\sigma$, γ is a coupling coefficient and the eigenvalues are

$$E_a = \frac{\hbar\omega}{2} - \sigma - \delta = E_o - \frac{1}{2}\sigma - \delta$$

$$E_b = \frac{\hbar\omega}{2} + \delta = E_o + \frac{1}{2}\sigma + \delta \ . \qquad (15)$$

δ is (in general) a complex quantity whose real part gives the energy shift of the level, while the imaginary part is related to the decay-rate.[2,3,6,7]

By substituting Eqs. (14) and (15) into Eq. (13), we obtain, neglecting small overlap terms,

$$K(0,T; 0, -T) = \left(\frac{M\omega}{\pi\hbar}\right)^{1/2} e^{-\omega T}\, e^{\sigma T/\hbar}\ \times$$

$$\times \left[\cosh \left(\frac{\sigma + 2\delta}{\hbar} T\right) - \cos 2\phi \ \sinh \left(\frac{\sigma + 2\delta}{\hbar} T\right) \right] \qquad (16)$$

We now consider two opposite situations:

(i) <u>The slightly asymmetric case:</u> $\sigma \ll \gamma$, $\cos 2\phi \simeq \sigma/2\gamma \simeq 0$. Under these conditions Eq. (16) becomes

$$K(0,T;0, -T = (\frac{M\omega}{\pi\hbar})^{1/2} \; e^{-\omega T} e^{\sigma T/\hbar} \; \cosh \; (\frac{\sigma + 2\delta}{\hbar} T) \tag{17}$$

$$\approx (\frac{M\omega}{\pi\hbar})^{1/2} \; e^{-\omega T} e^{\sigma T/\hbar} \left\{ \cosh\frac{\sigma T}{\hbar} \left[1 + \frac{1}{2} (\frac{2}{\hbar}\frac{\delta T}{\hbar})^2 \right] + (\frac{2}{\hbar}\frac{\delta T}{\hbar}) \sinh \frac{\sigma T}{\hbar} \right\},$$

since we have, typically, $2\delta T/\hbar \ll 1$.

In Eq. (17) the dominant term involving δ is either the linear term $2\delta T/\hbar$ or the quadratic one $(2\delta T/\hbar)^2$ depending on the relative importance of 2δ with respect to σ. When σ is small compared with 2δ (<u>quasi-symmetric case</u>), Eq. (17) can be approximated as

$$K_{0,2}(0,T;0,-T) \approx (\frac{M\omega}{\pi\hbar})^{1/2} \; e^{-\omega T} \left[1 + \frac{1}{2} (\frac{2\delta T}{\hbar})^2 \right] \tag{18}$$

which holds exactly for $\sigma = 0$ (perfectly symmetric case) in which $2\delta = \Delta E$ is the tunnelling splitting. The subscripts 0,2 in Eq. (18) mean that we consider in the kernel only the contributions arising from the purely stationary solution $x(t) \equiv 0$, which correspond to the first term in parentheses, and the double tunnelling event contributions, kink plus antikink, which correspond to the second term in parentheses. The latter can be compared with the corresponding expression of K_2 as deduced by the functional integral. According to the prescriptions of Ref. 15 we have

$$K_2(0,T;0, -T) = \left[K_1(x_0,0; 0 -T) \right]^2 2y =$$

$$= \frac{12 \, S_0}{\pi\hbar} \frac{(\omega \, aT)^2}{y} \; e^{-\omega T} \; e^{-2S_0/\hbar}$$

Assuming for the parameter a the estimate $a \sim 2\omega T \sim 1/2$ and for the width y of the path-beam the estimate in Ref. 15

$$y = (\frac{\pi\hbar aT}{2M})^{1/2} \sim (\frac{\pi\hbar}{16\omega M})^{1/2}$$

we obtain

$$K_2(0,T;0,-T) = \frac{12\,S_o}{\pi\hbar} \left(\frac{16\omega M}{\pi\hbar}\right)^{1/2} (\omega a T)^2 \; e^{-\omega T} e^{-2S_o/\hbar} \tag{19}$$

which can be used to make a comparison with the second term in Eq. (18) to give the tunnelling splitting

$$\Delta E = \hbar\omega \left(\frac{24\,S_o}{\pi\hbar}\right)^{1/2} e^{-S_o/\hbar} \tag{20}$$

in agreement with other analyses.[4,15]
 Still for moderate asymmetries, but in the opposite case of $\sigma \gg 2\delta$, Eq. (17) can be approximated as

$$K(0,T;0,-T) \simeq \left(\frac{M\omega}{\pi\hbar}\right)^{1/2} e^{-\omega T} e^{-\sigma T/\hbar} \left[\cosh\frac{\sigma T}{\hbar} + \frac{2\delta T}{\hbar} \sinh\frac{\sigma T}{\hbar} \right] \tag{21}$$

which for $\sigma T/\hbar$ sufficiently large, becomes

$$K_{0,1}(0,T;0,-T) \stackrel{\sim}{\sim} \frac{1}{2} \left(\frac{M\omega}{\pi\hbar}\right)^{1/2} e^{-\omega T} e^{2\sigma T/\hbar} \left(1 + \frac{2\delta T}{\hbar}\right) . \tag{22}$$

Here the subscripts 0,1 mean that we are considering contributions arising from zero-bounces ($\bar{x}(t) \equiv 0$) and single bounces. The implications of this result in connection with Eq. (12) will be discussed later.
(ii) Strongly asymmetric case: $\sigma \gg \gamma$, $\cos 2\phi = \sigma/\sqrt{\sigma^2+4\gamma^2} \simeq 1$. In this case Eq. (16) simply reduces to

$$K(0,T;0,-T) = \left(\frac{M\omega}{\pi\hbar}\right)^{1/2} e^{-\omega T} e^{\sigma T/\hbar} \exp\left(-\frac{\sigma+2\delta}{\hbar} T\right) = \left(\frac{M\omega}{\pi\hbar}\right)^{1/2} e^{-\omega T} e^{-2\delta T/\hbar}. \tag{23}$$

This expression, which is equivalent to that given by Callan and Coleman (Eq. (2.16) in Ref.6), is suitable to treat the complete asymmetric case, though it is not applicable to moderate asymmetries (e.g. potential like in Eq. (8), where $\gamma \sim 3V_o$, since then we cannot satisfy the

the condition $\gamma \ll \sigma$). By developing in Eq. (23) $e^{-2\delta T/\hbar} \simeq 1 - 2\delta T/\hbar$, comparing the linear term in T (corresponding to single bounces contributions) with "our" kernel, Eq. (7), and taking into account the factor $\frac{1}{2}$, we have

$$\text{Im}\delta = \mp \frac{1}{2} \hbar\omega \left(\frac{6S_B}{\pi \hbar} \right)^{1/2} e^{-S_B/\hbar} \qquad (24)$$

where we have assumed as before, $y = (\pi\hbar aT/2M)^{1/2}$ and $a = 2\omega T$. The result of Eq. (24) is exactly the same as reported by Schulman, [17] obtained by more rigorous but more complex procedures. Therefore, our approach to quantum fluctuations finds also in this case an exact confirmation. The decay rate of the unstable level is given by

$$\Gamma = 2 \frac{|\text{Im}\delta|}{\hbar} = \omega \left(\frac{6S_B}{\pi\hbar} \right)^{1/2} e^{-S_B/\hbar}$$

$$(25)$$

We note that the result of Eq. (24), or (25), has a structure analogous to that of the tunnelling splitting in a symmetrical double-well potential, Eq. (20). This resemblance is, of course, only formal since numerically the two results are very different: The classical action for a bounce is about twice the classical action for a kink ($S_B \sim 2S_0$), (typically the two quantities, Γ and $\Delta E/\hbar$, differ by some or several orders of magnitude.)

We ask now how is it possible to link these two extreme situations and to analyze, with the same procedure, the case of moderate asymmetries. This can be done by comparing the kernel as obtained by our method, Eq. (12), and that obtained by Feynman's sum, namely the second term in Eq. (21). Thus, taking a and y as before, we obtain for the tunnelling rate

$$\frac{\delta}{\hbar} = \omega\, e^{-\frac{\sigma}{4\hbar\omega}}\, \text{csch}\left(\frac{\sigma}{4\hbar\omega}\right) \cdot \left(\frac{12S_0}{\pi\hbar}\right)^{1/2}\left(\frac{8V_0}{\sigma}\right)^{1/2} e^{-2S_{as}^{(o)}/\hbar} \qquad (26)$$

(an analogous expression is obtained with the second term of Eq. (22)). As noted earlier δ will be, in general, a complex quantity whose real part is the energy shift of the level, while the imaginary one is the decay rate. In the limit of strong asymmetry we were able to identify the imaginary part (Eq. 24) and, as a consequence, to obtain the decay rate, Eq. (25). On the other hand, in the opposite limit of "zero-asymmetry" there is not any evidence of an imaginary part and the effect of the tunnelling is a "strong" splitting of the degenerate levels, ΔE in Eq. (20). In this (resonant) case we have a continuous "flip-flop" of the system between the two wells, that is we have a coherent tunnelling. In the other case (aperiodic case) the tunnelling causes, in a given time, an irreversible process, that is the decay, beyond the barrier, of the system initially prepared in the well (incoherent tunnelling). In the intermediate situations, corresponding to Eq. (26) we have presumably a combination of the two behaviours but we do not see how to discriminate between them. However, we note that Eq. (26) resembles Eq. (24) since $2S_{as}^{(o)} \sim 2S_o \sim S_B$; in Eq. (26) the enhancing factor

$$\frac{\delta}{\Gamma \hbar} = \left(\frac{8V_o}{\sigma}\right)^{1/2} e^{-\frac{\sigma}{4\hbar\omega}} \operatorname{csch}\left(\frac{\sigma}{4\hbar\omega}\right) \tag{27}$$

which takes here the place of $e^{S_o/\hbar}$, can be considered as a link between the resonant case ($\Delta E \propto e^{-S_o/\hbar}$) and the aperiodic one ($\delta \propto e^{-2S_o/\hbar}$). The graph of the ratio $(\delta/\Gamma\hbar)$ against $(\sigma/8V_o)$ is shown in Fig. 3 for typical parameter values. The curve is traced beyond the limit of validity of Eq. (26) up to values of $\sigma \sim 2V_o$ in order to cross the level given by Eq. (24) which represents equal contribution to the real and imaginary part of δ. The dashed line in Fig. 3 is expected to be an approximate representation of the imaginary part of δ. The inset of Fig. 3 shows the behaviour of Eq. (27) near the top of the resonance peak and the crossing of the curve obtained by comparing the quadratic term in Eq. (17) (more accurate for $\sigma \ll \delta$) with Eq. (19),

$$\frac{\delta}{\hbar\Gamma} = \frac{1}{\sqrt{2}} e^{-\frac{\sigma}{8\hbar\omega}} \operatorname{sech}^{1/2}\left(\frac{\sigma}{4\hbar\omega}\right) e^{S_o/\hbar} \tag{28}$$

For very small values of $\sigma (\ll 4\hbar\omega)$ this ratio has practically a constant value, corresponding to one half of the tunnelling splitting

$$\frac{\Delta E}{2\hbar\Gamma} = \frac{1}{\sqrt{2}} e^{S_o/\hbar} . \tag{29}$$

We can therefore conclude that our procedure of evaluating quantum fluctuations is capable of giving the right expressions for the tunnelling splitting in the symmetric case, Eq. (20), and for the decay -rate of a metastable level, Eq. (25). Moreover, it seems also to be

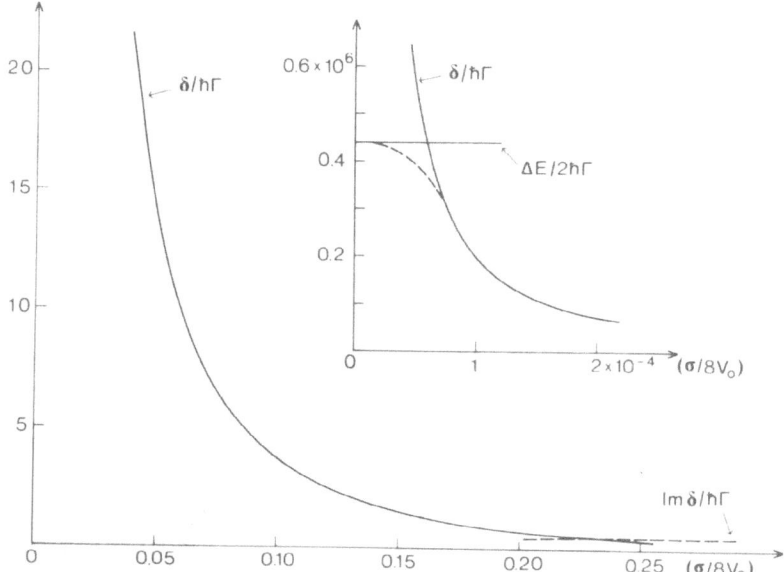

Fig. 3   Tunnelling rate $\delta/\hbar$ normalized to decay rate Γ, according to Eq. (27), as a function of the asymmetry factor $\sigma/8V_0$ for $\hbar\omega = \frac{2}{5} V_0$ (continuous line). The (presumed) imaginary part of δ is represented by dashed line. In the inset: detail of $\delta/\hbar\Gamma$ near the resonance peak ($\sigma \to 0$) where the curve corresponding to Eq. (27) crosses that corresponding to Eq. (28), practically coincident with one half of the tunnelling splitting, $\Delta E/2$. Dashed line represents a smooth connection between them.

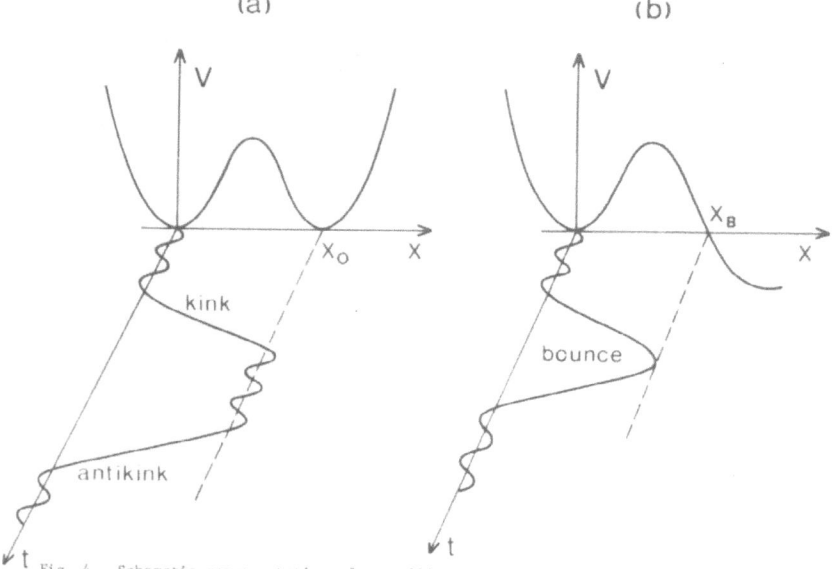

Fig. 4   Schematic representation of tunnelling events in symmetrical a) and asymmetrical b) potentials, according to the instanton method description. Nearly periodic oscillations in the well are before and after each tunnelling event: kink or antikink in a), bounce in b).

suitable to treat the intermediate case of asymmetrical double-well potentials, supplying a link between the two extreme situations. The tunnelling rate in Eq. (26) describes a quasi coherent behaviour, the imaginary part of δ being negligible, except in the limit of $\sigma \to \infty$. The
plausibility of the procedure, based on quasi-translational modes, finds
a qualitative confirmation as explained in the following discussion.

4. NEARLY PERIODIC OSCILLATION DESCRIPTION AND DISCUSSION

We have seen that the kernels for the bounce, Eqs. (7) and (12), contain
a time-dependent factor of the type $(aT/y)e^{-\omega T}$. The same time dependence is present in the kernel for the kink in the symmetric problem (Eq.
(20) in Ref. 15). Regarding y, the width of the path-beam, as a constant
quantity (in agreement with the large T-limit of Eq. 6. 42 in Ref. 3),
we arrive at the conclusion that the time-dependence, in either the kernel for the <u>bounce</u> or for the single symmetric <u>kink</u>, due to the fluctuation is (apart from the factor $e^{-\omega T}$)

$$\left(\frac{\delta^2 S}{\delta x^2}\right)_{\substack{\text{one bounce} \\ \text{or one kink}}} \propto aT. \qquad (30)$$

On the other hand, for the kernel of the kink plus antikink in the perfect symmetric case we have, according to Eq. (19) containing the same
factor $e^{-\omega T}$,

$$\left(\frac{\delta^2 S}{\delta x^2}\right)^{1/2}_{\substack{\text{kink} + \\ \text{antikink}}} \propto (aT)^2. \qquad (31)$$

We have also shown how both the kernels for the resonant case (single
kink or kink plus antikink, Eq. (8)) correctly give the tunnelling splitting, while the kernels for the bounce, Eqs. (7) and (12), give the decay-rate and the energy-shift in the asymmetric cases. The following
interpretation of these facts is suggested on the basis of an analogy
with Miller's description of periodic orbits.[18]
   The importance of neighbouring paths with time displacement aT is
due to the possibility of 0,1,2,... nearly periodic oscillations (NPO)
in the left (initial) well and in the right well, provided this is (nearly) identical with the left well, before and after each tunnelling
event. More precisely: NPO in the left well before the kink and NPO in
the right well before the antikink (see Fig. 4a). If there was no coupling between wells ($\gamma = 0$ and , as a consequence, $\Delta E = 0$) each oscillation would be exactly periodic inside each well, without contribution
to the phase of the kernel. However, since $\Delta E \neq 0$, one gets contribution to the phase proportional to $\Delta E.T$, see Eq. (13); this shows up in

$(\delta^2 S / \delta x^2)$.

If wells are perfectly symmetric one gets for a single tunnelling event (kink or antikink) a contribution proportional to T, or aT as in Eq. (30), while for a double tunnelling event (kink plus antikink) we have

$$
\left(\frac{\delta^2 S}{\delta_\chi^2} \right)^{1/2}_{\substack{kink+ \\ antikink}} \sim \left(\frac{\delta^2 S}{\delta x^2} \right)^{1/2}_{kink} \quad \left(\frac{\delta^2 S}{\delta x^2} \right)^{1/2}_{antikink} \quad \propto \quad T^2.
$$

$$(32)$$

The reason for this result is that in the path sum

$$
\int_{kink} e^{-S(kink)/\hbar} \, Dy(t)
$$

the phases due to the NPO in the right well are lost and therefore we can factorize the kernel as follows.

$$
\int_{\substack{kink+ \\ antik.}} e^{-S(kink+antik.)/\hbar} \, Dy(t) = \int_{kink} e^{-S(kink)/\hbar} \, Dy(t) \, x
$$

$$
x \int_{antikink} e^{-S(antik.)/\hbar} \, Dy(t).
$$

$$(33)$$

If the wells are (very) asymmetric there are no NPO in the right well at the level of the initial (left) well (see Fig. 4b). This is exactly true if we are considering a complete asymmetric case (see Eq. (3)),but also for an asymmetric double-well potential (see Eq. (8)) there is not any correspondence between the levels of the two separate wells. In these cases the bounce (Eq. (4), but also Eq. (9)) gets contribution only from the left well and, again, fluctuations are proportional to T, as in Eq. (30). Therefore, in the asymmetric case each <u>bounce</u> get a T factor just as in the symmetric case each <u>kink</u> gets a T factor. The generalization is clear. In the *symmetric* case the kernel for an even number 2n of events, n(kink + antikink), will be

$$
\int_{\substack{n(kink+antik.) \\ = 2n \, kink}} e^{-S(2nkink)/\hbar} \, Dy(t) \sim \left[\int_{kink} e^{-S(kink)/\hbar} Dy(t) \right]^{2n}
$$

(Eq. (33) is the particular case of n = 1). Analogously, in *asymmetric* case the kernel for any n number of bounces will be

$$\int_{n \text{ bounces}} e^{-S(n \text{ bounces})/\hbar} Dy(t) \sim \left[\int_{\text{bounce}} e^{-S(\text{bounce})/\hbar} Dy(t) \right]^{n}.$$

For strong asymmetry ($\gamma \ll \sigma \to \infty$) the bounce type behaviour (Fig. 4b) is completely adequate. The decay-rate of the metastable level is obtained independently of the potential beyond the bounce point x_B; in fact, the kernel in Eq. (23) is independent of σ and the bounce never actually probes the bottom of the lower well. [7] The origin of the imaginary part of the energy can be seen in the absence of a (nearby) barrier of the potential or, in other words, in the fact that the wavefunction beyond the barrier is outgoing.[8,19]

It is more difficult to explain on this basis the case of a slight asymmetry ($\sigma \ll \gamma$). Clearly, until $\sigma \ll 2\delta$, the kernel will be of the type T^2, Eqs. (31) and (32), while for $\sigma \gg 2\delta$, the kernel will be of the type T, Eq. (30). For $\sigma \sim 2\delta$ we have, presumably, a hybrid behaviour of the system, a sort of superposition of the two above mentioned cases, as can be seen from the kernel of Eq. (17). Still for slight asymmetry ($\sigma \ll \gamma$), even if the classical trajectory is a quasi-bounce (see Eq. 9 and the dashed curve in Fig. 2), the kernel (Eqs. (12), (21) and (22)) properly depends on σ; that, is on the shape of the potnetial beyond the barrier. In this case the quasi-bounce, in a sense, does probe the bottom of the lower well since the head of the bounce, x_B, , is very near to x_0, the coordinate of the lower minimum. In the language of the periodic orbit theory (which works at the energy levels and not at "zero-energy" as the instanton method does) this means that between each tunnelling event (kink or antikink) the system travels forth and back across each well, even if in a different way with respect to the resonant case. [18] Accordingly, the wavefunction in the lower well is no longer outgoing but stationary.

REFERENCES

1 - R.P. Feynman and A.R. Hibbs, *Quantum Mechanics and Path Integrals*, (Mc Graw-Hill, New York, 1965).

2 - S. Coleman, in *The Whys of Subnuclear Physics*, edited by A. Zichicchi (Plenum, New York, 1979).

3 - L.S. Schulman, *Techniques and Applications of Path Integration*, (Wiley-Interscience, New York, 1981).

4 - E. Gildener and A. Patrascioiu, Phys. Rev. D 16, 423 (1977).

5 - S. Coleman, Phys. Rev. D 15, 2929 (1977).

6 - C.G. Callan Jr. and S. Coleman, Phys. Rev. D 16, 1762 (1977).

7 - J.P. Sethna, Phys. Rev. B 24, 698 (1981); 25, 5050 (1982).

8 - U. Weiss and W. Haeffner, Phys. Rev. D 27, 2916 (1983).

9 - U. Weiss, Phys. Rev. A 25, 2444 (1982).

10- I. Affleck, Phys. Rev. Lett. 46, 388 (1981).

11- A.O. Caldeira and A.J. Leggett, Phys. Rev. Lett. 46, 211 (1981) D. Waxman and A.J. Leggett, Phys. Rev. B 32, 4450 (1985).

12- H. Grabert and U. Weiss, Phys. Rev. Lett. 52, 2193 (1984); 53, 1787 (1984); 54, 1605 (1985).

13- M.P.A. Fisher and A.T. Dorsey, Phys. Rev. Lett. 54, 1609 (1985).

14- S. Chakravarty and S. Kivelson, Phys. Rev. B 32, 76 (1985).

15- D. Mugnai and A. Ranfagni, Phys. Letters 109A, 219 (1985).

16- L.P. Felsen and N. Marcuvitz, *Radiation and Scattering of Waves*, (Prentice-Hall, Englewood Cliffs, N.J., 1973). Chap.4.

17- Ref. 3, Eq. 29.25, p. 283.

18- W.H. Miller, Adv. Chem. Phys. 25, 69 (1974); J. Chem. Phys. 63, 996 (1975); J. Phys. Chem. 83, 960 (1979).

19- U. Weiss, private communication.

TUNNELING - QUANTUM RESTORATION OF CLASSICALLY BROKEN SYMMETRIES

Yossef Dothan
Raymond and Beverly Sackler Faculty of Exact Sciences
School of Physics and Astronomy
Tel-Aviv University
Tel-Aviv 69978, ISRAEL

Introduction.

Quantum field theory combines the principles of quantum mechanics with the principles of special relativity and with the principles of causality and locality. Such a combination leads to an infinite number of degrees of freedom conveniently labeled e.g. by the points of space. The wave-function is replaced by a wave-functional supplying a probability amplitude for a field configuration. The last fifteen years brought about an understanding of quantum field theory in general and non-abelian gauge theories in particular that goes beyond the understanding gained through the application of simple perturbation theory. The most important achievement of this development is the so-called "standard model" of particle physics. As part of this development it was realized that there are tunneling phenomena in quantum field theory. It is thus not surprising that more than a decade ago a calculational scheme known as the "dilute instanton gas approximation" was developed by A.M. Polyakov and by G.t'Hooft to deal with tunneling phenomena [1].
The purpose of this talk is to review this scheme using two simple one dimensional quantum mechanical problems as examples and trying to emphasize the aspects of quantum barrier penetration. The moral of the story is that a practitioner of field theory is interested in a calculational scheme that is generalizable to any number of degrees of freedom and in particular an infinite number of degrees of freedom. It should be said explicitly that this generalizability to field theory is not an interesting property as far as the practitioner of quantum mechanics is concerned. On the contrary, the examples that I shall present may be dealt with more directly and the method I shall review may look cumbersome in the context of these examples.
The material presented is neither new nor original and it is presented for the purpose of facilitating communication among scientists from various disciplines.

J. Jortner and B. Pullman (eds.), Tunneling, 81–89.

The Double Well Potential.

Let us start with a simple quantum mechanical system of a non-relativistic particle of mass m moving on the line $-\infty < q < \infty$ under the influence of a non negative potential $V(q)$. The Hamiltonian of the system is given by

$$H = \frac{p^2}{2m} + V(q)$$

We further assume that if for a given energy there exist several disconnected regions of allowed classical motion they are related to each other by a symmetry of the Hamiltonian H. This means that if H has a minimum which is not invariant under G -the group of symmetries of H- this minimum must be energy degenerate with other minima of H so that the set of minima is invariant under the symmetry group G. Let us review two simple examples.

   I.consider the potential

$$V(q) = \frac{1}{2} m \omega^2 q^2$$

The symmetry group of the Hamiltonian is the two element group Z_2 generated by reflections at the origin or parity

$$P : q \dashrightarrow - q$$

Clearly the only invariant point is the origin $q = 0$. It is also clear that the potential V is parity invariant. Now V has a single minimum at the invariant point and the Hamiltonian has a single classical ground state namely $p = 0$ $q = 0$. This is the state of a particle sitting at rest at the bottom of the well.

   II. The second example involves the potential function

$$V(q) = \frac{\lambda}{4} (q^2 - \mu^2/\lambda)^2 \qquad \text{with } \lambda > 0 \text{ and } \mu > 0 .$$

The symmetry group is again Z_2 generated by parity. In this example the potential has two minima located at $q = \pm\sqrt{\mu^2/\lambda}$ and transforming into each other under parity. Thus there are two classical ground states corresponding to the particle being at rest at the bottom of either the left well or the right well. Clearly any classical motion with small enough energy breaks the left-right symmetry as it takes place in only one of the two wells.

   Quantization brings with it the possibility of tunnelling. It is clear that in this simple example the system will indeed tunnel and the two lowest eigenstates of the Hamiltonian H will have well defined parities. The question that needs an answer in the most general situation is similarly: does the system tunnel between degenerate ground states? or stated with an emphasis on the symmetry aspect: does symmetry breaking by a non-invariant degenerate ground state survive quantization?

The basic relation we use is the Feynman path-integral representation for matrix elements of the operator $\exp(-2HT/\hbar)$ namely

$$\langle q_f \mid \exp(-2HT/\hbar) \mid q_i \rangle = \int Dx \, \exp(-S/\hbar)$$

Let us explain the notation: the states $\mid q_i \rangle$ and $\mid q_f \rangle$ are eigenstates of the position operator, H is the Hamiltonian which by our assumptions is a non-negative operator while T is a positive c-number. Remembering that $\exp(-i2HT/\hbar)$ is the time translation operator by 2T we see that the time variable t was analytically continued to -it. This remark ties in with the right hand side where S is the Euclidean action defined as -i multiplying the action integral in which the time t is analytically continued to -it. The integral is over all paths starting at time -T from q_i and arriving at time T to q_f integrated with the measure Dx. The Lagrangian corresponding to the Hamiltonian H is

$$L = \frac{q^2}{2m} - V(q)$$

while the Euclidean action is given by

$$S = \int_{-T}^{T} dt \left[\frac{m}{2}\dot{q}^2 + V(q) \right]$$

To see that the basic formula may carry interesting information let us introduce in the left hand side a complete set of H eigenstates $\mid n \rangle$ with the corresponding energies E_n (ordered so that for $E_n \geq E_m$ for $n > m$). This gives

$$\sum_n \exp(-2E_n T/\hbar) \, \langle q_f \mid n \rangle \langle n \mid q_i \rangle$$

Thus for asymptotic values of $E_1 T/\hbar$ only the low energy eigenstates of H make a non-negligible contribution to the left hand side. We therefore may evaluate the right hand side for small values of $\hbar$ and large values of T by the method of steepest descent. The leading contributions to the path integral are from paths q(t) on which the Euclidean action S attains a local minimum, or more technically paths on which the first variation of S vanishes while the second variation is positive. The estimate can be improved by a Gaussian integration about the extremal paths.

The result is a sum of contributions which to leading order have the form

$$\left[\det \frac{\delta^2 S[q_c]}{\delta q(t)\, \delta q(t')} \right]^{-\frac{1}{2}} \exp(-S[q_c]/\hbar)$$

where the minimizing path is a solution of the Euclidean equations of motion

$$\left. \frac{\delta S}{\delta q(t)} \right|_{q=q_c} = 0$$

satisfying the boundary conditions $q_c(-T) = 0$ and $q_c(T) = 0$. The method is useful if $S[q_c]$ does not diverge as $T \longrightarrow \infty$. Thus we are looking for paths of classical motions under the influence of the "inverted" potential $-V(q)$ that start and end with vanishing velocity at minima of V or paths starting and ending at classical ground states of H. This remark is useful because it allows us to use qualitative and intuitive classical mechanics reasoning.

To see how this works consider first a generalization of example I where the potential V is any parity invariant potential with a single minimum at the invariant point q=0. As there is only one minimum we have to consider the matrix element for $q_f = q_i = 0$. The only path that minimizes the action S and fulfills the boundary conditions is clearly $q(t) = 0$. The eigenvalue equation for the second variation of the Euclidean action is

$$-m \frac{d^2 \eta_n}{dt^2} + V''(0)\, \eta_n = \lambda_n \eta_n$$

Defining ω by

$$V''(0) = m\omega^2$$

we get

$$< 0\ |\ \exp(-2HT/\hbar)\ |\ 0 > = \left[\det \left[-\frac{d^2}{dt^2} + m\omega^2 \right] \right]^{-\frac{1}{2}} =$$

$$= \sqrt{\frac{m\omega}{\pi\hbar}}\ e^{-\omega T}$$

Therefore

$$E_0 = \tfrac{1}{2}\hbar\omega[1 + O(\hbar)] \qquad \text{and} \qquad |< q=0\ |\ n=0 >|^2 = \sqrt{\frac{m\omega}{\pi\hbar}}\ [1 + O(\hbar)]$$

Thus in this approximation the quantum particle is in a harmonic oscillator ground state wave function centered at the minimum of the potential and with a width matching $V''(0)$.

We now turn to the double well potential in example II and denote by q_0 the location of the minimum of the right well. There are four matrix elements that we consider but because of the parity symmetry there are two relations among them namely

$$< q_0 \mid \exp(-2TH/\hbar) \mid q_0 > \; = \; < -q_0 \mid \exp(-2TH/\hbar) \mid -q_0 >$$

and

$$< -q_0 \mid \exp(-2TH/\hbar) \mid q_0 > \; = \; < q_0 \mid \exp(-2TH/\hbar) \mid -q_0 >$$

In addition to the two paths $q_c(t) = q_0$ and $q_c(t) = -q_0$ the following paths are approximate solutions that become exact in the limit
$T \longrightarrow \infty$

$$q_c(t) = q_0 \tanh \left[\sqrt{\frac{\mu^2}{2m}} \, (t - t_1) \right]$$

and

$$q_c(t) = q_0 \tanh \left[-\sqrt{\frac{\mu^2}{2m}} \, (t - t_1) \right]$$

The first is referred to as an instanton centered at t_1 while the second is an anti-instanton centered at t_1. The name instanton coined by t'Hooft (Polyakov used the name pseudo-particle) signifies a solution of the Euclidean equations of motion starting at one classical ground state and ending at another and giving rise to a finite action. The instanton exhibited above is localized around t_1 in the sense that most of the contribution to the Euclidean action comes from t values near t_1. It should be stressed that an instanton is an approximate solution for any value of the parameter t_1 in the range $-T < t_1 < T$ not too close to the end points. Once we have it at our disposal we can construct other approximate solutions by stringing together many instantons and anti-instantons with centers at $t_1,....,t_n$ subject to the conditions

$$-T < t_1 < t_2 < ... < t_n < T$$

which are not too close to each other. The boundary conditions and the symmetry of the problem dictate that such multi-instanton approximate solutions have instantons and anti-instantons in alternating order. For a path from $-q_0$ to q_0 the string starts and ends with an instanton while for a path from $-q_0$ back to $-q_0$

the string starts with an instanton and ends with an anti-instanton.
This gives rise to the "instanton gas" referred to in the
introduction. Summing over all of these paths one gets for the two
lowest lying energy eigenstates $| + >$ and $| - >$ with energies
E_+ and E_- and with the indicated parities the following results

$$| < \pm q_0 | + > |^2 = | < \pm q_0 | - > |^2 = \frac{1}{2} \sqrt{\frac{m \omega}{\pi \hbar}}$$

$$\frac{1}{2} (E_+ + E_-) = \frac{1}{2} \hbar \omega$$

$$\Delta E = E_- - E_+ = \frac{1}{2} \hbar \omega \left[64 \frac{m \omega q_0^2}{\pi \hbar} \right]^{1/2} \exp - \frac{2}{3} \frac{m \omega q_0^2}{\pi \hbar}$$

where the frequency ω is given by

$$\omega = \sqrt{2 \mu^2 / m}$$

Note that the expression for ΔE considered as a function of $\hbar$ has
an essential singularity at $\hbar = 0$ and therefore does not admit a
power-series expansion in $\hbar$.

From the discussion it is now clear that the existence of
instanton solutions signals the occurrence of tunneling. In other
words if the barrier between the two minima was such that no
instanton solutions existed then the lowest energy state would have
been doubly degenerate representing the system on either side of the
barrier. Thus the answer to the physical question we posed namely
does the system tunnel between degenerate classical ground states is
answered positively if we can find instanton solutions to the
Euclidean equations of motion.

The Periodic Potential

A periodic potential is one that satisfies

$$V(q + a) = V(q)$$

The symmetry group is Z the group of translations by integral
multiples of the basic translation a. If V is not a constant then a
minimum at q_0 is degenerate with minima at $q_0 + na$ for any integer
n. A simple example of such a potential is

$$V(q) = m \frac{\alpha^2}{\beta^2} (1 - \cos \beta q)$$

for which $a = 2\pi / \beta$ and the minima are located at $q_n = n a$.
There is an instanton solution given by

$$q_c(t) = 4/\beta \, \arctan [\, \exp \alpha(t - t_1) \,]$$

which connects adjacent minima. In this case the Hamiltonian H has a band structure and the lowest (valence) band gives the leading contribution to the matrix elements of $\exp(-2HT/\hbar)$. Let $U(a)$ be the unitary operator of translation by a. Then $U(a)$ commutes with the Hamiltonian and may be diagonalized simultaneously with it. Since there is no non-zero integer n for which $[\, U(a) \,]^n = I$, the spectrum of $U(a)$ is given by $e^{-\theta}$ with θ in the range $-\pi < \theta \leq \pi$. Let $|\, \theta >$ be a simultaneous eigenstate of $U(a)$ and H with energy $\varepsilon(\theta)$ in the valence band. Then

$$< q_m \mid \exp(-2HT/\hbar) \mid q_n > =$$

$$= \int_{-\pi}^{\pi} \frac{d\theta}{2\pi} < q_m \mid \theta > < \theta \mid q_n > \exp -\frac{2\varepsilon(\theta)T}{\hbar}$$

With the relation

$$< q_{n+1} \mid \theta > = e^{i\theta} < q_n \mid \theta >$$

it is possible to summarize the results of the calculations in the form

$$|< q_0 \mid \theta >|^2 = \sqrt{\frac{m\alpha}{\pi\hbar}}$$

$$\varepsilon(\theta) = \hbar(\alpha/2 - \delta \cos \theta)$$

where

$$\delta = 2\alpha \left[\frac{16 \, m \, \alpha}{\pi \, \hbar \, \beta^2} \right]^{1/2} \exp -\frac{8 \, m \, \alpha}{\hbar \, \beta^2}$$

Note that $|< q_0 \mid \theta >|$ is independent of θ as in the tight binding approximation. Actually the results reproduce the leading asymptotic expression for the band width [2].

The Vacuum of Quantum-Chromo-Dynamics(Q.C.D.)

The problem of the vacuum of a non-Abelian gauge theory bears a formal resemblance to the problem of the periodic potential. The calculations are much more involved and use some algebraic topology to prove the existence of a countable infinity of degenerate classical minima of the energy and to calculate the instantons that signal tunneling. The physical system involves not only the gauge potentials but also the fermions that couple to them. The following is a short discussion of a simplified system that involves only the gauge potentials. In 1954 Yang and Mills generalized the principle of gauge invariance familiar from electromagnetism. For the case of the gauge group SU(n) these generalized gauge transformations are

$$A_\mu(x) \dashrightarrow M^\dagger(x)\, A_\mu(x)\, M(x) + \frac{g}{i}\, M^\dagger(x)\, \partial_\mu M(x)$$

$$F_{\mu\nu}(x) \dashrightarrow M^\dagger(x)\, F_{\mu\nu}(x)\, M(x)$$

where A_μ -the gauge potentials and $F_{\mu\nu}$ -the gauge field strengths are hermitian trace-less nxn matrices and M is a unitary nxn matrix with determinant 1. The constant g is referred to as the gauge coupling constant. We expand the A_μ and the $F_{\mu\nu}$ in a basis of traceless hermitian nxn matrices λ_k normalized so that the trace of their square is 2.

$$A_\mu = A_{\mu k}\, \tfrac{1}{2}\, \lambda_k$$

$$F_{\mu\nu} = F_{\mu\nu k}\, \tfrac{1}{2}\, \lambda_k$$

and write the Hamiltonian in the gauge invariant form

$$H = \int d^3x \, \tfrac{1}{2}\, [\, \overline{E}_k{}^2 + \overline{B}_k{}^2 \,]$$

Thus the energy is minimized for

$$\overline{E}_k = \overline{B}_k = 0$$

or remembering that $\overline{E}_k$ and $\overline{B}_k$ are the space-time and space-space component of $F_{\mu\nu k}$

$$F_{\mu\nu} = 0$$

which means that A_μ is a gauge transform of the zero potential, namely

$$A_\mu = \frac{g}{i}\, M^\dagger\, \partial_\mu M$$

It is at this point in the discussion that algebraic topology is needed to show that there is a countable infinity of classes of mappings M characterized by an integer n. Each class gives rise to a classical ground state. Upon quantization this situation would lead to a countable infinity of degenerate vacua denoted by | n >. By now it should be clear that the question that arises is whether there is tunneling between different | n > states. Once again the existence of instanton solutions signals tunneling. As in the case of the periodic potential tunneling gives rise to states | θ > characterized by an angle θ $-\pi \leq \theta < \pi$ known as the vacuum angle.

Once the fermions are coupled the picture becomes much more involved however as long as all fermions are massive the phenomenon of θ-vacua persists. In particular the vacuum angle is measurable by measuring the neutron's electric dipole moment. However experimentally the neutron's electric dipole moment is bounded from above by an extremely small number causing the vacuum angle to be smaller than about 10^{-8}. The smallness of the vacuum angle is still a puzzle in particle physics as there is no known physical principle that will cause it to be either extremely small or exactly zero.

Acknowledgement
It is a pleasure to thank professor J. Jortner for the invitation to take part in this interesting meeting. I am grateful to Herbert Neuberger from whom I learned many things about the dilute instanton gas approximation.

References

1)For excellent reviews and a complete list of references see:
     S.Coleman International School of Subnuclear Physics, Erice, Italy, 1977.   A.Zichichi editor, pp. 805-916, Plenum press, New-york 1978.
     S.Rajaraman, Solitons and Instantons, North-Holand, Amsterdam 1982.
2)J.Meixner,F.W.Schäfke, Mathieusche Funktionen und Spheroid-funktionen, Springer Verlag (1954).

REACTION PATHS AND SURFACES FOR HYDROGEN ATOM TRANSFER REACTIONS IN POLYATOMIC MOLECULES

William H. Miller
Department of Chemistry, University of California,
and Materials and Molecular Research Division,
Lawrence Berkeley Laboratory,
Berkeley, California 94720

ABSTRACT. The reaction path/surface Hamiltonian model for polyatomic reaction dynamics is reviewed. Applications to formaldehyde decomposition ($H_2CO \rightarrow H_2 + CO$), vinylidene isomerization ($H_2C=C: \rightarrow HC\equiv CH$), and intramolecular H-atom transfer in malonaldehyde show that tunneling effects play an important role in these processes.

1. INTRODUCTION

Over the last 5-6 years my co-workers and I have developed a series of theoretical methodologies for describing reaction dynamics in polyatomic molecular systems from first principles, namely reaction path and reaction surface Hamiltonian models.[1] This paper reviews these approaches very briefly and discusses their application to several hydrogen atom transfer processes. The effect of tunneling is very significant in these examples and is, in fact, the most interesting feature of them.

2. REACTION PATH/SURFACE HAMILTONIAN AND TUNNELING MODELS

The reaction path Hamiltonian for a general polyatomic reaction, and various applications of it, has been reviewed several times recently,[1h,1k] and the reader should see these for a more detailed presentation. The basic idea[2,3] is that one describes the reaction as motion along the steepest descent path (in mass-weighted cartesian coordinates) that passes through the transition state from reactants to products, plus local harmonic motion away from this reaction path.

It is possible to apply such an approach in an <u>ab initio</u> framework since quantum chemists[4] have developed efficient algorithms for computing the gradients of the Born-Oppenheimer potential surface $V(x)$ (x = all 3N mass-weighted cartesian coordinates). The reaction path $x_0(s)$, as a parametric function of the reaction coordinate s (the cartesian distance along the reaction path), is then determined by the gradient-following prescription

91

$$\frac{d}{ds} \underset{\sim}{x}(s) = - \frac{\partial V(\underset{\sim}{x})}{\partial \underset{\sim}{x}} \Big/ \Big| \frac{\partial V}{\partial \underset{\sim}{x}} \Big| \; . \tag{1}$$

$$\rightarrow \underset{\sim}{x}_0(s) \; .$$

One also needs the force constant matrix along the reaction path $\underset{\approx}{K}(s)$,

$$\underset{\approx}{K}(s) = \Big(\frac{\partial^2 V}{\partial \underset{\sim}{x} \partial \underset{\sim}{x}} \Big)_{\underset{\sim}{x} = \underset{\sim}{x}_0(s)} \; , \tag{2}$$

which is diagonalized (for each value of s) to determine the eigenvectors $\{\underset{\sim}{L}_k(s)\}$ and frequencies $\{\omega_k(s)\}$ for local harmonic motion orthogonal to the reaction path. The reaction path coordinates $s, \{Q_k\}$, $k=1, \ldots, F-1$ ($F=3N-6$ since we consider only total angular momentum $J=0$ in this simplified presentation) are then the reaction coordinate s and the normal coordinates $\{Q_k\}$ for vibrational motion perpendicular to the reaction path. The "old" cartesian coordinates $\underset{\sim}{x}$ are related to these "new" coordinates by

$$\underset{\sim}{x} = \underset{\sim}{x}_0(s) + \sum_k Q_k \underset{\sim}{L}_k(s) \; . \tag{3}$$

The Hamiltonian for the molecular system in terms of these new coordinates, and their conjugate momenta, is (for $J=0$),

$$H(p_s, s, \underset{\sim}{P}, \underset{\sim}{Q}) = H_0 + H_1 + H_2 + \ldots \tag{4}$$

where

$$H_0 = \tfrac{1}{2} p_s^2 + V_0(s) + \sum_{k=1}^{F-1} \Big(\tfrac{1}{2} P_k^2 + \tfrac{1}{2} \omega_k(s)^2 Q_k^2 \Big) \tag{5a}$$

$$H_1 = - \sum_{k=1}^{F-1} Q_k B_{k,F}(s) p_s^2 \; , \tag{5b}$$

etc. $V_0(s)$ is the potential energy along the reaction path, and the functions $\{B_{k,F}(s)\}$, $k=1, \ldots, F-1$ describe coupling of the vibrational modes to the reaction coordinate s. These coupling functions are related to the curvature of the reaction path. The total curvature $\kappa(s)$ of the reaction path in the F-dimensional space at distance s along it is

$$\kappa(s) = \left[\sum_{k=1}^{F-1} B_{k,F}(s)^2 \right]^{\frac{1}{2}} ; \qquad (6)$$

the individual coupling functions $B_{k,F}(s)$ are a measure of how the total curvature of the reaction path projects locally onto the various modes k orthogonal to it. It is this curvature of the reaction path which causes the coupling. Synonymous with the effect of non-separability of the reaction coordinate on tunneling, therefore, is the effect of reaction path curvature on tunneling.

2.1 Tunneling Models

The simplest description of tunneling effects on the rate constant for the reaction is obtained by neglecting the coupling between modes (i.e., using only H_0 for the Hamiltonian Eq. (4)) and then using a statistical approximation - i.e., transition state theory, RRKM theory, etc. - to determine the average rate constant. The unimolecular rate constant, for example, for a molecule with total energy E and total angular momentum J is given by[5,6]

$$k(E,J) = [2\pi\hbar\rho(E,J)]^{-1} \sum_{\underset{\sim}{n},K} P(E-\epsilon^{\ddagger}_{\underset{\sim}{n}JK}) , \qquad (7a)$$

where $\{\epsilon^{\ddagger}_{\underset{\sim}{n}JK}\}$ are the vibrational and rotational energy levels of the "activated complex" in terms of the (F-1) vibrational quantum numbers $\underset{\sim}{n} = (n_1, n_2, \ldots, n_{F-1})$ and angular momentum quantum numbers (J,K). (J is the conserved total angular momentum quantum number and K is the non-conserved projection of total angular momentum onto a body-fixed axis.) $P(E_F)$ is the one-dimensional tunneling probability as a function of the energy E_F along the reactant coordinate at the transition state. ρ is the density of reactant states per unit energy

$$\rho(E,J) = \sum_{\underset{\sim}{n},k} \delta(E-\epsilon_{\underset{\sim}{n}JK}) , \qquad (7b)$$

where $\{\epsilon_{\underset{\sim}{n}JK}\}$ are the energy levels of the reactant molecule in terms of the F vibrational quantum numbers $\underset{\sim}{n} = (n_1, \ldots, n_F)$ and angular momentum quantum numbers (J,K).

Coupling of the reaction coordinate to the transverse vibrational modes - via the curvature coupling in the perturbative H_1, Eq. (5b) - however, can have a significant effect on tunneling probabilities. A very simple model for describing these effects, which is usually semi-quantitative, is the semiclassical infinite order sudden approximation.[7,1c] Here the tunneling probability is

$$P_{\underset{\sim}{n}}(E) = e^{-2\theta_0} \prod_{k=1}^{F-1} I_0(\theta_k) , \qquad (8a)$$

where θ_0 is the usual barrier penetration integral

$$\theta_0 = \int_{s_<}^{s_>} ds\ \sqrt{2[V_n(s)-E]}\ , \tag{8b}$$

and V_n is the vibrationally adiabatic potential,

$$V_n(s) = V_0(s) + \sum_{k=1}^{F-1} \hbar\omega_k(s)(n_k+\tfrac{1}{2})\ ; \tag{8c}$$

i.e., the factor $e^{-2\theta_0}$ in Eq. (8a) is the zeroth order tunneling probability. The action integrals $\{\theta_k\}$ contain the influence of reaction path curvature,

$$\theta_k = \Big|\ \int_s^s ds\ \sqrt{2[V_n(s)-E]}\ [\frac{2n_k+1}{\omega_k(s)}]^{1/2}\ e^{i\delta_k(s)}\ B_{k,F}(s)\Big|\ ; \tag{8d}$$

Since the Bessel function I_0 in Eq. (8a) is an increasing function of its argument (and $I_0(0) = 1$), one sees that reaction path curvature in general increases the tunneling probability.

For the well-studied test case $H + H_2 \to H_2 + H$, this simple model [Eqs. (8a-d)] corrects the tunneling probability from being a factor of 50-100 too small if reaction path curvature is ignored to within a factor of 2 of the correct value.[1c] If the curvature correction of the tunneling probability is smaller than this, then one has some confidence that this model will be able to described it reasonably well.[8-10] There are other such simple models that do an equally good job.

2.2 Reaction Surface Hamiltonian

When the curvature coupling elements become too large, the reaction path description of the dynamics becomes poor. Often, though, it is only one of the coupling elements $\{B_{k,F}(s)\}$ that becomes large - i.e., only one mode k is strongly coupled to the reaction coordinate motion. In such cases it is therefore useful to allow two degrees of freedom to undergo arbitrarily large amplitude motion, by introducing two reaction-like coordinates, with all other degrees of freedom described as local harmonic motion away from this two-dimensional "reaction surface". Since two degrees of freedom are allowed arbitrarily large amplitude motion, all dynamics that involves strong coupling between only two degrees of freedom can be described exactly via this approach. Since two degrees of freedom is dynamically equivalent to a collinear atom-diatom system, all dynamical phenomena that appear in collinear A+BC reactions can be described accurately by this reaction surface model. The present case of H-atom transfer typically involves only two degrees of freedom in an intimate way and should thus be described well.

The same method used to derive the reaction <u>path</u> Hamiltonian[1a] can be generalized to construct the reaction <u>surface</u> Hamiltonian.[1j,1ℓ] It has the same general structure as Eqs. (4) and (5); e.g., H_0 is

$$H_0 = \tfrac{1}{2}\, \underline{P}_r \cdot \underset{\approx}{S}(r_1,r_2)^{-1} \cdot \underline{P}_r + V_0(\underline{r})$$

$$+ \sum_{k=1}^{F-2} \tfrac{1}{2}\, P_k^2 + \tfrac{1}{2}\, \omega_k(\underline{r})^2 Q_k^2 \,, \qquad (9)$$

where $\underline{r} = (r_1, r_2)$ and $\underline{P}_r = (P_{r_1}, P_{r_2})$ are the two coordinates and momenta for motion on the reaction surface, and $\underset{\approx}{S}^{-1}$ is the 2×2 Wilson G-matrix for these two degrees of freedom. $\{Q_k\}$ are the local normal coordinates for harmonic vibrational motion away from the reaction surface. H_1 for the reaction surface Hamiltonian involves a generalization of the curvature couplings of Eq. (5b) – i.e., if the two-dimensional surface were a <u>plane</u>, these couplings would vanish – and couples motion <u>on</u> the reaction surface, i.e., coordinates (r_1,r_2), to motion away from it, i.e., coordinates $\{Q_k\}$. At the level of H_0, therefore, one solves accurately for dynamics on the two-dimensional reaction surface – a very manageable problem – and then coupling of this motion to the vibrational modes k is treated perturbatively.

3.   APPLICATIONS TO HYDROGEN ATOM TRANSFER

3.1   Unimolecular Decomposition of Formaldehyde

The photodissociation of formaldehyde has been studied in

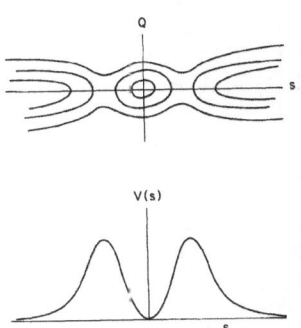

FIGURE 1.   Sketch of the potential energy surface for the formaldehyde dissociation $H_2CO \rightarrow H_2 + CO$. The upper figure is a contour plot of the potential as a function of the reaction coordinate s and the out-of-plane bend coordinate Q. The lower figure is the potential along the reaction path, i.e., along the line Q=0 of the upper figure. The barrier height for this reaction is ~80-90 kcal/mole.

considerable detail experimentally, and there have also been very
elaborate quantum chemical calculations for the potential energy
surface.[11]   Fig. 1 shows a sketch of the ground electronic state
potential energy surface.

   The rate constant k(E,J) [from Eq. (7)] for the unimolecular
decomposition

$$H_2CO \rightarrow H_2 + CO$$

is shown in Fig. 2 for J=0, as a function of energy relative to the
classical threshold. The dashed line is the classical RRKM result
that neglects tunneling. Since the radiative decay rate of excited
formaldehyde is ~10^5 sec^{-1}, unimolecular decomposition will be the
major decay pathway if this rate is greater than 10^5 sec^{-1}. From Fig.
2 one sees that this will be the case for energies as much as 10
kcal/mole below the classical threshold. Tunneling thus plays a large
role in the unimolecular dissociation of formaldehyde into molecular
products H_2 + CO.

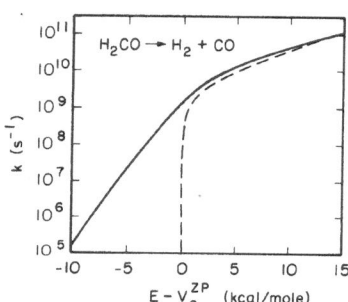

FIGURE 2.   The microcanonical unimolecular rate constant for the
formaldehyde dissociation (J=0) as a function of energy relative to
the classical threshold. The solid line includes the effects of
tunneling, and the broken line is the classical result that neglects
tunneling.

   k(E) in Fig. 2 is the <u>average</u> unimolecular rate for this reaction
as a function of total energy in the molecular, but it is of
considerable interest to determine the <u>state-specific</u> unimolecular
rate constants, i.e., the decay rates of individual quantum states of
formaldehyde. Experimental results for these quantities are beginning
to become available.[12-15]   Calculations have been carried out for a
two-mode model of formaldehyde,[18] the two modes being the reaction
coordinate and the out-of-plane bend mode (this being chosen since it
was considered to be the one most weakly coupled to the reaction
coordinate are thus most likely to show state-specific effects).

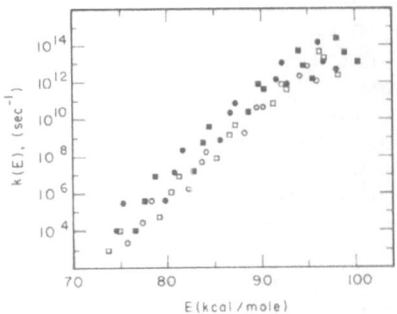

FIGURE 3. State-specific unimolecular rate constants for the two-mode model of the formaldehyde dissociation.

Fig. 3 shows these state-specific rate constants for the two-mode model of the present reaction, and one does indeed see a significant degree of state-specificity. That is, the decay rates for different states with essentially the same total energy can differ by as much as two orders of magnitude. This is approximately the same degree of state-specificity that has been seen experimentally.[14-15]

3.2 Isomerization of Vinylidene

Vinylidene ($H_2C=C:$) is a very shallow minimum on the ground electronic state potential energy surface of H_2C_2.[1f] Acetylene ($HC\equiv CH$), of course, is the much more stable, absolute minimum. Fig. 4 shows the average (microcanomical) rate constant k(E) (for J=0) for isomerization of vinylidene to acetylene,

$$H_2C=C: \rightarrow HC\equiv CH$$

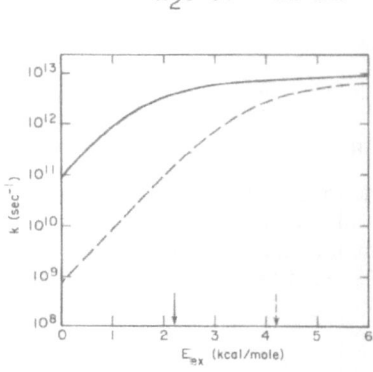

FIGURE 4. Microcanonical rate constant for the vinylidene → acetylene isomerization, as a function of vibrational excitation energy in vinylidene. The solid curve is for a barrier height of 4 kcal/mole and the dashed curve for 6 kcal/mole.

as a function of the excitation energy of vinylidene above its ground
vibrational state. This is shown for two different values of the
assumed barrier height; if the barrier is scaled down to be only 2
kcal/mole, the ground state $(E_{ex}=0)$ rate increases to $\approx 3 \times 10^{12}$ sec^{-1}.

Even with a lifetime $\tau \equiv k^{-1}_{ex}$ of less than a picosecond, however,
vinylidene lives long enough to have a physically meaningful
existence, i.e., it has a well-defined vibrational spectrum.
Moreover, this vibrational spectrum has recently been observed in
photodetachment experiments by Lineberger's group.[16] The observed
vibrational frequencies (and isotopic shifts) are in excellent
agreement with the ab initio quantum chemistry values.

Lineberger's[16] experiments show that not only does vinylidene
exist, but so do its excited vibrational states. Motivated by this,
more detailed quantum mechanical calculations were carried out[11] to
determine the energies and isomerization rates for individual states
of vinylidene. The results are consistent with all experimental
observations to date and show an interesting degree of mode-
specificity. For example, the isomerization rate is increased by a
factor of 6 if the CH_2 rock mode (~1.3 kcal/mole of excitation energy)
is excited; this is the mode which evolves into the reaction
coordinate and is thus expected to be the most efficient in promoting
the reaction. Exciting the CH_2 scissors mode (~3.3 kcal/mole
excitation energy) also increases the rate, but only by a factor of
~2. Exciting the C-C stretch mode (~4.7 kcal/mole excitation energy)
is predicted to decrease the rate by a factor of ~2; this is an
adiabatic effect due to the C-C bond becoming stiffer in going from
vinylidene to acetylene.

These calculations thus provide a rather detailed prediction of
mode-specific features in this reaction.

3.3 Malonaldehyde

Hydrogen atom transfer in malonaldehyde,

is a classic example of a double-well potential; the potential along
the reaction coordinate is shown in Fig. 5. The tunneling splitting
in the ground vibrational state has been observed in the microwave
spectrum[17] and thus provides a direct measure of tunneling in this
polyatomic system.

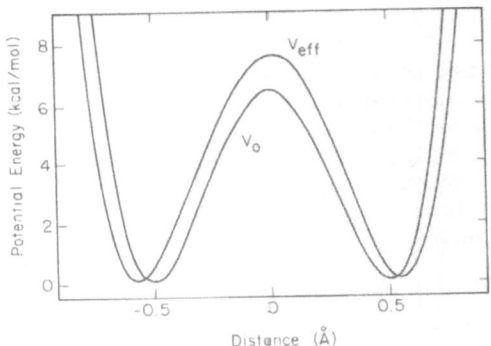

FIGURE 5. Potential energy along the reaction path for intramolecular
H-atom transfer in malonaldehyde. V_o is the "bare" potential and V_{eff}
the vibrationally adiabatic potential, shifted so that the local
minima both are at 0.

Because this reaction path is very strongly curved in this case,
the reaction surface model noted in Section 2.2 was used for this
application.[11] The two reaction surface variables were chosen to be
the two OH bond lengths, r_1 and r_2; Fig. 6 shows the 2-dimensional
surface for this reaction.

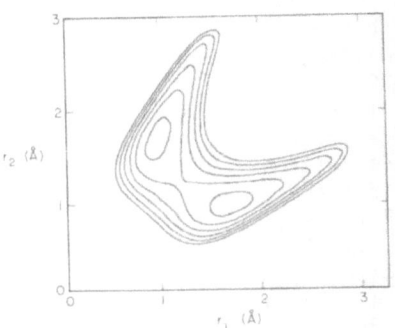

FIGURE 6. Contour plot of the potential energy for malonaldehyde on
the 2-dimensional (r_1,r_2) reaction surface.

Calculations of the tunneling splitting were carried out[11] by
diagonalizing the reaction surface Hamiltonian, Eq. (9), in a 2-
dimensional basis set and then including all the other degrees of
freedom by perturbation theory. It is found that the experimental
tunneling splitting (21cm^{-1}) can be obtained if the barrier height is
adjusted to be 6 to 7 kcal/mole. Within ± 2 kcal/mole, this is in
agreement with various levels of <u>ab initio</u> quantum chemistry
calculations.[18]

4. CONCLUDING REMARKS

As is well-recognized nowadays, tunneling can be a significant aspect of chemical dynamics when the motion is primarily that of hydrogen atoms. This paper has reviewed a methodology, and some of its applications, for carrying out ab initio calculations for such processes in polyatomic molecular systems. Further developments in the methodology and continued applications to various chemical processes can be expected.

ACKNOWLEDGMENTS
This work has been supported by the Director, Office of Energy Research, Office of Basic Energy Sciences, Chemical Sciences Division of the U.S. Department of Energy under Contract Number DE-AC03-76SF00098 and also in part by the National Science Foundation Grant CHE84-16345.

REFERENCES

1. (a) W. H. Miller, N. C. Handy, and J. E. Adams, J. Chem. Phys. 72, 99 (1980); (b) W. H. Miller, in Potential Energy Surfaces and Dynamics Calculations, edited by D. G. Truhlar (Plenum, New York, 1981), p. 265; (c) C. J. Cerjan, S.-h. Shi, and W. H. Miller, J. Phys. Chem. 86, 2244 (1982); (d) S. K. Gray, W. H. Miller, Y. Yamaguchi, and H. F. Schaefer, J. Chem. Phys. 73, 2733 (1980); (e) S. K. Gray, W. H. Miller, Y. Yamaguchi and H. F. Schaefer, J. Am. Chem. Soc. 103, 1900 (1981); (f) Y. Osamura, H. F. Schaefer, S. K. Gray, and W. H. Miller, ibid, 103, 1094 (1981); (g) B. A. Waite, S. K. Gray, and W. H. Miller, J. Chem. Phys. 78, 259 (1983); (h) W. H. Miller, J. Phys. Chem. 87, 3811 (1983); (i) T. Carrington, Jr., L. M. Hubbard, H. F. Schaefer, and W. H. Miller, J. Chem. Phys. 80, 4347 (1984); (j) T. Carrington, Jr. and W. H. Miller, J. Chem. Phys. 81, 3573 (1984); (k) W. H. Miller, in The Theory of Chemical Reaction Dynamics, edited by D. C. Clary (D. Reidel, Boston, 1986), p. 27; (ℓ) T. Carrington, Jr. and W. H. Miller, J. Chem. Phys. 84, 4364 (1986).
2. For early work on reaction paths and reaction coordinates, see (a) S. Glasstone, K. J. Laidler, and H. Eyring, The Theory of Rate Processes (McGraw-Hill, New York, 1941); (b) R. A. Marcus, J. Chem. Phys. 45, 4493, 4500 (1966); 49, 2610 (1968); (c) G. L. Hofacker, Z. Naturforsch. Teil A. 18, 607 (1963); (d) S. F. Fischer, G. L. Hofacker, and R. Seiler, J. Chem. Phys. 51, 3941 (1969).
3. (a) K. Fukui, J. Phys. Chem. 74, 4161 (1970); (b) K. Fukui, S. Kato, and H. Fujimoto, J. Am. Chem. Soc. 97, 1 (1975); (c) K. Yamashita, T. Yamabe, and K. Fukui, Chem. Phys. Lett., 84, 123 (1981); (d) A. K. Fukui, Acc. Chem. Res., 14, 363 (1981); (e) K. Ishida, K. Morokuma, and A. Komornicki, J. Chem. Phys. 66, 2153 (1977).

4. (a) P. Pulay in Application of Electronic Structure, edited by H. F. Schaefer (Plenum, New York, 1977), p. 153; (b) J. W. McIver, Jr., and A. Komornicki, J. Am. Chem. Soc. 94, 2625 (1972); (c) B. R. Brooks, W. E. Laidig, P. Saxe, J. D. Goddard, Y. Yamaguchi, and H. F. Schaefer, J. Chem. Phys., 72, 4652 (1980); (d) J. A. Pople, R. Krishnan, H. B. Schlegel, and J. S. Binkley, Int. J. Quantum Chem. Symp., 13, 225 (1970); (e) Y. Osamura, Y. Yamaguchi, P. Saxe, M. A. Vincent, J. F. Gaw, and H. F. Schaefer, Chem. Phys. 131 (1982); Y. Yamaguchi, Y. Osamura, G. Fitzgerald, H. F. Schaefer, J. Chem. Phys. 78, 1607 (1983); (f) P. Pulay, ibid, 78, 5043 (1983).

5. R. A. Marcus, J. Chem. Phys. 45, 2138 (1966).

6. W. H. Miller, J. Amer. Chem. Soc. 101, 6810 (1979).

7. W. H. Miller and S.-H. Shi, J. Chem. Phys. 75, 2258 (1981).

8. R. A. Marcus and M. E. Coltrin, J. Chem. Phys. 67, 2609 (1977).

9. R. T. Skodje and D. G. Truhlar, J. Chem. Phys. 79, 4882 (1983).

10. B. C. Garrett and D. G. Truhlar, J. Chem. Phys. 79, 4931 (1983).

11. C. B. Moore and J. C. Weisshaar, Ann. Rev. Phys. Chem. 34, 525 (1983).

12. J. C. Weisshaar and C. B. Moore, J. Chem. Phys. 72, 2875, 5415 (1980).

13. W. E. Henke, H. L. Selzle, T. R. Hays, E. W. Schlag, and S. H. Lin, J. Chem. Phys. 76, 1335 (1982).

14. H. L. Dai, R. W. Field, and J. L. Kinsey, J. Chem. Phys. 82, 1606 (1985).

15. D. R. Guyer, W. F. Polik, and C. B. Moore, J. Chem. Phys.

16. S. M. Burnett, A. E. Stevens, C. S. Feigerle, and W. C. Lineberger. Chem. Phys. Lett. 100, 124 (1983).

17. S. L. Baughcum, Z. Smith, E. B. Wilson, Jr., and R. W. Duerst, J. Am. Chem. Soc. 106, 2265 (1984).

18. M. J. Frisch, A. C. Scheiner, H. F. Schaefer, and J. S. Binkley, J. Chem. Phys. 82, 4194 (1985).

OPTICAL STUDIES OF PROTON TRANSFER PROCESSES AT LOW TEMPERATURES IN
HYDROGEN BONDED MOLECULAR CRYSTALS

H.P Trommsdorff
Laboratoire de Spectrométrie Physique associé au C.N.R.S.
Université Scientifique, Technique et Médicale de Grenoble
B.P. 87, 38402 Saint-Martin d'Hères Cedex, France

ABSTRACT. The tautomerization of carboxylic acid dimers at low temperatures is an example of a translational tunneling system. It provides
a model for the study of the interplay of tunneling and relaxation in a
condensed phase environment. A new experimental approach for the study
of such systems was made possible by the discovery of dilute mixed crystal systems of indigo dyes in benzoic acid. The spectral changes of the
dye can be used as very sensitive probes of the structure and the dynamics of neighboring benzoic acid dimers. The measured rates of tautomerization and the deuteration effect are discussed in relation to previous measurements by NMR at higher temperatures and with regard to the
reaction coordinate.

1. INTRODUCTION

Tunneling phenomena reveal in a spectacular manner the quantum mechanical behaviour of a chemical system, they are invoked to explain the
occurence of chemical reactions at 0 K, and are associated with the
fact that the wave function of any particle is more or less delocalized.
This delocalization increases as the mass becomes smaller and tunneling
phenomena are therefore most easily observed for electrons where they
frequently dominate thermally activated processes even at and above
room temperature.
    For havy nuclei, tunneling becomes observable at very low tempera-
tures only ; reactions involving the transfer of hydrogen atoms, on the
other hand, usually show classical behaviour at room temperature and exhibit quantum processes over a reasonable range of low temperatures and
are therefore best suited to study the transition from one regime to the
other. This special role of hydrogen atoms appears also in the hydrogen
bond, which again has to be linked with the delocalization of the proton
between two havier nuclei. Chemical reactions at 0 K correspond to the
transition of a system from one quantum state to an other of lower energy and are therefore a special case of nonradiative relaxation processes : they are special in the sense that the change of geometry between
the two states is sufficiently large such that the two states of the

J. Jortner and B. Pullman (eds.), Tunneling, 103–115.

system can be recognized as different chemical species.

Frequently the dominant geometry change is associated with the coordinate of a single particle, as is the case, for example, in hydrogen transfer reactions. The effective mass of the hydrogen along the reaction path is of course influenced and determined by the geometry changes of the rest of the molecule, that is by the structural relaxation that occurs during the motion of the hydrogen. At higher temperatures the two species exist in a thermal distribution over the energetically accessible states, and depending upon their energy difference, the speed of the reaction, and the timescale of the observation, may no longer be recognized as being different. The existence as different species at low temperatures is linked to the existence of an energy barrier of sufficient height and width separating the two species, such that the lifetime of the unstable species becomes long enough to make it observable.

When the different species corresponding to the minima in the potential energy surface are indistinguishable and therefore isoenergetic, the delocalization of the wavefunctions in the two wells leads to a splitting of the otherwise degenerate states. The textbook example of a tunneling splitting of this nature is ammonia. In condensed phases the environment usually lifts the degeneracy of the two species and at low temperatures (when the thermal energy is smaller than the energy difference produced by the environment), only the most stable species exists.

Most studies of tunneling systems at low temperatures have therefore been directed toward systems which remain symmetric even in a condensed phase environment : this is the case for crystals of light symmetric molecules such as hydrogen and methane, for example, or for molecules containing a methyl group. Even though the environment of the methylated molecule may be, and usually is, asymmetric, the potential for the rotation of the methyl group is exactly threefold degenerate because the three hydrogen atoms of the methyl group are identical. The tunneling splitting of such systems is measured directly by high resolution inelastic neutron scattering /1,2/ and can also be determined by electron spin resonance techniques /3,4/, when the methyl group is attached to a radical ; it has also been shown, that an indirect determination of this splitting can be obtained from the temperature dependence of the proton T_1 relaxation in NMR /5,6/.

Because of their intrinsic high symmetry all these systems are fundamentally different from systems where the hydrogen motion is translational, as in hydrogen transfer reactions. In particular the relaxation behaviour differs because in a rotational tunneling system selection rules are imposed on the nuclear spin functions of the protons in the different tunneling levels. Only one system, which exhibits translational tunneling between symmetric wells has been studied so far in detail, namely hydrogen atoms in niobium metal /7/. In this system the hydrogen atoms can occupy two wells in the lattice, and, at high dilutions, the energies of these two wells are identical. In this paper an other example of a translational tunneling system will be discussed : the tautomerization by double proton transfer of benzoic acid dimers.

2. TAUTOMERIZATION OF BENZOIC ACID

Many carboxylic acids form dimers linked by two hydrogen bonds in both
vapor and condensed phases. There exist two tautomer forms which are
interconverted by simultaneous two proton exchange.

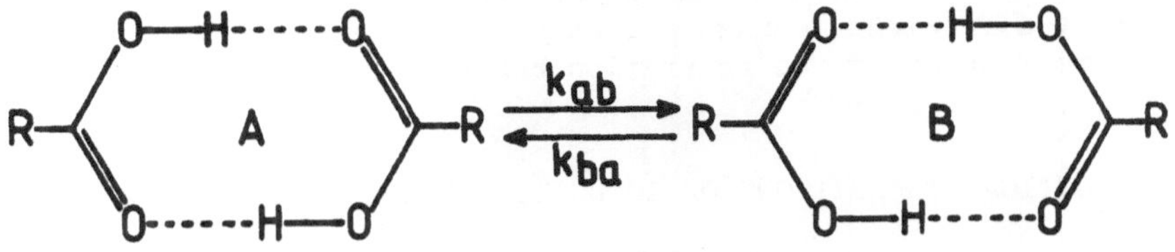

Figure 1. Tautomerization of carboxylic acid dimers.

The double well potential and the dynamics of the interconversion has
recently been the focus of numerous experimental /8-12/ and theoretical
/13-15/ studies. While the two tautomer forms of an isolated symmetric
dimer are identical, they become distinguishable in a condensed phase
environment. In crystals, therefore, the acid protons are more or less
ordered and x-ray determinations of the structure show the C-O double
and single bonds to be more or less well defined depending upon the
temperature and the energy difference of the two tautomers, which in
turn is determined by the packing of the crystal /16,17/.
     The dynamics of tautomerization by double proton transfer in car-
boxylic acid dimer crystals has been studied by nuclear magnetic resonance (NMR) /8-10/ and inelastic neutron scattering (INS) /11,12/ techniques. These studies have established that the tautomerization at room
temperature occurs as a thermally activated process above the barrier
separating the two tautomer forms. In benzoic acid these studies established that the height of this barrier is about 400 cm^{-1} and that the
preexponential factor is $2.7*10^{-1}$ sec^{-1} /8,9,12/. The energy difference
of the two tautomer forms in the crystal is determined by multiple van
der Waals contacts with neighboring molecules and was found to be 35
cm^{-1}, in agreement with previous evaluations based upon the temperature
dependence of infrared bands /18,19/. At lower temperatures the rate of
tautomerization slows down and the deviations of the measured proton T_1
relaxation curves from the theoretical model below about 120 K were
attributed to the onset of tunneling, but no quantitative evaluation of
the speed of this process was made. At very low temperatures (< 10 K)
both NMR and INS measurements fail because the thermal population of
the unstable tautomer form becomes negligeable ; the protons are ordered
in the most stable configuration and all dynamics is frozen out.
     In order to study the dynamics of tautomerization at very low tem-
peratures it is therefore necessary to either find a system where the
energy difference of the two tautomer forms is not large compared to
kT, or an other mechanism to populate the unstable tautomer form of
higher energy. Of all simple carboxylic acids studied so far, the smallest energy difference of the two tautomers was found for benzoic acid
(50 K). Trimesic acid, with a more complicated structure at room temperature containing 48 molecules per unit cell, contains also dimers which

are equivalent by symmetry /20/, but at low temperatures the structure
is unknown and the search for a tunneling splitting by INS has been un-
successful /12/. In systems were the two tautomers have different energies, the overlap of the wavefunctions corresponding to the two forms
becomes very small and the INS intensity becomes too low for useful measurements ; in addition the energy resolution of state of art INS is
hardly sufficient to make a lineshape analysis and to obtain dynamical
information about the relaxation behaviour of such a system.

3. OPTICAL TRANSITIONS OF IMPURITY MOLECULES IN BENZOIC ACID

We were confronted with the problem concerning the proton dynamics in
benzoic acid when we used this material as a host matrix for a dye molecule (pentacene) and used these mixed crystals as model systems for non-
linear optical studies /21/ : at low temperatures, when the protons are
ordered, the observed spectra exhibit inhomogeneous linewidths of about
0.5 cm^{-1}, which are typical for a "normal" mixed crystal system. As the
temperature is raised, however, the spectral lines broaden /22/ : as in
"normal" crystals the modulation of the transition energies by phonons
leads to a homogeneous broadening, but in the case of benzoic acid an
additional factor has to be considered. At higher temperatures the less
stable tautomer form becomes populated and as the energy of a guest molecule depends upon the structure of the environment, the tautomerization leads to an additional broadening of the spectral lines.
      The question therefore arises as to whether this additional broade-
ning is inhomogeneous or homogeneous, that is whether the proton disorder is static or dynamic on the timescale of the experiment. The timescale of the fluctuations of the environment of an impurity molecule
(the dye) depends upon the distance up to which the dye "senses" the
tautomerization of a benzoic acid dimer i.e. the number of dimers to
which it is sensitive, and the speed of the interconversion process.
The first NMR measurements of the high temperature (100–300 K) behaviour
of carboxylic acid crystals became available at this time only /8/, and
we therefore considered ways to obtain information about the speed of
tautomerization at lower temperatures from optical measurements.
      This objective could be attained by the discovery of other suitable
dopants of benzoic acid, namely thioindigo (TI) and selenoindigo (SI)
/23,24/. The doping by impurity molecules alters slightly the energies
of neighboring benzoic acid dimers : the energy difference of the two
tautomer forms of dimers next to the impurity molecule is therefore not
the same as in a pure crystal (35 cm^{-1}), but may be larger or smaller.
These variations depend upon the relative position of the neighbor, the
chemical nature and also the electronic state of the impurity molecule.
For TI et SI there are two physically acceptable ways in which a guest
molecule can replace substitutionally a host dimer of BA, and two origin
regions, separated by about 500 cm^{-1}, are indeed observed for both molecules. For the sites absorbing at higher energy (which in the following
we shall call blue sites) of both TI and SI it happens that the energy
difference of the two tautomer forms of a given BA dimer next to the
impurity in its ground electronic state becomes sufficiently small so

that both forms are populated even at liquid helium temperatures. When
the impurity is in its first excited electronic state, these energy
differences become much larger and in thermal equilibrium the more stable tautomer would be populated only at these temperatures. For the sites absorbing at lower energy (red sites) the tautomer form which is
most stable when the impurity molecule is in its ground electronic state
becomes unstable when the neighboring impurity is brought into an excited electronic state. Optical techniques which monitor the spectral
changes of the impurity (a dye with very strong electronic transitions),
are very sensitive at low temperatures, and can therefore be used to
study proton tunneling processes.

4. SPECTRAL OBSERVATIONS

In crystals at low temperatures, when the protons are ordered, the BA
dimers are centrosymmetric. The indigo dyes replacing a BA dimer also
have a center of inversion, and there exist therefore pairs of equivalent BA dimers, separated by two lattice spacings, surrounding the indigo dye. In order to discuss the optical transitions we need to consider
the one pair of dimers only for which the energy difference of the two
tautomer forms was altered by the impurity in the aforementioned manner, for all other pairs the energy difference of the two tautomer forms
is higher and only the most stable tautomer is populated in the temperature range of our experiments.
     For one pair of dimers caging the guest molecule there exist four
possible configurations of the acid protons. If the two tautomer forms
of one benzoic acid dimer are labelled a and b, the four configurations
are a1a2, a1b2, b1a2 and b1b2 (see figure 2). In a centrosymmetric lattice field only three separate energy states result, because the a1b2
and b1a2 configurations have the same energy. In the absorption and
emission spectra of TI and SI in BA we observe indeed triplets of lines
which can be assigned to these three states of different energy ; these
triplets of lines will be labelled R1, R2, R3 and B1, B2, B3, for the
"red" and "blue" sites respectively. The line corresponding to a1b2
and b1a2 is identified by measurements in an applied electric field
which lifts the energy degeneracy of these two configurations and leads
to a linear splitting of this line which is proportional to the change
upon excitation of the field induced dipole moment of the dye. The lines assigned to the centrosymmetric configurations a1a2 and b1b2, in
contrast, only show a quadratic shift due to the change of polarizability.
     In the ground electronic state the energies of these configura-
tions or sites are determined by measuring their relative population
as a function of the temperature ; the energies in the excited state
are then obtained from the differences in transition energies. The resulting energy level diagram for TI in BA is given in Fig.2. Note that
the energy difference of the B1 and B2 sites is small as compared to kT
even at the lowest temperatures used in our experiments (1.6K) : this is
a particularly interesting situation, because the double well potential
for the tautomerization of one of the neighbors of TI is in this case

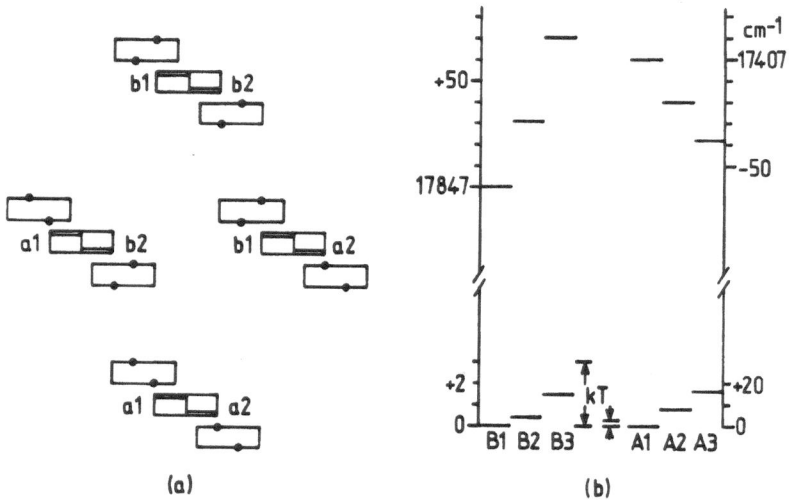

Figure 2.a) Schematic drawing of the four configurations of two benzoic acid dimers caging one dye molecule. b) Resulting energy level diagram for thioindigo in benzoic acid. For the temperature of 4.2 K the value of kT = 3 cm^{-1} is indicated.

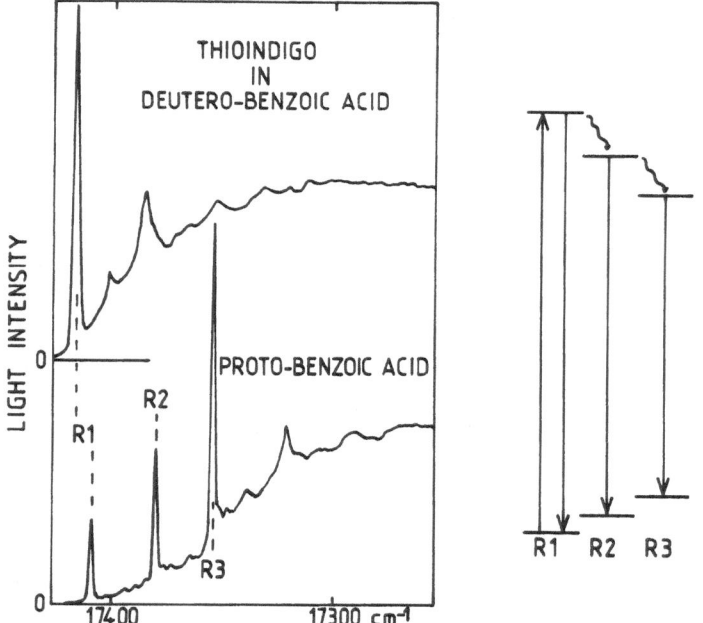

Figure 3. Fluorescence spectra of thioindigo in proto and deutero benzoic acid at 1.6 K for the origin region of the "red" sites, obtained under cw laser excitation of the site having its 0-0 band at R1.

nearly symmetric and the tunneling splitting may therefore contribute
to the observed energy difference. Measurements at lower temperatures
are needed and are presently underway in order to establish whether the
tunneling leads in this case to a partial delocalization of the wavefunction over both potential wells.

Among the red sites only R1, which has the lowest energy, is sufficiently populated at 1.6 K to be observable in the absorption spectra ; in the excited state, however, the relative energies of the three
sites are reversed and the excitation of R1 leads subsequently to the
population of R2 and R3, as is demonstrated in Fig.3. As the oscillator
strength of the TI transition is expected to be largely insensitive to
the environment and as the phonon sidebands of the three red sites are
very similar, the relative populations of the three sites can be evaluated from the intensities of the R1, R2 and R3 lines, and, knowing the
excited state lifetime, can in turn be used to evaluate the rate at
which the site interconversion, that is the tautomerization, occurs.
As will be discussed below, these rates can also be measured more directly by monitoring the rise and decay of the fluorescence from each
site subsequent to the pulsed excitation of R1.

Fig.3 also demonstrates the effect of substituting the acid protons by deuterium : the rate of tautomerization becomes much slower
than the rate of decay of the excited state of TI and almost all of the
emission arises from the initially excited site R1. From spectra of this
type the rate of tautomerization of deutero-BA is evaluated to be at
least three orders of magnitude smaller than for proto-BA.

5. DYNAMICS OF TAUTOMERIZATION

Measurements of the population dynamics, following selective pumping of
one site, such as described above were made, using time correlated single photon counting techniques, for all sites of TI and SI in BA /23,
25/. These measurements, together with the data obtained from the cw
excited fluorescence spectra, support the model proposed in Fig.2 : for
the red sites the experimental data are the most precise and it is found
that the rate of pupolation transfer from R1 to R2, k_{12}, equals twice
the rate k_{23} from R2 to R3, while the rate for direct transitions from
R1 to R3, k_{13} is, within experimental accuracy, equal to 0. These relations correspond precisely to the model : the population transfers between the different sites occur via the tautomerization of a neighboring
BA dimer and $k_{23} = k_{12}/2$ represents the rate of tautomerization. The relative energies of the red sites of TI and SI are very similar, and it
is therefore gratifying to find that the rate of tautomerization measured for the two systems is within 5% the same : it is a property of the
BA host and is not specific for the guest molecule.

In addition to these fluorescence measurements, which monitor the
tautomerization of a BA dimer next to an electronically excited molecule, some measurements have been performed, which indicate that the rates
of tautomerization near a ground state impurity molecule have similar
values. These measurements involve the determination of the homogeneous

linewidth from holeburning experiments. In the absence of pure dephasing in a crystal at very low temperatures, the damping parameter equals one half of the sum of the inverse of the lifetimes of the two states involved in the transition. Knowing the lifetime of the excited state from the fluorescence decay measurements a finite lifetime of the ground state was obtained for the blue sites of TI in BA which exist in thermal equilibrium ; this finite value is ascribed to the tautomerization process and measures its rate /23/.

The tautomerization near a ground state impurity could also be measured by monitoring the changes of the ground state populations of the different sites, after having pumped a significant fraction of the population of one site into an excited state. These experiments are technically more demanding byt very recent experiments using both pump probe and transient grating techniques, indicate again that the site interconversion process in the ground state occurs on a similar timescale /26/. This is consistent with the evaluations via holeburning, but more precise measurements are required to substantiate these findings and to determine values for the rate constants. The energy difference of the tautomer forms are in some of these situations quite small (1 to 5 cm^{-1}) and a measure of the tautomerization rate is particularly important is this case in order to learn how these rates vary as a fonction of the asymmetry of the double well (see below).

6. DISCUSSION

6.1. Comparison with NMR measurements

Fig.4 compares the rate of tautomerization by thermal activation as determined from NMR with the values found in our measurements. Over the limited temperature range of the optical measurements (12 K) these rates are temperature independent. The rates determined at 1.6 to 12 K correspond to the situation where the tautomerization occurs no longer as a thermally activated process but by tunneling. The values are consistent with the proposition /8/ that tunneling becomes important below about 120 K as it appears from Fig.4 that at this temperature the tunneling rate amounts to about 10% of the thermally activated rate and that it would dominate at temperatures below 70 K.

In NMR, the rate of tautomerization is obtained by fitting the variation of the measured proton T_1 relaxation time as a function of the temperature to a two site jump model in which the correlation time τ_c is given by :

$$1/\tau_c = k_{ab} + k_{ba} \tag{1}$$

and k_{ba} is represented as a thermally activated rate :

$$k_{ba} = k_o * \exp(-V/kT) \tag{2}$$

$$k_{ab} = k_{ba} * \exp(-2A/kT) \tag{3}$$

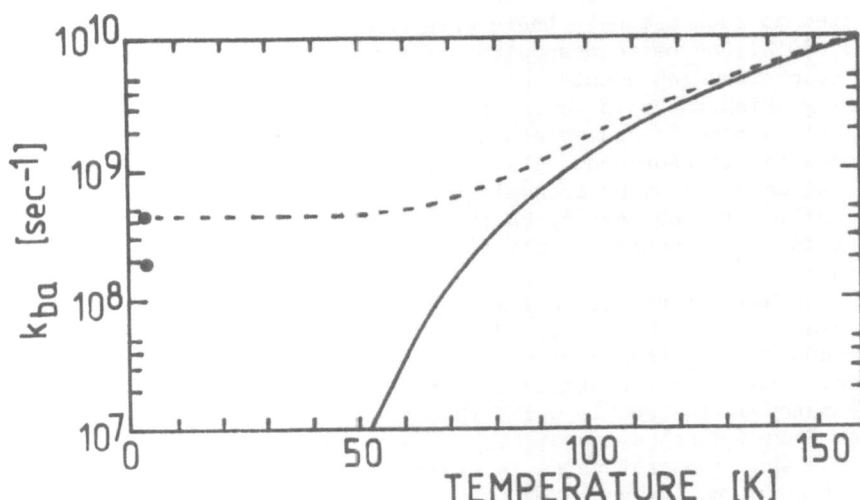

Figure 4. Rate of tautomerization of benzoic acid dimers as a function of temperature. The continuous line represents the thermally activated rate k_{ba} measured by NMR /8,9/. The points give the tunneling rate at low temperatures obtained in the optical measurements and the dotted line gives the sum of the two.

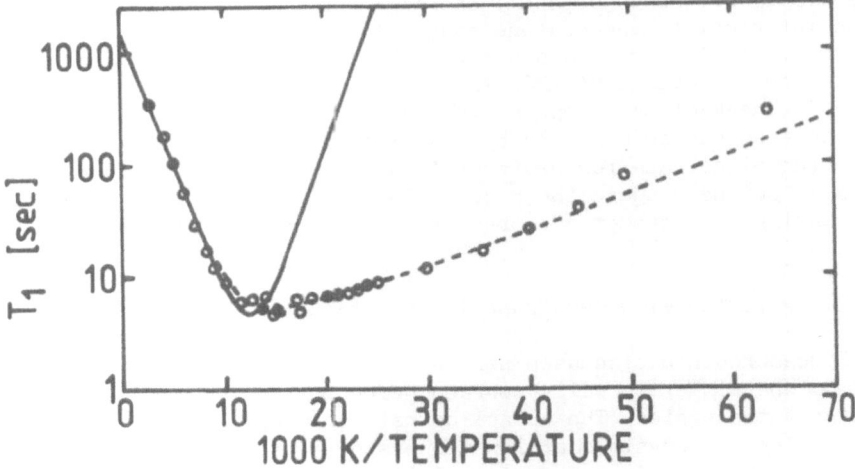

Figure 5. Proton spin lattice relaxation time T1 as a function of temperature taken from the measurements of Nagaoka et al. /9/. The full line gives the fit to a two site jump model including the thermally activated rate only ; the dotted line gives a fit including the tunneling rate.

V is the height of the barrier and 2A the energy difference of the two tautomers. While at high temperatures the experimental data can be well fitted to such a model, there are large deviations at temperatures below 120 K. One basic assumption of the model is that the thermally activated process represents incoherent jumps from one well to the other during which the spin memory is lost. We propose that the rate of the tunneling process can be added to the rate of the thermally activated process as it represents also an incoherent process as long as the tunneling matrix element is small as compared to the asymmetry of the double well : in the case of BA this is certainly true as all measurements show that the tunneling splitting of the symmetric well is smaller than 1 cm^{-1}.

We have attempted to mimic the experimentally observed proton T_1 relaxation curve by simply adding the rate determined in our optical measurements to the previously determined rate for the activated process. Without any adjustments of the parameters (barrier height and asymmetry of the double well) this gives a fit which does show a transition from a steep slope at high to a smaller slope at lower temperatures at the same temperature as is experimentally observed, but in order to mimic quantitatively the experimental data the parameter 2A, representing the energy difference of the two wells has to be increased from its value of about 35 cm^{-1}, as determined previously /8,17,18/, by about 60% (see Fig.5). It should be noted however that these determinations were made at somewhat higher temperatures and that the value of 2A, i.e. the asymmetry of the double well, could quite conceivably increase somewhat at lower temperatures and that it is also not obvious that the tautomerization rate can be represented over the whole temperature range as simply as was done here : one might on the contrary expect that in the intermediate temperature range of about 20 to 120 K thermally activated tunneling processes do contribute to the tautomerization rate /27/.

For deuterated BA /28/ the NMR data of the deuterium T_1 relaxation show a break of the slope of the low temperature branch of the curve, which also can be mimicked by adding to the thermally activated rate the tunneling rate for deutero-BA as determined from optical measurements, but here again the exact value of the slope is only obtained by adjusting the parameter representing the energy difference of the two wells.

6.2. Relation with the tunneling splitting

All measurements discussed here have been made on condensed phase systems, where the two wells corresponding to the two tautomer forms have different energies. The tunneling rate from one well to the other, as determined in crystals at low temperatures, does not represent the frequency at which, in a symmetric well, the system oscillates between the two forms after having been prepared in one well. Yet it is interesting to describe the transition from a symmetric to an asymmetric double well and to relate the properties of the two (there has been considerable theoretical interest in the low temperature dynamics of a tunneling system coupled to a heat bath, see ref.27 and references therein).

For our purpose it is sufficient to consider the lowest two states

of the double well only, and it is convenient to use localized wave-
functions, denoted by $|a>$ and $|b>$, which describe the system in one or
the other tautomer form. With these basisfunctions, and a convenient
choice of the zero of energy, the diagonal matrix elements of this
simple 2x2 Hamiltonian are :

$$<a|H|a> = -<b|H|b> = -A. \qquad (4)$$

2A measures the static asymmetry of the double well. The offdiagonal
matrix elements are :

$$<a|H|b> = <b|H|a> = T \qquad (5)$$

2T, in the symmetric well, equals the tunneling splitting. The wavefunctions of the upper and lower levels are then simply given by :

$$|u> = \sin\phi|a> + \cos\phi|b> \text{ and } |l> = \cos\phi|a> - \sin\phi|b> \qquad (6)$$

where $\tan2\phi = T/A$; the energy separation of these two levels is :

$$\Delta E = 2\sqrt{A^2+T^2} \qquad (7)$$

Using Fermi's golden rule to express the rate of relaxation from the
upper to the lower well we have :

$$k_{ul} = 2\pi/\hbar|<u|R|l>|^2\rho(\Delta E) \qquad (8)$$

In a crystal the phonon density of states $\rho(\Delta E)$ in this energy range is
proportional to the square of the energy. The matrix element of the relaxation matrix R can be expressed in the localized wavefunctions as :

$$<u|R|l> = \sin 2\phi(<a|R|a> - <b|R|b>)/2 + \cos2\phi<a|R|b> \qquad (9)$$

For the rate of relaxation one obtains therefore :

$$k_{ul} \propto 2\pi/\hbar[T(<a|R|a> - <b|R|b>)/2 + A<a|R|b>]^2 \qquad (10)$$

In the symmetric double well the second term equals zero and the first
term, which represents the modulation by phonons of the relative energies of the two wells is responsible for the population relaxation ;
$<a|R|b>$ represents the modulation of the tunneling matrix element and
leads to phase relaxation in the symmetric double well. In the asymmetric well both terms, in principle, can contribute to the population
relaxation between the two levels : if the first term dominates it is
found that the rate of relaxation is proportional to the square of the
tunneling matrix element T multiplied by the modulations of the static
asymmetry of the double well. If the second term is largest, the rate
of pupulation relaxation between the two levels is predicted to be pro-
portional to the square of the product of A and $<a|R|b>$, that is propor-
tional to the square of the modulations of the tunneling matrix element.
        Precise measurements of the tautomerization rate were made on sys-

tems with a static asymmetry ranging from about 17 to 35 cm^{-1} and the tautomerization rate was found to increase by a factor of 2.7 : this is not sufficient to conclusively decide upon the relaxation mechanism and further measurements on systems with much smaller asymmetry, such as were discussed above, are required.

6.3. Deuteration effect and the reaction coordinate

The description of the potential is one of the key issues for the under-standing of the tautomerization reaction, and the height of the poten-tial barrier separating the two tautomers is one convenient criterium for the comparison of theoretical and experimental investigations. It has become clear that, in order to calculate realistic barriers, it is necessary to take into account the structural relaxation of the heavy atom skeleton during the motion of the proton (for a discussion see Hayashi et al. /15/). On the experimental side it is not clear in how far the activation energies measured for the thermally activated rates represent barrier heights : as the barrier becomes more transparent at its top the measured values may underestimate the real height.

When structural relaxation is taken into account, not only the calculated barriers are lowered, but also the relative importance of the motion of the proton along the reaction coordinate is decreased : in other words the effective mass of the proton is increased. As a con-sequence the isotope effect is predicted to be much smaller. For the tunneling splitting, for example, the calculations of Graf et al. on the formic acid dimer /13/ show that not only the absolute values are reduced by 3 to 5 orders of magnitude but also that the ratios of the tunneling splittings (proto/deutero) become smaller than three for bar-rier heights approaching the experimental values.

Such a small deuteration effect seems difficult to reconcile with the large effect found here for the tunneling rates ($\sim 10^3$). Part of this discrepancy may stem from the fact that the above calculation was done for a onedimensional cut of the potential surface representing the minimum energy path connecting the two minima, and it should be interes-ting to apply more refined calculations taking into account more coor-dinates, such as developped currently /29/, to this problem.

ACKNOWLEDGEMENTS

The work discussed in this paper has been performed in collaboration with J.M. Clemens, G.R. Holtom and R.M. Hochstrasser at the University of Pennsylvania in Philadelphia, and with M.Pierre at the University of Grenoble ; a fuller account of it is found in ref.23 to 26. Many discus-sions with R. Silbey concerning tunneling and relaxation have been very enlightening. Discussions with S. Hayashi and N. Hirota and their groups at Kyoto have also been very useful and the permission to quote and dis-cuss their partially-unpublished work is gratefully acknowledged.

REFERENCES

/1/ B. Alefeld and A. Kollmar, Phys. Lett. L57 (1976) 289
/2/ S. Clough, A. Heidemann and M.N.J. Pawley, J. Phys. C13 (1980) 4009
/3/ S. Clough and T. Hubson, J. Phys. C7 (1947) 3387
/4/ S. Clough, A.J. Horsewill and M.N.J. Pawley, J.Phys. C15 (1982) 3803
/5/ S. Clough, A. Heidemann, A.J. Horsewill, J. Lewis and M.N.J. Pawley, J. Phys. C14 (1981) L525
/6/ S. Clough and P.J. McDonald, J. Phys. C15 (1982) L1139
/7/ A. Magerl, A.J. Dianoux, H. Wipf, K. Neumaier and I.S. Anderson, Phys.Rev. Lett. 56 (1986) 156, and references therein
/8/ S. Nagaoka, T. Terao, F. Imashiro, A. Saika, N. Hirota and S. Hayashi, Chem. Phys. Lett. 80 (1981) 580
/9/ S. Nagaoka, T. Terao, F. Imashiro, A. Saika, N. Hirota and S. Hayashi, J. Chem. Phys. 79 (1983) 4694
/10/ B.H. Meier, F. Graf and R.R. Ernst, J. Chem. Phys. 76 (1982) 767
/11/ B.H. Meier, R. Meyer, R.R. Ernst, A. Stöckli, A. Furrer, W. Hälg and I. Anderson, Chem. Phys. Lett. 108 (1984) 522
/12/ B.H. Meier, Ph. D. Thesis, ETH Nr. 7620, Zürich 1984
/13/ F. Graf, R. Meyer, T.-K. Ha and R.R. Ernst, J. Chem. Phys. 75 (1981) 2914
/14/ S. Nagaoka, N. Hirota, T. Matsushita and K. Nishimoto, Chem. Phys. Lett. 92 (1982) 498
/15/ S. Hayashi, J. Umemura, S. Kato and K. Morokuma, J. Phys. Chem. 88 (1984) 1330
/16/ L. Leiserovitz, Acta Cryst. B32 (1976) 775
/17/ Z. Berkovich-Yellin and L. Leiserovitz, J. Am. Chem. Soc. 104 (1982) 4052
/18/ S. Hayashi and N. Kimura, Bull. Inst. Chem. Res., Kyoto Univ. 44 (1966) 335
/19/ S. Hayashi and J.Umemura, J. Chem. Phys. 60 (1974) 2630
/20/ D.J. Duchamp and R.E. Marsh, Acta Cryst. B25 (1969) 5
/21/ P.L. Decola, J.R. Andrews, R.M. Hochstrasser and H.P. Trommsdorff, J. Chem. Phys. 73 (1980) 4695
/22/ K. Duppen, L.W. Molenkamp, J.B.W. Morsink, D.A. Wiersma and H.P. Trommsdorff, Chem. Phys. Lett. 84 (1981) 421
/23/ J.M. Clemens, R.M. Hochstrasser and H.P. Trommsdorff, J. Chem. Phys. 80 (1984) 1744
/24/ H.P. Trommsdorff, R. Casalegno, R.J.D. Miller, J.M. Clemens and R.M. Hochstrasser, J. Luminescence 31/32 (1984) 517
/25/ G.R. Holtom, R.M. Hochstrasser and H.P. Trommsdorff, to be published
/26/ M. Pierre and H.P. Trommsdorff, to be published
/27/ P.E. Parris and R. Silbey, J. Chem. Phys. 83 (1985) 5619
/28/ From T. Terao, Thesis, Kyoto 1985, S. Hayashi, private communication
/29/ T. Carrington, Jr. and W.H. Miller, J. Chem. Phys. 84 (1986) 4364 and references therein.

HYDROGEN TUNNELING IN CHEMICAL REACTIONS. COMPARISON OF THE GOLDEN-
RULE APPROACH WITH TRANSITION-STATE THEORY*

Philip D. Pacey
Department of Chemistry, Dalhousie University,
Halifax, Nova Scotia, Canada   B3H 4J3

Willem Siebrand and Timothy A. Wildman
Division of Chemistry, National Research Council of Canada,
Ottawa, Ontario, Canada   K1A 0R6

A recent treatment of hydrogen-tunneling reactions based on the Golden
Rule of time-dependent perturbation theory is compared with the conventional treatment based on transition-state theory. As a specific
example, hydrogen abstraction by methyl radicals from methyl groups is
studied in detail. Empirical barriers derived by the Golden Rule and
by transition-state theory are compared with a theoretical barrier calculated by ab initio methods for the symmetric reaction of $^\bullet CH_3$ with
CH_4. It is concluded that the methods yield barriers of essentially
the same height and width. This agreement, however, does not extend to
the dynamics derived from the barrier shape. The reason for the
discrepancy is discussed as a first step towards unification of the two
methods.

1. INTRODUCTION

   Tunneling, defined as transfer through a potential-energy barrier,
is a quantum effect (Bell, 1980). Nevertheless, most treatments of
hydrogen tunneling use a semi-classical approach. This may reflect the
desire to treat this reaction in the same way as classically allowed
chemical reactions, namely by transition-state theory (Robinson and
Holbrook, 1972; Forst, 1973) which is essentially semi-classical. It
also serves to demonstrate explicitly the breakdown of Newtonian
mechanics and thus to focus attention on the special character of this
transfer. Whether or not that is desirable remains a question of
taste. There is much to be said for an uncompromising quantum-
mechanical approach both from a conceptual and a practical point of
view. Such an approach was presented in two recent papers (Siebrand,
Wildman and Zgierski, 1984a,b), hereafter referred to as SWZ1 and SWZ2.

---

*Issued as NRCC No. 25531

117

J. Jortner and B. Pullman (eds.), Tunneling, 117–137.
© *1986 by D. Reidel Publishing Company.*

It is based on the Golden Rule of time-dependent perturbation theory and accounts with reasonable accuracy for a representative series of hydrogen transfer reactions in condensed phases. Nevertheless, it would be useful and instructive to consider a treatment closely linked to transition-state theory which is by far the most popular approach to reaction kinetics. Numerous references to this theory, as it applies to tunneling reactions, can be found in recent reviews (Garrett and Truhlar, 1984; Truhlar, Hase and Hynes, 1983) and in an earlier paper by one of us (Furue and Pacey, 1986), hereafter referred to as FP.

In this contribution, we explore the relationship between the two approaches with the ultimate aim of integrating them into a unified theory of hydrogen-transfer reactions. At first sight, they may seem totally different. Transition-state theory describes the process in terms of the path of minimum energy between reactants and products, the so-called reaction coordinate, and focuses on the highest-energy configuration along this path. The properties of this transition state relative to the reactant state are expressed statistically in terms of partition functions, the dynamics of the process being formulated entirely in terms of motion along the reaction coordinate. The Golden Rule, on the other hand, describes the transfer as due to coupling between the initial (reactant) and final (product) states and makes no use of intermediate configurations. However, on closer inspection, connections appear. Although transition-state theory is not concerned with products, the rate of tunneling transfer depends on the height and width of the barrier which are implicitly governed by the nature of the products. In the Golden-Rule approach, degrees of freedom participating in the transfer are treated dynamically rather than statistically. However, in practice the number that can be handled in this manner is limited, so that a statistical treatment of the other degrees of freedom is indicated.

Since tunneling is defined as transfer through a barrier, the nature of the barrier and of the penetrating particle, are of special interest. In the Golden-Rule approach, the barrier does not appear explicitly, but can be constructed from the states involved in the transfer and their coupling. In transition-state theory, the barrier can be deduced empirically from the observed temperature dependence of the transfer. Adiabatic barriers can also be calculated quantum-chemically. The shape and width of the barrier will of course be a function of the coordinate(s) used. In transition-state theory one uses the reaction coordinate, in the Golden-Rule approach the relevant dynamical degrees of freedom. Thus before we can compare these barriers, we have to find the proper relationship between the coordinate systems.

As a concrete example, we consider hydrogen abstraction by methyl radicals from molecules CH_3X which has been studied extensively both in the gas phase and in condensed phases. For data and references, we refer to our earlier papers (FP; Doba et al., 1984a,b). We compare the methods used in these papers to analyze the data and relate the results to new ab initio calculations of the potential-energy surface for the case X=H, as a first step towards a more general description of non-classical transfer reactions.

2. QUANTUM-CHEMICAL CALCULATIONS

In this article we consider bimolecular reactions of the form

$$AH + B \rightarrow A + HB \tag{1}$$

where H is a hydrogen isotope. Specifically, we choose the reaction

$$CH_3X + {}^{\bullet}CH_3 \rightarrow {}^{\bullet}CH_2X + CH_4 \tag{2}$$

which has been studied under a wide variety of conditions ranging from low-temperature glasses to high-temperature gases. In its simplest form, i.e., with X=H, it is amenable to quantum-chemical calculations. We shall use the theoretical adiabatic potential for the system CH_4 + ${}^{\bullet}CH_3$, i.e., the minimum energy path from reactants to products, as a benchmark for comparison of potentials derived empirically from kinetic data. Early calculations based on semi-empirical methods produced a wide range of barrier heights. INDO calculations indicated that the transition state is 169 kJ·mol^{-1} more stable than the reactants. (Rayez-Meaume, Dannenberg and Whitten, 1978). MINDO methods are more successful, but yield barriers lower than the experimental activation energy, which is probably a lower bound on the barrier height (Dewar and Haselbach, 1970; Gilliom, 1934; Canadell, Olivella and Poblet, 1984). The best ab initio calculations to date yield a barrier of about 104 kJ·mol^{-1} (Wildman, 1986).

The preferred geometry of the methyl-methane complex has C_{3v} symmetry with the transferred hydrogen on the CC axis and the methyl groups staggered. The transition state symmetry is D_{3d}. Calculations were performed for a series of hydrogen positions along the CC axis with full geometry optimization.

As described elsewhere (Wildman, 1986), the computational method was based on a modified version (Peterson and Poirier, 1981) of the GAUSSIAN80 program package (Binkley et al., 1980) and the standard [4s2p/2s] double-zeta basis (Huzinaga, 1965; Dunning, 1970) plus p polarization functions on the transferred hydrogen. CI calculations included all single and double replacements relative to the restricted SCF wavefunction (SDCI), excluding the C1s cores and core complements. A size-consistency correction (SCC) was applied uniformly (Davidson and Silver, 1977).

Total energies are collected in table 1; they define the adiabatic barrier along the minimum energy path. Electron correlation has a significant effect on the barrier. Earlier calculations including d polarization functions suggest that the CI result should be regarded as an upper limit to the actual barrier (Wildman, 1986).

Table 1. Total energy and relative energy along the minimum energy
path for $^\bullet CH_3$ + CH_4 with the [4s2p/2s(1p)] basis.

r/pm		SCF		SCF + SDCI[a] + SCC	
CH_3–H	H---$^\bullet CH_3$	E/au	V/kJ•mol^{-1}	E/au	V/kJ•mol^{-1}
133.45	133.45	−79.67482	149.9	−79.91027	98.4
117.90	140.00	−79.68059	134.8	−79.91107	96.3
113.83	150.00	−79.69038	109.1	−79.91685	81.1
111.40	162.00	−79.70092	81.4	−79.92404	62.2
109.98	175.00	−79.70997	57.6	−79.93073	44.7
108.72	200.00	−79.72120	28.1	−79.93954	21.5
108.08	250.00	−79.72978	5.6	−79.94642	3.5
108.02	300.00	−79.73172	0.5	−79.94774	0
107.91	∞	−79.73192	0	−79.94769	0.1

[a]Including 11279 configurations. The coefficient of the root configuration never fell below 0.95.

        To calculate the reaction rate, we also require vibrational poten-
tials. Vibrational frequencies and, to a lesser extent, anharmonici-
ties of the $^\bullet CH_3$ and CH_4 moieties are known spectroscopically. Those
of the $[CH_3$---H---$CH_3]^\bullet$ complex required in the transition state
approach are not known from experiment, but may be calculated quantum-
chemically. Results of such calculations will be reported elsewhere.

3. GOLDEN-RULE APPROACH

        According to the Golden Rule of time-dependent perturbation
theory, the rate constant for the transition from an initially prepared
state $|i>$ to a continuum of final states $|\{f\}>$ driven by a time-
independent coupling operator U is (Merzbacher, 1970)

$$k_{fi} = (2\pi/\hbar)|<\{f\}|U|i>|^2 \, \delta(\varepsilon_i-\varepsilon_f) \, , \tag{3}$$

where the delta function imposes energy conservation. This rule
applies if the final states $|f>$ form a smooth continuum of density ρ_f
in an interval $\Delta\varepsilon_f$ such that

$$\Delta\varepsilon_f \gg |U_{fi}|^2\rho_f \gg |U_{fi}| \, . \tag{4}$$

This condition implies weak coupling through U and sufficiently strong
coupling to a thermal heat bath to generate the desired continuum
$|\{f\}>$. Hence eq. (3) is particularly appropriate for slow transfer
processes in condensed phases.

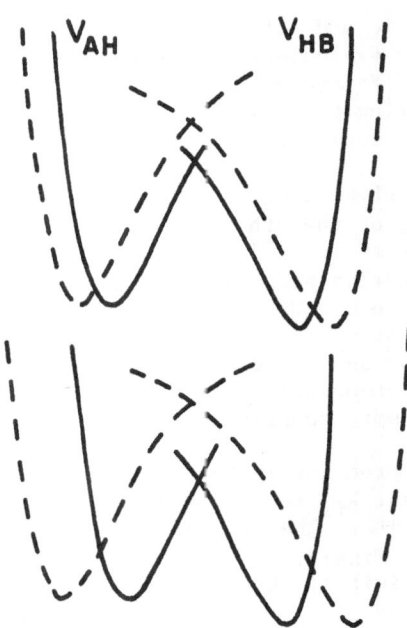

Fig. 1. Model for hydrogen abstraction by a radical B from a molecule
AH. The intersecting solid or dashed potentials represent AH and HB
stretching vibrations, subject to the A---B stretching vibration.
Solid and dashed lines depict potentials appropriate to the inner and
outer turning points, respectively, of the AB vibration. Top and
bottom graphs correspond to a low and high degree of excitation,
respectively, of this vibration.

The application to hydrogen transfer is illustrated in fig. 1.
The hydrogen motion in the reactants before reaction and in the
products after reaction can be visualized in terms of the appropriate
potential-energy surfaces. In the absence of coupling, these surfaces
intersect freely: this is the zeroth-order "diabatic" description
which defines the states $|i>$ and $|f>$. If the coupling is turned on,
the lowest-order result is transition from $|i>$ to $|f>$. For weak
coupling, higher-order contributions, which would renormalize these
wavefunctions, can be neglected.

For a detailed calculation of U_{fi}, we refer to SWZl. Here we note
simply that $|i>$ and $|f>$ should contain all the hydrogen motions rele-
vant to the transfer, thus not only AH and BH stretching and bending,
but also the motions of the atoms in A and B that carry the hydrogen
atoms. As illustrated in fig. 1, the latter motions will strongly
affect the height and width of the effective barrier. This barrier is
diabatic in the present approach; for comparative purposes we can
transform it into the usual adiabatic barrier by applying the coupling
to the zeroth-order surfaces. Note, however, that $|i>$ and $|f>$ are

eigenstates of the diabatic and not of the adiabatic potentials; the eigenstates of the latter are delocalized over the two minima. In condensed phases $|i>$ and $|f>$ are products of vibrational and rotational wavefunctions. The coupling U is essentially an electronic integral involving orbitals of the atoms between which H is transferred.

To turn the states $|\{f\}>$ into a smooth continuum, we couple them to a heat bath consisting of the other degrees of freedom of the molecules A and B and the medium. This coupling is taken so large that neighboring levels $|f>$ overlap strongly. Formally this is achieved by broadening each level ε_f into a Gaussian with a width equal to the level spacing of the lowest-energy oscillator considered explicitly in $|i>$ and $|f>$. Such an approach is justified a posteriori by the observed exponential (first-order) nature of the transition. Choppy continua would lead to a more complicated time dependence.

The calculation now proceeds in three elementary stages. First we calculate the rate constant k_{fi} for every contributing pair of initial and final states $|i>$ and $|f>$. Then we sum over all final states $|f>$, leading to $k(i) = \Sigma_f k_{fi}$. Finally, we calculate the Boltzmann populations $\exp(-\varepsilon_i/kT)$ of all initial state levels $|i>$ and sum over the resulting contributions:

$$k(T) = \sum_i k(i)\exp(-\varepsilon_i/kT)/ \sum_i \exp(-\varepsilon_i/kT) \ . \tag{5}$$

This final sum governs the temperature dependence of the observed rate constant.

In this approach, many of the input parameters, such as effective masses, moments of inertia, interatomic distances and vibrational frequencies can be derived from spectral observations or are available from other sources. The coupling can be calculated quantum-chemically. This makes the treatment tractable and also provides a consistency test.

In SWZ2 the method has been applied to a number of condensed-phase hydrogen-transfer reactions, selected on the basis of the availability of a sufficiently complete set of experimental data. Most of these reactions involve large molecules for which some of the input parameters can only be determined empirically. It is therefore important that the kinetic data cover a wide range of temperatures and include both hydrogen and deuterium transfer. Although the two-oscillator model used in SWZ2 could not deal adequately with all systems selected, in general the method was found to compare favorably with experiment.

In another report (Doba et al., 1984a), the same method has been applied to the reactions

$$CH_3OH + {}^\bullet CH_3 \rightarrow {}^\bullet CH_2OH + CH_4$$

$$CD_3OD + {}^\bullet CH_3 \rightarrow {}^\bullet CD_2OD + CH_3D \tag{6}$$

in a methanol glass in the range $4K \leqslant T \leqslant 100K$. Although the rate constants show dispersion in this medium, a unique rate constant associated with the average tunneling distance could be deduced from the data. The results and their analysis in terms of the Golden-Rule treatment are displayed in fig. 2.

In this analysis two vibrational degrees of freedom are treated explicitly: the CH-stretching vibration in the initial and final state, and the CC-stretching vibration taken to be the highest frequency "lattice" mode in the glass. The two oscillators are taken colinear and the rotations of the methyl groups (including the tumbling of the methyl radical) are assumed to be fast on the timescale of the reaction whose observed half-life varies from minutes to weeks depending on temperature and isotope. All relevant input parameters are known or can be reliably estimated. An excellent fit to the data is obtained with minor adjustments of these estimates. The parameter values are taken from Doba et al. (1984a) with a minor adjustment of J.

In this approach the barrier does not enter explicitly, but can be derived from the input parameters. To allow comparison with other treatments and with the quantum-chemical barrier, we now carry out such a derivation. Since our model contains two degrees of freedom, it leads to a two-dimensional potential-energy surface. If we denote the $^{\bullet}CH_2OH$ and $^{\bullet}CH_3$ radicals by A and B, respectively, and approximate the CH-stretch potential by a Morse function, the AH and HB motions will be governed by potentials

$$V_A(r) = D\{1-\exp[-(r-\bar{r}_A)(m\omega^2/2D)^{\frac{1}{2}}]\}^2$$

$$V_B(r) = D\{1-\exp[(r-\bar{r}_B)(m\omega^2/2D)^{\frac{1}{2}}]\}^2 \ , \tag{7}$$

where the reduced mass m, the frequency ω and the dissociation energy D are taken to be the same for the two oscillators. In the solid environment, the separation R of the atoms A and B is fixed except for fluctuations associated with the A---B (stretching) vibration. This vibration is assumed harmonic and governed by a potential

$$V(R) = \tfrac{1}{2}M\Omega^2(R-\bar{R})^2 \ , \tag{8}$$

where

$$R = \bar{r}_B - \bar{r}_A + 2\xi \ , \tag{9}$$

2ξ being the sum of the equilibrium AH and BH separations. Choosing the point halfway between A and B as the origin, i.e., setting $\bar{r}_A = -\bar{r}_B$, we obtain from (7)

$$V_A(r) = D\{1-\exp[(-\tfrac{1}{2}R+\xi-r)(m\omega^2/2D)^{\frac{1}{2}}]\}^2$$

$$V_B(r) = D\{1-\exp[(-\tfrac{1}{2}R+\xi+r)(m\omega^2/2D)^{\frac{1}{2}}]\}^2 \ . \tag{10}$$

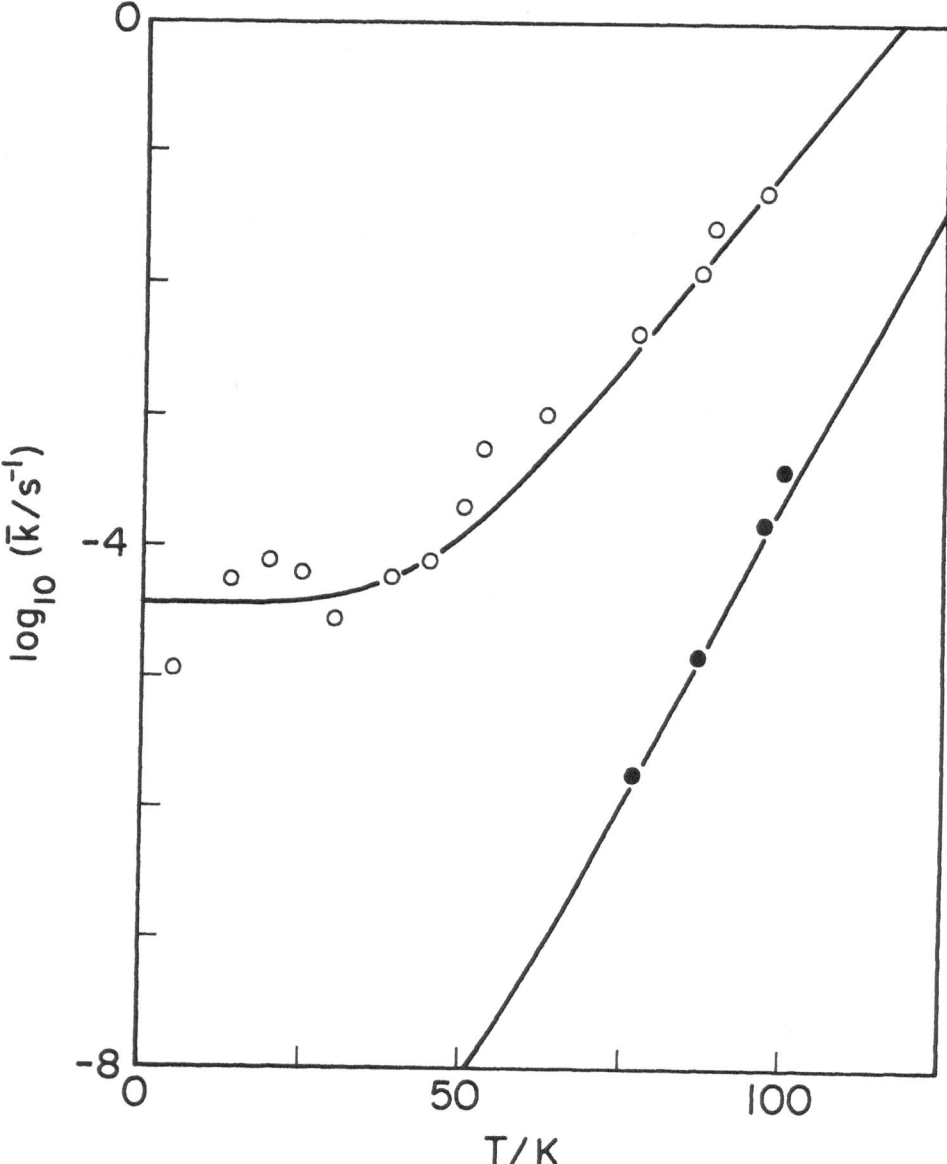

Fig. 2. Semilogarithmic plot of rate constants for methyl radicals in a methanol glass against temperature, taken from Doba et al. (1984a) with minor additions and corrections. Open and solid symbols refer to rate constants $\bar{k}$ corresponding to the average tunneling distance in CH_3OH and CD_3OD, respectively. Solid curves are calculated by the Golden-Rule method.

Combination with (8) yields the diabatic potentials

$$E_A^0(R,r) = V(R) + V_A(r) + \Delta E$$

$$E_B^0(R,r) = V(R) + V_B(r) \ , \tag{11}$$

where ΔE is the exothermicity of the reaction. The corresponding adiabatic potential (lower surface), we obtain by including the mixing of the diabatic surfaces by the coupling $J(R,r)$:

$$E(R,r) = \tfrac{1}{2}\left(E_A^0 + E_B^0\right) - \tfrac{1}{2}\left[\left(E_A^0 - E_B^0\right)^2 + 4J^2\right]^{\frac{1}{2}} \ . \tag{12}$$

For simplicity, we take J to be independent of R and r; in SWZ2, this was shown to introduce only a small error in the analysis of the data. In terms of the present adiabatic surface it means that this surface may be in error for extreme values of R and r, but these regions are unimportant for low-temperature transfer. The resulting surface, calculated for parameter values appropriate to methyl in methanol, is depicted in fig. 3. For clarity, a perpendicular axis system has been chosen. The surface shows the properties of a weakly coupled system: the mixing of the two diabatic surfaces is very small outside the crossing region. Since it is deduced from the experimental data, it will reproduce these data if the calculation is reversed. In this sense, the surface gives a realistic picture of the dynamics of the reaction.

However, for a direct comparison with conventional transition-state results, we need to reduce this surface to a one-dimensional barrier along the reaction coordinate, s. For a given energy E, equally distributed between the two oscillators, the classical path s is determined by the turning points

$$R_{\pm} = \bar{R} \pm (E/M\Omega^2)^{\frac{1}{2}}$$

$$r_{A\pm} = \bar{r}_A - (2D/m\omega^2)^{\frac{1}{2}} \ln\left[1 \mp (E/2D)^{\frac{1}{2}}\right]$$

$$r_{B\pm} = \bar{r}_B + (2D/m\omega^2)^{\frac{1}{2}} \ln\left\{1 \mp \left[(\tfrac{1}{2}E+\Delta E)/D\right]^{\frac{1}{2}}\right\} \tag{13}$$

such that $\bar{r}_B - \bar{r}_A = R_- - 2\xi$. Thus we have

$$s(E) = r_{B+} - r_{A+}$$

$$= \bar{R} - 2\xi - \left(\frac{E}{M\Omega^2}\right)^{\frac{1}{2}} + \left(\frac{2D}{m\omega^2}\right)^{\frac{1}{2}}\left\{\ln\left[1 - \left(\frac{\tfrac{1}{2}E+\Delta E}{D}\right)^{\frac{1}{2}}\right] + \ln\left[1 - \left(\frac{E}{2D}\right)^{\frac{1}{2}}\right]\right\}, \tag{14}$$

which can be inverted to yield the diabatic barrier $E^0(s)$. This barrier consists of two intersecting curves $E_A^0(s)$ and $E_B^0(s)$ from which we can obtain the adiabatic barrier $E(s)$ by means of eq. (12).

The resulting diabatic and adiabatic barriers are depicted in fig. 4. To facilitate comparison with the quantum-chemical barrier, we

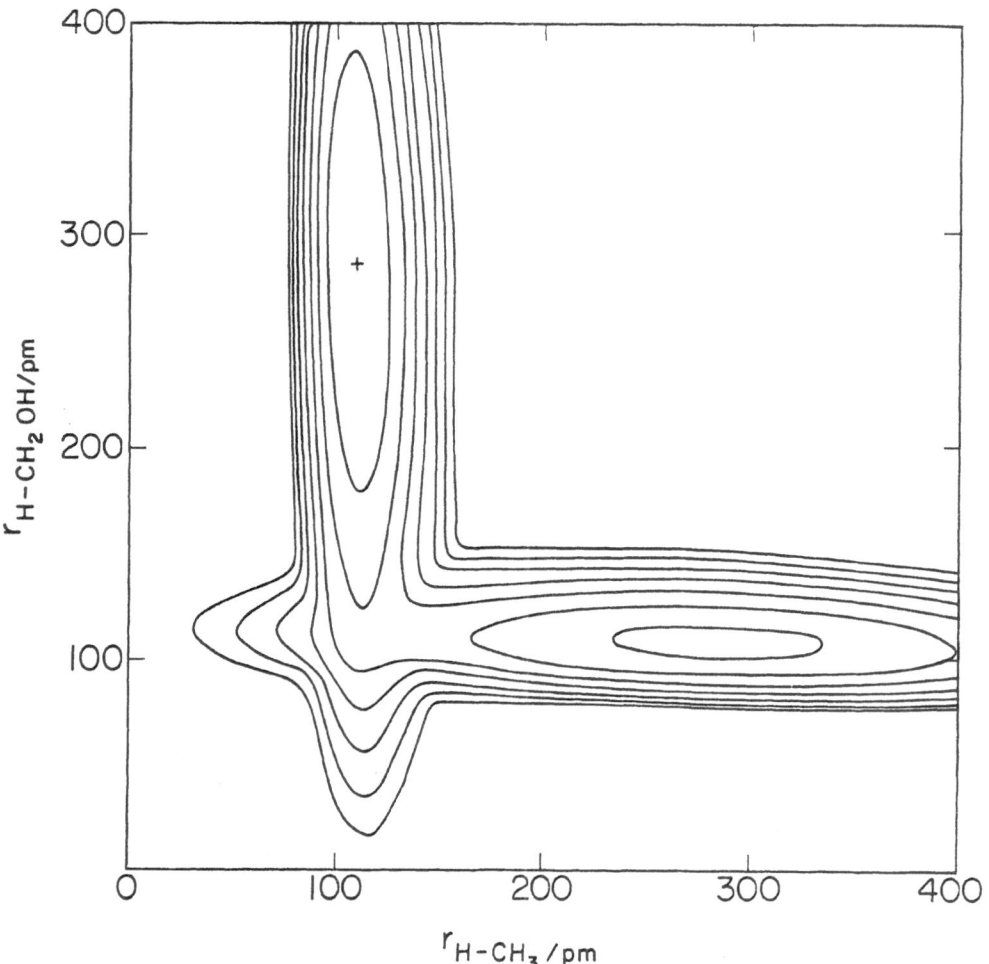

Fig. 3. Adiabatic potential-energy surface derived from the Golden-
Rule treatment of the data displayed in fig. 2 (see text). Parameter
values are those given by Doba et al. (1984a) except that J has been
taken constant. Contours are drawn at intervals of 20 $\hbar\Omega$ = 31 kJ/mol
and + denotes the zero of energy.

also depict a symmetric adiabatic barrier, derived from the same parameters as the asymmetric barrier except for ΔE = 0 (zero exothermicity).
These empirical barriers are thus the "best" approximation the Golden
Rule can provide for a one-dimensional barrier governing hydrogen
transfer in the system $^{\bullet}CH_3$ + CH_3OH. Before probing whether this
barrier can regenerate the data from which it is derived, we consider
the problem from the viewpoint of transition-state theory.

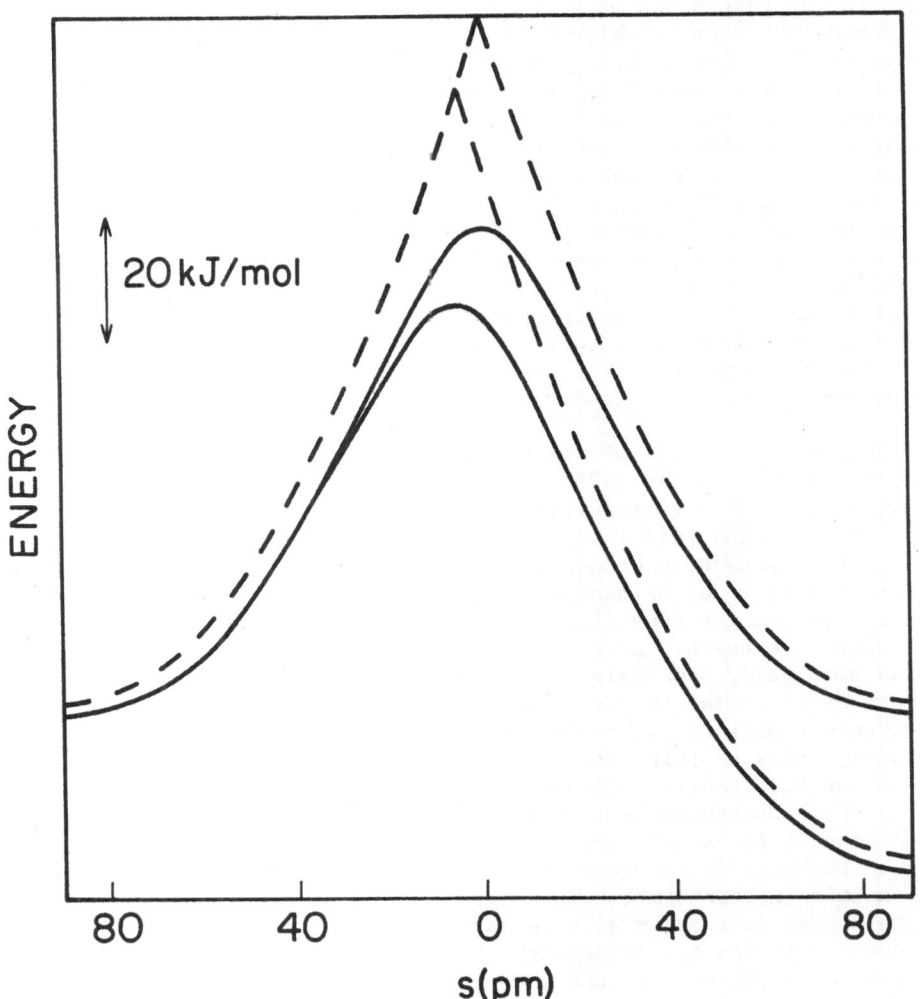

Fig. 4. Potential-energy barriers derived from the Golden-Rule analysis for the hydrogen abstraction rate constants observed for methyl radicals in a methanol glass. The broken lines depict diabatic barriers, the solid lines adiabatic barriers calculated for a constant coupling J between the diabatic potentials. The lower diabatic and adiabatic barriers are derived from the experimental data on the assumption that the exothermicity ΔE = 26 kJ/mol. The upper barriers are derived from the same parameter values except for ΔE = 0; hence they are symmetric.

4. TRANSITION-STATE THEORY

Classically, a reaction leading from one (meta)stable state to another requires passage over a potential-energy barrier. Hence

reaction can take place only if the system has enough energy to reach
the top of the barrier which thus acts as the transition state, i.e.,
the critical dividing surface between reactants and products. Transition-state theory describes the reaction in terms of this surface. It
is based on two general assumptions: (i) the reactant states are
completely equilibrated even though they are not in equilibrium with
the products; and (ii) the reaction rate is given by the unidirectional
flux through the dividing surface, without recrossing. Situations are
known where these assumption break down. For instance, assumption (i)
does not hold for low-pressure unimolecular reactions in the gas phase
and diffusion-limited reactions in condensed phases; assumption (ii)
may fail for highly energetic systems and for condensed-phase reactions
in viscous media, subject to friction. However, similar limitations
apply to the Golden Rule, so that they do not interfere with our
comparison of the two methods.

The current status of transition-state theory has been reviewed by
Truhlar, Hase and Hynes (1983), who cite numerous references. The
theory is essentially classical in its conception and yields a rigorous
upper bound to the rate in classical mechanics. For practical applications, it is generally necessary to introduce further assumptions,
including some based on quantum-mechanical concepts. This is
especially true for tunneling reactions since they are classically
forbidden. To begin, one assumes that electronic and nuclear motions
are adiabatically separable. The reaction is then described in terms
of motions of nuclei in potentials that are eigenvalues of the
electronic Hamiltonian. Reactants (and products) are uniquely defined
as local minima on this potential surface, but there is more than one
choice for the transition state. In conventional transition-state
theory it is chosen so as to contain the saddle point between reactants
and products; in variational transition-state theory, one chooses
instead the dividing surface that results in the minimum transfer rate.
The reaction coordinate running perpendicular to the dividing surface
is the degree of freedom singled out for special treatment. The other
degrees of freedom (translations, rotations and vibrations) can
contribute to the extent that their properties change on passing from
the reactants to the transition state. Usually they are assumed
adiabatically separable and treated by standard thermodynamic methods.
This means that the rate is modulated by the ratio of the partition
functions Q^S and Q^R appropriate to the dividing surface (transition
state) and the reactants, respectively:

$$k(T) = L^{\#}A(T)\left(Q^S_{3N-1}/Q^R_{3N}\right)\exp\left(-E_e/k_BT\right) \ , \tag{15}$$

where N denotes the number of nuclei (including solvent nuclei where
appropriate) to be included explicitly. The degree of freedom excluded
from Q^S is the reaction coordinate. For separable degrees of freedom
j, the partition functions reduce to products

$$Q^S_{3N-1} = \prod_{j=2}^{3N} q^S_j \ , \qquad Q^R_{3N} = \prod_{j=1}^{3N} q^R_j \tag{16}$$

where

$$q_j = \sum_n \exp\left[-\left(\varepsilon_{jn}-\varepsilon_{j0}\right)/k_BT\right] \; , \tag{17}$$

ε_{jn} being the energy of the nth level of the jth degree of freedom, and hence ε_{j0} that of the zero-point level. Hence the activation energy E_e is measured between the zero-point levels of the transition state and the reactants. The factor $L^{\ddagger}$ in (17) is of statistical origin:

$$L^{\ddagger} = g_e^S \prod_j \sigma_j^R / g_e^R \prod_j \sigma_j^S \tag{18}$$

where g_e and σ_j are electronic degeneracies and symmetry numbers, respectively, not contained in (17).

The dynamics of the reaction is represented by the frequency factor $A(T)$ which describes the motion along the reaction coordinate. For classically allowed reactions, one sets $A(T) = k_BT/h$, but for tunneling reactions, a more elaborate treatment is necessary. It is usually assumed that the tunneling can be represented in terms of barrier penetration by a free particle. This allows a standard quantum-mechanical calculation of the probability $\gamma_m(E)$ that a particle of mass m and energy E will reach the other side of a one-dimensional barrier $V(s)$. However, this approach is not without problems. Away from the dividing surface in the reactions discussed here, the reaction coordinate corresponds not only to the motion of the hydrogen atom: according to the analysis of section 3, it also involves the relative motions of the two carbon atoms between which the hydrogen is transferred. Moreover, in condensed phases, the energies before and after reaction will be discrete rather than continuous. In view of these problems, we expect the simple barrier-penetration picture to work best near the top of the barrier where large quantum numbers lead to near-classical behavior. In that case we can write

$$A(T) = \left(k_BT/h\right)\exp\left(E_e/k_BT\right)\int_0^{\infty} \gamma_m(E)\exp\left(-E/k_BT\right)d\left(E/k_BT\right) \; . \tag{19}$$

Substitution into (15) yields

$$k(T) = (L^{\ddagger}/h)\left(Q_{3N-1}^S/Q_{3N}^R\right)\int_0^{\infty} \gamma_m(E)\exp\left(-E/k_BT\right)dE \; . \tag{20}$$

In practice, the barrier is often approximated by a function for which $\gamma_m(E)$ can be evaluated analytically, e.g., an Eckart barrier.

In FP this model has been used to analyze a series by hydrogen abstraction reactions by methyl radicals in the gas phase. These reactions are thermally activated, but if the logarithm of the rate constant is plotted against the reciprocal of the temperature, the result is usually not a straight line. Such curved Arrhenius plots

have long been recognized as an indication that the reaction proceeds
by tunneling. They can be formally written as

$$\ln k = \ln\alpha + \ln\kappa + n \ln T - E_e/k_B T \qquad (21)$$

where α includes k_B/h and the temperature-independent part of the
partition functions, κ is the "tunneling factor"

$$\kappa(T) = \exp(-E_e/k_B T) \int_0^\infty \gamma_m(E)\exp(-E/k_B T)d(E/k_B T) , \qquad (22)$$

and n is the contribution from sources other than κ.

Figure 5, taken from FP, shows a typical fit to eq. (21). The
parameters derived from this analysis for a series of related gas-phase
reactions are collected in table 2, where they are compared with
barrier parameters derived from the quantum-chemical calculations of
section 2 and the Golden-Rule analysis of section 3. The similarity of
these parameters indicates that transition-state theory can account
satisfactorily for high-temperature hydrogen transfer in the gas phase.
Unfortunately, no deuterium transfer data are available for these

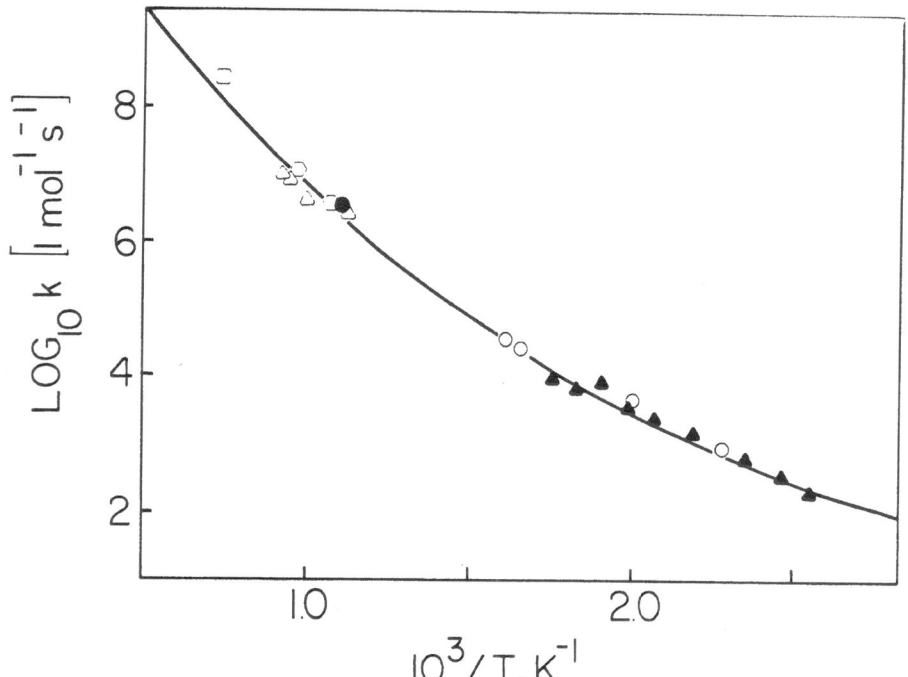

Fig. 5. Arrhenius plots for the reaction $^\bullet CH_3 + C_2H_6 \rightarrow CH_4 + {}^\bullet C_2H_5$,
taken from FP, who give detailed references to the data. The curve is
a least-squares fit of eq. (21).

Table 2. Parameter values derived for one-dimensional adiabatic barriers governing hydrogen transfer from the methyl group in CH_3X to a methyl radical.

CH_3X	Phase	Method[a]	$V_{max}(s)$[b] kJ/mol	E_a[b] kJ/mol	$\Delta s_{\frac{1}{2}}$[c] (pm)
C_2H_6	gas	TST	(87)	74±6	38±3(54)
$C(CH_3)_4$	gas	TST	(79)	66±2	41±2(58)
$(CH_3)_2O$	gas	TST	(78)	65±3	37±2(52)
CH_4	–	QC	98.4	(85)	76
CH_3OH	glass	GRT	77.3	(64)	68

[a]TST = transition-state theory (FP), QC = quantum chemistry (this work), GRT = Golden-Rule treatment (this work).
[b]numbers in brackets are corrected for zero-point energy differences.
[c]for $m = m_H$; numbers in brackets correspond to $m \simeq \frac{1}{2}m_H$ (see FP).

reactions, so that it remains to be established whether this treatment yields the correct isotope effect.

5. DISCUSSION

We are now ready to compare the two kinetic approaches with each other and with the quantum-chemical results. First we compare the one-dimensional adiabatic barriers. In fig. 6 the solid curves marked 1, 2 and 3 refer, respectively, to the quantum-chemical barrier of table 1, the symmetric Eckart barrier derived from fig. 5, and the symmetric adiabatic barrier of fig. 4. They all have roughly the same shape with minor differences in height and somewhat larger differences in width. One reason for these differences is that we are dealing with different systems. The presence of a group X≠H in CH_3X will affect the size of the barrier. In the molecules studied here, the CH bonds are all weaker than those in CH_4, so that the barriers will tend to be lower than the symmetric barrier calculated quantum-chemically. This calculation, on the other hand, is likely to be an overestimate due to the limited basis set used. Note that curve 2 in fig. 6 is corrected for differences in zero-point energies between the reactants and the transition state. These enter implicitly in the analysis of the gas-phase data by means of transition-state theory. In the analysis leading to the barriers in table 2, as reported in FP, four bending modes were treated classically. The difference in zero-point energy for the remaining modes can be estimated indirectly from the quantum-chemical force-field calculations on $[CH_3-H-CH_3]^{\bullet}$ mentioned in section 2. The result for that system, namely 13.2 kJ/mol, should be added to the barrier heights for the gas-phase reactions; the corrected values are shown in brackets in table 2.

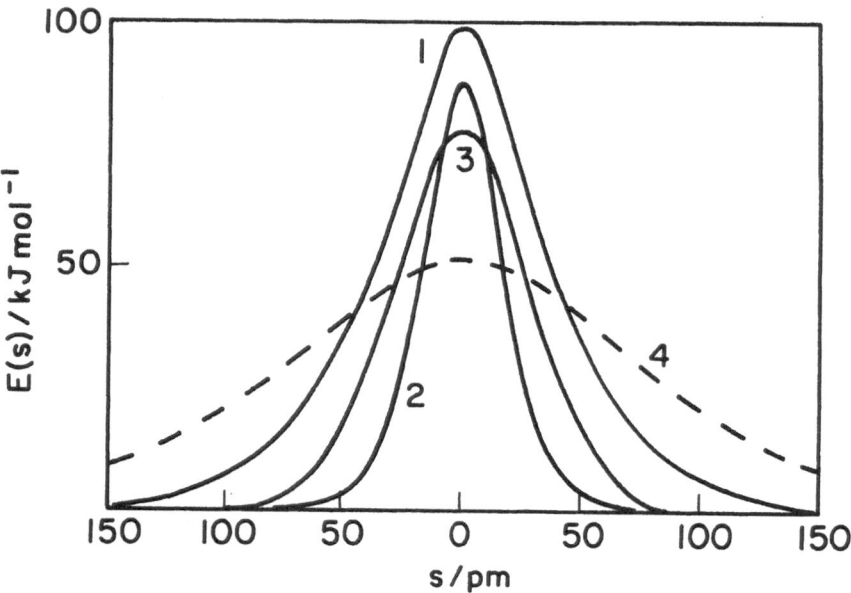

Fig. 6. Comparison of one-dimensional adiabatic barriers for the reaction $^•CH_3 + CH_3X$. Curve 1 corresponds to the CI results of table 1, curve 2 is a symmetric Eckart barrier derived from the fit of fig. 5 for $m = m_H$, curve 3 is the symmetric adiabatic potential displayed in fig. 4 and curve 4 is a symmetric Eckart barrier taken from Le Roy, Murai and Williams (1980) for the reaction $^•CH_3 + CH_3OH$ (glass).

The width of the barriers is expressed as $\Delta s_{\frac{1}{2}}$, the full width at half-maximum along the reaction coordinate. In the quantum-chemical calculations and in the Golden-Rule analysis, this coordinate is not mass-weighted. In the transition-state analysis, however, $\Delta s_{\frac{1}{2}}$ is proportional to $m^{-\frac{1}{2}}$. For purposes of comparison we have set $m = m_H$, the proton mass, to calculate $\Delta s_{\frac{1}{2}}$. The alternative assumption used in FP, amounting to $m \simeq \frac{1}{2}m_H$, would have increased the $\Delta s_{\frac{1}{2}}$ values of the gas-phase reactions, including curve 2 in fig. 6, by a factor $\sim 2^{\frac{1}{2}}$.

As mentioned in section 2, semi-empirical quantum-chemical methods have tended to yield much lower barriers. The empirical barriers derived by Le Roy, Murai and Williams (1980) from low-temperature kinetic data in glasses are lower as well as broader than our results. A typical example, relating to hydrogen abstraction by methyl in a methanol glass is shown as a broken line in fig. 6. This curve, marked 4, is a symmetric Eckart barrier. Although its derivation is based on kinetic data that were not properly corrected for the inhomogeneity of the glass (Doba et al., 1984a), this affects the barrier only slightly. Its shape results from the use of a one-dimensional model that breaks down at low temperatures, as we shall show presently. The remaining

three barriers in fig. 6 are in substantial agreement, since the
residual discrepancies might well be due to differences in the
substrates and uncertainties in the calculations. Hence we conclude
that three completely different methods, namely quantum chemistry,
transition-state theory and the Golden-Rule method yield essentially
the same barrier for hydrogen abstraction by methyl radicals from
methyl groups. Apparently this barrier is not a model-dependent
parameter, but represents a characteristic property of the reaction.

Knowing the barrier, we can calculate high-temperature rate
constants for the reaction by transition-state theory generalized to
tunneling reactions. It remains to be investigated whether this
barrier can also yield reliable low-temperature results. From earlier
work (Laplante and Siebrand, 1978; SWZ2), we know that a one-dimensional Golden-Rule treatment will not reproduce the low-temperature kinetic
data measured in glassy methanol. We therefore try transition-state
theory. To simplify the arithmetic, we approximate the asymmetrical
adiabatic barrier of fig. 4 by an asymmetrical Eckart barrier (Bell,
1980)

$$V(b) = fy/(1+y)^2 + gy/(1+y) \; ; \qquad y = \exp(g/b) \; . \tag{23}$$

Here g is the endothermicity and the barrier height (activation energy)
is given by

$$E_e = V_{max}(b) - E_0 = (f+g)^2/4f - E_0 \; , \tag{24}$$

E_0 being the difference in zero-point energy between transition state
and reactants. The permeability of this barrier for a particle of mass
m and energy E is given by

$$\gamma_m(E) = \frac{\sinh^2[\pi\kappa b(1+\Delta)] - \sinh^2[\pi\kappa b(1-\Delta)]}{\sinh^2[\pi\kappa b(1+\Delta)] + \cosh^2[\pi(mb^2 f/\hbar^2 - \frac{1}{4})^{\frac{1}{2}}]} \tag{25}$$

where

$$K = (2mE)^{\frac{1}{2}}/\hbar \; ; \qquad \Delta = [(E-g)/E]^{\frac{1}{2}} \; . \tag{26}$$

The parameters b, f and g are chosen so as to reproduce the adiabatic
potential of fig. 4 as accurately as possible. The mass m is taken
equal to the mass of the free particle (H or D atom) since there is no
unambiguous way to determine a reduced or effective mass in this
framework. Substitution in eq. (20) yields

$$k(T) = Y(T) \int_0^\infty \gamma_m(E) \exp(-E/k_B T) d(E/k_B T) \; , \tag{27}$$

where $Y(T) = Y_n T^n$ according to eq. (21). The integral is evaluated
numerically and Y_n is treated as an adjustable parameter, chosen so as
to yield the observed rate constant for T = 77K.

The results for H and D transfer are depicted in fig. 7, where the solid and dashed curves refer to n = 0 and 2.5, respectively; the latter value was adopted in FP. It is seen that eq. (27) does not compare well with the observed rate constants. To appreciate the extent to which eq. (27) fails, one should compare fig. 7 with fig. 2 which relates the same data to the results of a Golden-Rule calculation based on the same parameter values. Equation (27) overestimates the deuterium isotope effect by a factor $\geqslant 100$, underestimates the temperature dependence at the high end of the experimental range and grossly overestimates it at the low end. These discrepancies cannot be removed by minor adjustment of the barrier parameters (Pacey, 1979). Although a much better fit to the temperature dependence can be obtained if these parameters are freely varied (Le Roy, Murai and Williams, 1980), the resulting barrier, marked 4 in fig. 6, is physically unacceptable: compared to the quantum-chemical result it is too low and too broad. Moreover it overestimates the observed deuterium isotope effect by a factor of order 10^9. If we restrict ourselves to barriers that can be obtained by independent methods, i.e., the solid-curve barriers of fig. 6, we must conclude that eq. (27) fails to account for low-temperature hydrogen transfer.

Evidently this failure is due to its neglect of quantum effects. If the temperature falls below $\hbar\Omega/k_B$, a limiting rate constant $k(0)$ is approached which is governed by the zero-point states. Since these states cannot be treated classically, the method breaks down. Classically, the rate constant would be expected to vanish for $T \to 0$. The existence of a nonzero limit $k(0)$ is supported by the data on hydrogen abstraction in a methanol glass and is consistent with the observation of a "tunnel" splitting of the zero-point levels of NH_3 and ND_3, associated with the inversion ("umbrella") vibration (Maessen et al., 1984). Such transitions cannot be described in terms of a reaction coordinate but require the explicit treatment of all relevant degrees of freedom, i.e., all those that contribute to the hydrogen motion. This obviously includes modes of the atoms carrying the hydrogen: the resulting potential will thus in general be multi-dimensional and never one-dimensional.

To calculate low-temperature hydrogen transfer rates, it is therefore not sufficient to know the one-dimensional potential energy along the reaction coordinate. Interaction with other relevant degrees of freedom must be included explicitly. It is doubtful whether this can be done adequately by the usual method in which this interaction is introduced in the form of curvature of the reaction path. Such an approach remains essentially classical and is expected to break down at the zero-point level. Our final conclusion is therefore that for low-temperature hydrogen transfer a full quantum treatment is necessary. If the experimental results are compatible with first-order kinetics, a Golden-Rule treatment will do. Since at high temperatures, transition-state theory corrected for tunneling is expected to be adequate, a unified treatment should include elements of both approaches. The present analysis is intended as a first step towards such a unification.

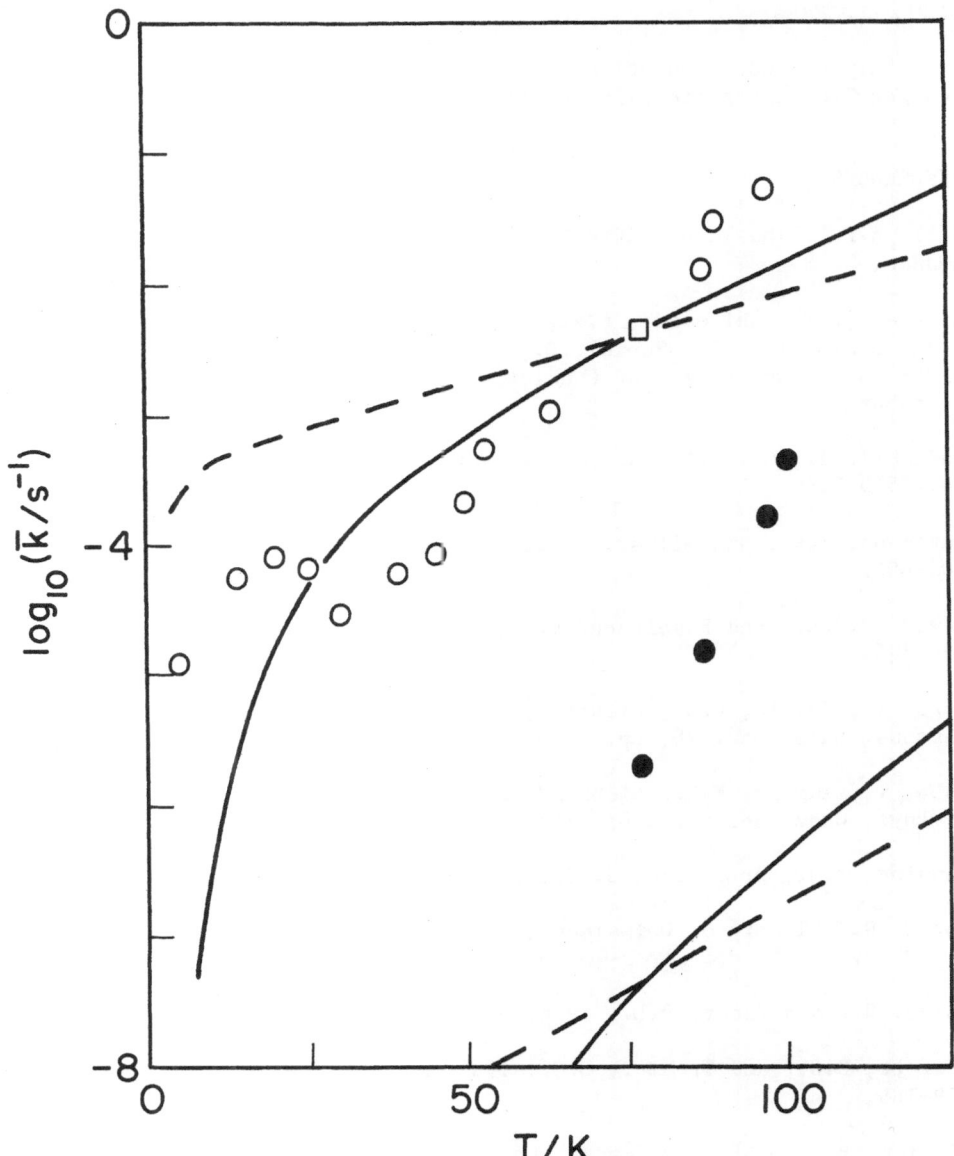

Fig. 7. Same plot as fig. 2 but with theoretical curves calculated from eq. (27) for the same parameter values used before to calculate the Golden-Rule results displayed as solid curves in fig. 2. Solid and broken curves refer to n = 2.5 and 0, respectively, where n is the exponent of T in the pre-exponential factor.

ACKNOWLEDGEMENT

   Helpful and stimulation conversations with Karl Grellmann and
Juergen Troe are gratefully acknowledged.

REFERENCES

Bell, R.P., "The Tunnel Effect in Chemistry", 1980, Chapman and Hall,
London.

Binkley, J.S., Whiteside, R.A., Krishnan, R., Seeger, R., DeFrees,
S.J., Schlegel, H.B., Topiol, S., Kahn, L.R., and Pople, J.A., 1980,
GAUSSIAN80. Department of Chemistry, Carnegie-Mellon University,
Pittsburgh.

Canadell, E., Olivella, S., and Poblet, J.M., 1984, J. Phys. Chem. 88,
pp. 3545-3549.

Davidson, E.R., and Silver, D.W., 1977, Chem. Phys. Letters 52, pp.
403-406.

Dewar, M.J.S., and Haselbach, E., 1970, J. Am. Chem. Soc. 92, pp.
590-598.

Doba, T., Ingold, K.U., Siebrand, W., and Wildman, T.A., 1984a, Faraday
Discuss. Chem. Soc. 78, pp. 175-191.

Doba, T., Ingold, K.U., Siebrand, W., and Wildman, T.A., 1984b,
J. Phys. Chem. 88, pp. 3165-3167.

Dunning, T.H., Jr., 1970, J. Chem. Phys. 53, pp. 2823-2833.

Forst, W., "Theory of Unimolecular Reactions", 1973, Academic Press,
New York.

Furue, H., and Pacey, P.D., 1986, J. Phys. Chem. 90, pp. 397-400.

Garrett, B.C., and Truhlar, D.G., 1984, Ann. Rev. Phys. Chem. 35, pp.
159-189.

Gilliom, R.D., 1984, J. Comput. Chem. 5, pp. 237-240.

Huzinaga, S., 1965, J. Chem. Phys. 42, pp. 1293-1302.

Laplante, J.P., and Siebrand W., 1978, Chem. Phys. Letters 59, pp.
433-436.

Le Roy, R.J., Murai, H., and Williams, F., 1980, J. Am. Chem. Soc. 102,
pp. 2325-2334.

Maessen, B., Bopp, P., McLaughlin, D.R., and Wolfsberg, M., 1984, Z. Naturforsch. 39a, pp. 1005-1006.

Merzbacher, E., 1970, "Quantum Mechanics", 2nd Edition, Wiley, New York, pp. 475-486.

Pacey, P.D., 1979, J. Chem. Phys. 71, pp. 2966-2969.

Peterson, M.R., and Poirier, R.A., 1981, MONSTERGAUSS. Department of Chemistry, University of Toronto, Toronto.

Rayez-Meaume, M.T., Dannenberg, J.J., and Whitten, J.L., 1978, J. Am. Chem. Soc. 100, pp. 747-749.

Robinson, P.J., and Holbrook, K.A., 1972, "Unimolecular Reactions", Wiley, New York.

Siebrand, W., Wildman, T.A., and Zgierski, M.Z., 1984a, J. Am. Chem. Soc. 106, pp. 4083-4088.

Siebrand, W., Wildman, T.A., and Zgierski, M.Z., 1984b, J. Am. Chem. Soc. 106, pp. 4089-4096.

Truhlar, D.G., Hase, W.L., and Hynes, J.T., 1983, J. Phys. Chem. 87, pp. 2664-2682.

Wildman, T.A., 1986, Chem. Phys. Letters, in press.

TUNNELING MEDIATED BY HEAVY LARGE AMPLITUDE MOTION

P. F. Barbara
Department of Chemistry
University of Minnesota
Minneapolis, MN 55455

ABSTRACT. The excited state intramolecular proton transfer kinetics
of a variety of organic molecules have been studied by picosecond
fluorescence spectroscopy. The temperature, solvent, and molecular
structure dependence of the excited state dynamics can be rationalized
by invoking proton tunneling. The kinetic studies suggest that the
proton tunneling, in some of the molecules studied, is strongly
influenced by vibrational motion of the heavy atoms (carbon and oxygen)
of the reactant.

1. INTRODUCTION

Recent picosecond laser studies on molecules undergoing ultrafast
excited-state-intramolecular-proton-transfer (ESIPT) have significantly
advanced the understanding of small barrier (< 5 kcal/mol) proton
transfer reactions of polyatomics. The present paper summarizes a few
of the principles established in these studies with an emphasis on
experiments from our laboratory on the molecules 3-hydroxyflavone
(3HF) [1-4], 2-(2'-hydroxyphenyl)benzothiazole (HBT) [5], 2-(2'-hydroxy-5'-
methylphenyl)-benzotriazole (HMPB), [6], 1,5-dihydroxyanthraquinone
(DHAQ) [7] and related compounds.
    The driving force for the ESIPT reactions that are discussed in
this paper is believed to be associated with the well-known acid/base
properties of aromatic molecules in S_0 and S_1, the ground and first
excited singlet states [8]. It has been observed that aromatic alcohols
and amines are more acidic in S_1 than S_0, and aromatically conjugated
carbonyls and carbon nitrogen double bonds are more basic in S_1. For
example, consider the ESIPT molecule HBT which is represented as
follows. The equilibrium geometry in S_0 is the enol tantomer and the
$S_0 \rightarrow S_1$ absorption is due entirely to enol absorption. The relaxed $S_1 \rightarrow S_0$
fluorescence is due to the keto form. Consequently, the emission is
strongly Stokes-shifted from the $S_0 \rightarrow S_1$ absorption. The spectroscopic
data on HBT [5] and related compounds [9-10] strongly suggests that
the proton transfer reaction coordinate is a double-minimum potential
with enol and keto minima, as discussed below.

139

J. Jortner and B. Pullman (eds.), Tunneling, 139–148.
© *1986 by D. Reidel Publishing Company.*

S₀ enol   **S₁ enol**   **S₁ keto**

Picosecond time and wavelength resolved fluorescence measurements have
shown that the proton transfer rate constant of a variety of ESIPT
molecules in non-hydrogen-bonding solvents is unresolvably rapid
($k_{PT} > 10^{11} s^{-1}$) even at crogenic temperatures [1-7,9-12]. However, in
hydrogen-bonding solvents, k_{PT} can be reduced by orders-of-magnitude.
In most ESIPT molecules studied to date, k_{PT} is so slow in hydrogen-
bonding environments that it does not effectively compete with other
processes that dissipate the enol form of S_1. Consequently, it is
impossible in these cases to obtain data on k_{PT} in hydrogen-bonding
solvents for most known ESIPT molecules. An important exception is the
compound 3HF, which exhibits an ESIPT reaction whose kinetics can be
accurately studied in a variety of hydrogen-bonding solvents [1-4],
including hydrogen-bond donating solvents such as alcohols and accepting solvents such as ethers.

The ESIPT reaction of 3HF was discovered by Sengupta and Kasha
[13], and subsequently studied by several researchers, for example,
see [1-4,14-16] and references therein. The stable S_0 isomer of 3HF
is the normal N form.

N   **T**

S_1 molecules of 3HF in the N form (λ_{F1}^{max} = 413 nm) undergo a rapid
proton transfer isomerization (N→T) to yield a tantomer T form
(λ_{F1}^{max} = 543 nm). The proton transfer (N→T) kinetics are strongly
dependent on solvent, isotope substitution and temperature.

Much of the recent research on ESIPT molecules has been concerned
with the underlying physical basis of the proton transfer processes.
In the case of non-hydrogen-bonding solvents, one of the key issues is
whether the unresolvably rapid proton transfer rate constant is a

simple consequence of the absence of a energy barrier to proton transfer or, alternatively, whether proton tunneling through a barrier is
actually responsible for the rapid rate. For ESIPT molecules in
hydrogen-bonding solvents, recent research has emphasized the study of
how hydrogen-bonding interactions modulate the proton transfer process.
    In particular, the dynamic participation of solute-solvent
hydrogen-bonding coordinate in the proton transfer reaction coordinate
has been discussed extensively for 3HF and derivatives of this compound
[3-4,17-18].
    The present understanding of these isssues will be discussed in
the following sections of this paper which deal with specific examples
of ESIPT molecules.

2. THE PHOTODYNAMICS OF HBT and HMPB

The fluorescence spectrum and the fluorescence excitation spectrum of
HBT in an argon matrix at 12 k is shown in Figure 1 [5]. The observed
emission is due to the keto form as shown in Eq. 1. The excitation
band, in contrast, is assigned to absorption of enol, i.e., $S_o \rightarrow S_1$. The
absence of detectable enol fluorescence is due to the rapid proton
transfer process in S_1.

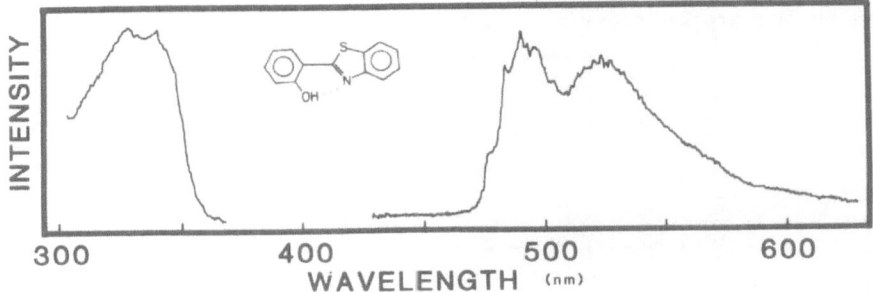

Figure 1. All fluorescence measurements were made with front-surface
collection at normal (90°) incidence and excitation at 30° incidence.
For the fluorescence excitation experiments (∿2nm resolution) (Figure
1), a 250-W xenon lamp (Hanovia) coupled to a 0.25-m Jarrel Ash monochromator was used for excitation. The fluorescence was monitored (521
nm) with a 25-m HR320 monochromator (Instruments SA) coupled to a 1P28
(RCA) photomultiplier tube (PMT). The photon flux of the excitation
source was monitored with a "quantum counter", i.e., an optically dense
dye solution viewed by a 1P28 PMT. The excitation spectrum in Figure 1
represents a ratio of fluorescence output to excitation photon flux
determined by an analog ratiometer. The nontime-resolved fluorescence
spectra in the figure was recorded with an HR320 polychromator coupled
to a photodiode array detector, PDA (Model 1412, Princeton Applied
Research, Photocathode S20). The sample was excited at 355 nm with a
Nd:YAG laser, see below.

Picosecond measurements [5,9-10] and static spectroscopic data show that the proton transfer rate constant actually exceeds $10^{11}s^{-1}$. This rapid proton transfer rate has been interpreted by a model that assumes no barrier exists for the enol→keto tantomerization in S_1. However, the static spectroscopic data indicates that the enol form of S_1 actually is metastable and lives for a few vibrational periods of spectroscopically relevant vibrational modes [5,9-10].

We have previously suggested that the proton transfer process of S_1 HBT might resemble an intramolecular vibrational redistribution (ivr) process. In other words, the ivr process would be the rate-limiting step to proton transfer which would imply that the "barrier" to proton transfer is zero or at least smaller than the vibrational zero point energy of the S_1 enol. On the other hand, the proton transfer may more accurately be represented by a model in which the enol→keto interconversion is viewed as a non-thermally-activated tunneling process through a small (∿1 kcal/mol) barrier. This latter model has been suggested for several symmetrical ESIPT molecules, for which tunneling splittings have actually been observed [19-20].

Another interesting feature of the photodynamics of HBT is portrayed in Fig. 2. The short wavelength band edge of the keto emission exhibits a rapid evolution at early times immediately following proton transfer. This effect has been ascribed [5,9,10] to vibrational relaxation of the S_1 keto form which presumably is vibrational energy excited by the proton transfer reaction.

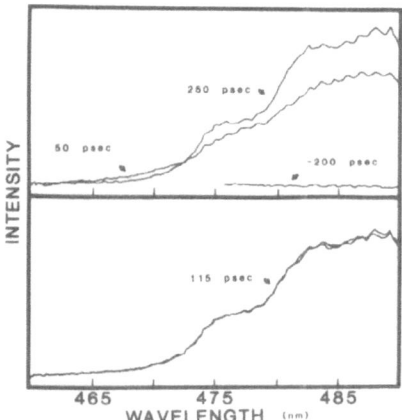

Figure 2. HBT emission spectra in argon at 12 K at different time regions with respect to excitation at 355 nm. The experimental details may be found in the text. Each spectrum represents data averaging for ∿5 min.

The proton transfer mechanism we have just outlined for HBT is generally relevant to a variety of PT molecules, including the widely studied methyl salicylate [11], the photochromic anils [9-10], and the polymer photostabilizer HMPB[7]. In the last case the excited state proton transfer processes is a critical element of the photo-dynamic scheme that is responsible for the extremely rapid and

efficient electronic radiationless decay of 3HMPB. In turn, the rapid
radiationless decay mechanism of HMPB is the source of the extra-
ordinary photostability of HMPB which is extensively used commercially
as a polymer photostabilizer.

Figure 3 portrays a photodynamic scheme for HMPB. The central
structure in the figure is the stable S_o isomer in non-hydrogen bonding solvents. It is analogous to the enol form of HBT. The structure
on the right hand side is analogous to keto HBT. The structure on the
left is relevant for hydrogen bonding solvents where the solvent is
represented by (:B). In environments of this type proton transfer is
too slow to observe experimentally. This effect has been interpreted
by assuming that intermolecular hydrogen-bond for HMPB "breaks" the
intramolecular hydrogen-bond. Presumably, the intramolecular hydrogen-
bond is necessary for ultrafast proton transfer [5,12,21].

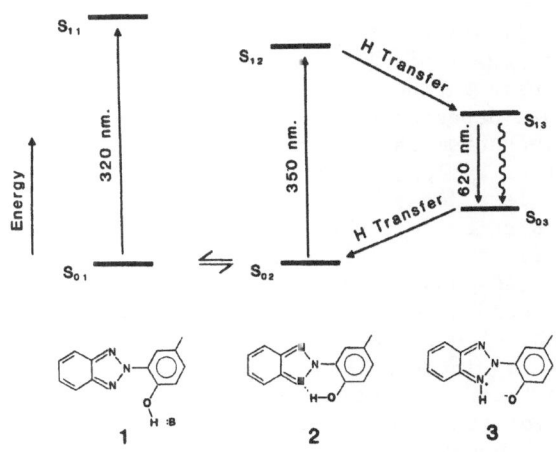

Figure 3. A schematic representation of the photodynamics of HMPB in
the form of a Jablonski diagram.

3. THE ULTRAFAST PROTON TRANSFER OF DHAQ

Certain ESIPT molecules exhibit enol and keto type fluorescence
simultaneously from the relaxed form of S_1. In these cases the vibrational zero point energy of the S_1 molecule is apparently greater than
the enol-keto energy gap, so in the relaxed form of S_1 both tantomers
are apparent. We have previously denoted this type of dual fluorescence
as intrinsic dual fluorescence to distinguish it from the more common
source of dual fluorescence for ESIPT molecules, i.e., a distribution
of slowly interconverting solute/solvent complexes and/or intramolecular rotamers [7]. The first report of intrinsic dual fluorescence
was made by Brus and coworkers for methyl salicylate [23].

A particularly interesting example of a molecule with intrinsic dual fluorescence is DHAQ which has been studied by static [22] and dynamic [7] methods. Picosecond data on DHAQ show that the ESIPT process of this molecule is unresolvably rapid ($k > 10^{11} s^{-1}$). Static, high-resolution spectroscopy on DHAQ indicates that a barrier actually exists [22] in S_1 between enol and keto. It seems reasonable, therefore, to interpret the rapid proton transfer as non-thermally-activated tunneling through a barrier between the enol and keto forms of S_1.

S_1 enol   **S_1 keto**

The fluorescence lifetime τ_{fl}^{obs} of DHAQ is only slightly temperature dependent in various types of solvent. The lifetime increases from 375 psec to 470 psec in the range 100–300 k. The absence of a significant temperature dependence of τ_{fl}^{obs} demonstrates that the radiationless decay mechanism is not thermally activated. We have interpret this result as evidence that the keto form is the direct precursor to radiationless decay of S_1 DHAQ.

The dynamic and static fluorescence spectroscopic properties of DHAQ are extraordinarily insensitive to whether the solvent is non-polar (hexane), polar (ACN and ethanol), hydrogen bond donating (ethanol) or hydrogen bond accepting (2MTHF). Apparently the intramolecular hydrogen bond of both the S_0 and S_1 molecules is extremely strong and intermolecular interactions are only a minor perturbation. The absence of a significant solvent effect on τ_{fl}^{obs} strongly suggests that intramolecular hydrogen bonding and ESIPT are associated with the rapid radiationless decay rate of DHAQ in all of the solvents that we have studied.

The driving force for ESIPT in DHAQ can be associated with the intramolecular charge transfer character of S_1 anthraquinones with electron-donating substituents. Charge-transfer in such compounds is believed to occur in the lowest excited states between the electron-donating substituents and the quinoid-oxygens. The negative charge should be localized on the oxygen.

4. PROTON TRANSFER KINETICS OF 3HF

The excited-state dynamics of 3HF, and derivatives of this compound, are considerably more complex than that for the other molecules mentioned above. In non-hydrogen bonding solvents the proton transfer kinetics are unresolvably rapid [3,4,17] in analogy with HBT, HMPB, and DHAQ. However, in hydrogen-bonding solvents k_{PT} varies from unresolvably rapid ($k_{PT} > 10^{11} s^{-1}$) to too slow to measure ($k_{PT} < 10^6 s^{-1}$) depending on the hydrogen-bonding ability of the solvent, the temperature, isotopic substitution, and the structure of the particular derivative of <u>3HF</u> being considered.

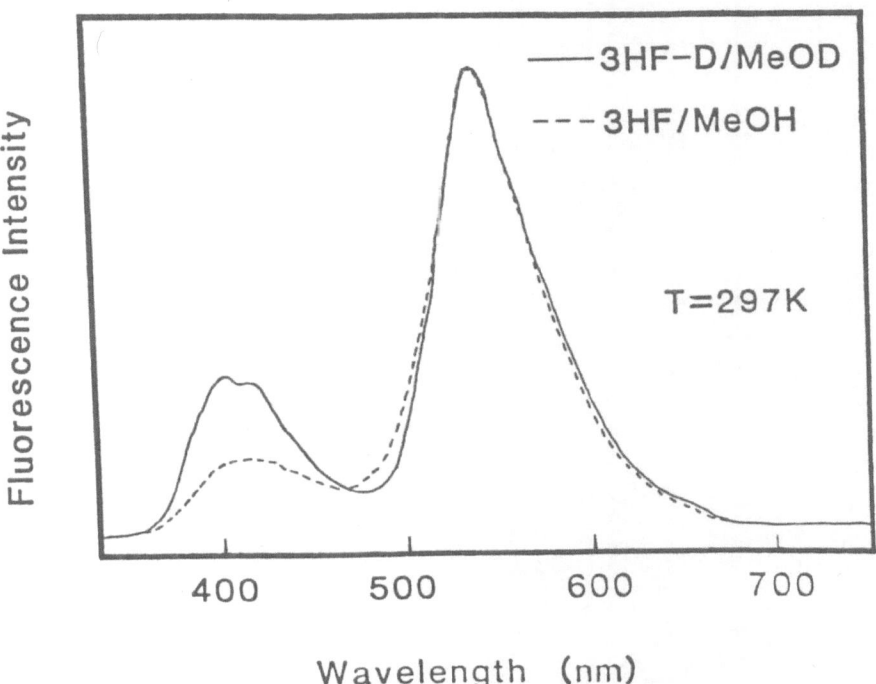

Figure 4. Static emission spectra of 3HF (...) and 3HF-D (-) in methanol . The fluorescence was induced by 355 nm, 30 ps light pulses. The spectra are not corrected for the wavelength sensitivity of our photodiode array spectrometer.

In a recent paper [1] we showed that the species responsible for the green (543 nm) fluorescence of 3HF was formed kinetically from the blue (413 nm) emitting species by two kinetic components. We rationalized this complex behavior by a model in which there are two pathways for ESIPT, i.e., a slow and a fast component. The present paper will deal with the slow component for PT.

The proton-transfer process of 3HF is responsible for the dual emissions of this compound which are portrayed in Fig. 4 under different environmental conditions. The proton transfer kinetics of 3HF have been characterized by a combination of static and dynamic measurements. Examples of the latter type of data for 3HF in methanol are shown in Fig. 5. The upper panel shows kinetic traces for the N form of 3HF and for a monodeuterated derivative of this compound. The lower panel portrays the proton transfer build-up of the T^* emission, which eventually decays at a slower timescale under the environmental conditions that apply to Fig. 5.

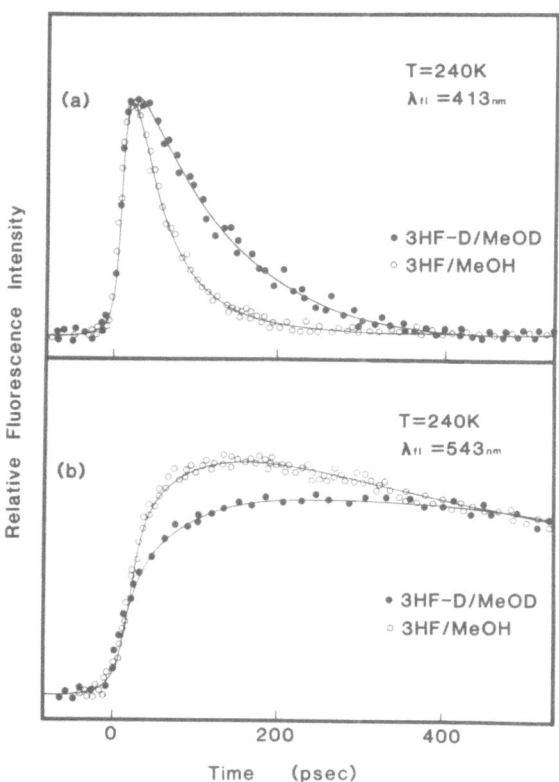

Figure 5. Time-resolved emission traces at 405 nm (a) and 530 nm (b) for 3HF in methanol (0) and 3HF in methanol-d$_1$ (●).

Solvent effects on the proton transfer kinetics have been extensively investigated [3]. The thermal activation parameter (E_a and A) for k_{PT} systematically increase with the hydrogen-bonding ability of the solvent, suggesting that the intermolecular solute/solvent hydrogen/bond is participating in the proton transfer coordinate of the 3HF/solvent complex. The solvent effect on k_{PT} does not appear to be a simple consequence of a viscosity or "friction" effect.

A model that can rationalize the various static and spectroscopic data on 3HF is represented by the following [5].

"tight complex"   "loose complex"

In this model the immediate precursor to proton transfer is a "loose complex" which resembles an unsolvated molecule. The proton transfer kinetics of the "loose" complex should be extremely rapid because the intramolecular hydrogen bond is stretched and almost broken. The "loose" complex is a thermally activated form of the "tight" complex. The equilibrium geometry of the solvated form of 3HF is assumed to be the "tight" complex. The temperature dependence of k_{PT} in this model is ascribed to thermal excitation of coordinates that are associated with the intermolecular hydrogen bond of the complex, i.e., the "tight" complex to "loose" complex interconversion.

An additional feature of the model is that the actual microscopic proton transfer step, i.e., the conversion of S_1 "tight" N to T is assumed to occur by a non-thermally activated tunneling mechanism in analogy to the proton transfer mechanism of unsolvated ESIPT molecules, see the discussion above of DHAQ and HBT. Thus in our model, thermal excitations of "heavy" vibrations of the solute/solvent complex mediate the tunneling process. The strongest evidence for the tunneling component of the model is that hydrogen/deuterium kinetic isotope effect ($k_H/k_D \approx 2$) for 3HF is not measurably temperature dependent over a large range [3].

A detailed study of 3HF derivatives show a clear correlation for the different derivatives between E_a and a variety of spectroscopic and chemical properties, including NMR chemical shifts and pK_a values. These trends generally support the model mentioned above because the trends verify that the proton transfer process is associated with energetic factors of the solute/solvent configuration [4]. The derivative study also suggests that torsional motion of the phenyl group about the bond joining it to the main ring of 3HF can mediate proton transfer by lowering the intrinsic barrier [4,17-18].

5. ACKNOWLEDGEMENTS

Acknowledgement is made to the donors of the Petroleum Research Fund, administered by the American Chemical Society, to the National Science Foundation (Grant No. CHE-8251158), and the Alfred P. Sloan Foundation.

6. REFERENCES

1. Strandjord, A.J.G.; Courtney, S. H.; Friedrich, D. M.; Barbara, P. F. *J. Phys. Chem.* 1983, **87**, 1125.
2. Strandjord, A.J.G.; Barbara, P. F. *Chem. Phys. Lett.* 1983, 9821.
3. Strandjord, A.J.G.; Barbara, P. F. *J. Phys. Chem.* 1985, **89**, 2355.
4. Strandjord, A.J.G.; Smith, D. E.; Barbara, P. F. *J. Phys. Chem.* 1985, **89**, 2366.
5. Ding, K.; Courtney, S. H.; Strandjord, A.J.G.; Flom, S. R.; Barbara, P. F. *J. Phys. Chem.* 1983, **87**, 1125.
6. Flom, S. R.; Barbara, P. F. *Chem. Phys. Lett.* 1983, **94**, 488.
7. Flom, S. R.; Barbara, P. F. *J. Phys. Chem.* 1985, **89**, 4489.
8. Ireland, J. F.; Wyatt, P. H. *Adv. Phys. Org. Chem.* 1976, **12**, 643.
9. Barbara, P. F.; Brus, L. E.; Rentzepis, P. M. *J. Amer. Chem. Soc.* 1980, **102**, 5631.
10. Barbara, P. F.; Rentzepis, P. M.; Brus, L. E., *J. Amer. Chem. Soc.*, 1980, **102**, 2786.
11. Smith, K. K.; Kaufman, K. J. *J. Phys. Chem.* 1978, **82**, 2286.
12. Houston, A.; Scott, G. W.; Gupta, A. *J. Chem. Phys.* 1982, **76**, 4978.
13. Sengupta, P. K.; Kasha, M. *Chem. Phys. Lett.* 1979, **68**, 382.
14. Woolfe, G. J.; Thistlethwaite, *J. Amer. Chem. Soc.* 1981, **103**, 6919.
15. Itoh, M.; Tokumara, K., Tanimoto, Y.; Okada, H.; Obi, K.; Tanaka, I. *J. Amer. Chem. Soc.* 1982, **104**, 4146.
16. McMorrow, D.; Dzygan, T.; Aartsma, T. J. *Chem. Phys. Lett.* 1984, **103**, 492.
17. McMorrow, D.; Kasha, M. *J. Phys. Chem.* 1984, **88**, 2235.
18. McMorrow, D.; Kasha, M. *J. Amer. Chem. Soc.* 1983, **105**, 3339.
19. Rossetti, R.; Brus, L. E. *J. Chem. Phys.* 1980, **73**, 1547.
20. Van Benthem, M. H.; Gillispie, G. D.; Haddon, R. C. *J. Phys. Chem.* 1982, **86**, 4281.
21. Werner, T.; Kramer, H.E.A. *Eur. Polym. J.* 1977, **13**, 501.
22. Van Benthem, M. H.; Gillispie, G. D.; *J. Phys. Chem.* 1984, **88**, 2954.
23. Goodman, J.; Brus, L. E.; *J. Am. Chem. Soc.* 1980, **100**, 7472.

SOME PRACTICAL ASPECTS OF TUNNELING IN UNIMOLECULAR REACTIONS

J. TROE
Institut für Physikalische Chemie der Universität
Göttingen, Tammannstraße 6, D-3400 Göttingen, Germany

ABSTRACT. Practical aspects of implementing tunneling contributions into unimolecular reactions of polyatomic molecules are discussed. One-dimensional expressions of transmission coefficients for allowed and forbidden reactions are combined with statistical unimolecular rate expressions. Specific rate constants k(E), thermally averaged high pressure rate constants k_∞ and low pressure fall-off rate coefficients are elaborated. The dissociation reactions of H_2CO and N_2O and intramolecular hydrogen shifts in large polyatomic molecules are used as experimental examples. Identifying the allowed or forbidden character and separating tunneling from statistical factors presents a most difficult problem in most practical cases.

1. INTRODUCTION

Although the basic principles of tunneling in chemical reactions are understood relatively well, the explicit treatment in the majority of practical applications is still on an unsatisfactory level. The present article considers some practical problems of implementing tunneling contributions in unimolecular reactions. In this situation the following points are of particular importance:
(i) The dynamics of unimolecular reactions is determined by contributions from often many coordinates. Therefore, the well known treatments of one-dimensional tunneling have to be generalized and applied to multi-dimensional cases.
(ii) The reaction may be an adiabatic,"allowed", or a non-adiabatic, "forbidden" process. In both cases tunneling can occur; however, different formalisms apply. In practice, it is often difficult to distinguish between strong- and weak-coupling situations.
(iii) Unimolecular reaction rates are influenced by a variety of molecular parameters, such as threshold energies, "activated complex" geometries and frequencies, multi-dimensional shapes of energy barriers, multi-dimensional crossing geometries and interaction matrix elements

149

J. Jortner and B. Pullman (eds.), Tunneling, 149–164.

in non-adiabatic reactions. It is, therefore, often difficult to relate individual of these parameters to measured quantities.
(iv) Unimolecular reaction rates of isolated molecules in well selected states so far have been measured only in exceptional cases. Normally, competition with collisional processes and/or averaging over many states was involved. It is, therefore, of importance to analyze the effects of tunneling under the variety of reaction conditions encountered in practical experiments.

It is the aim of the following article to discuss these various aspects of tunneling in unimolecular reactions. The topic is by far too complicated to allow for a detailed quantitative treatment. However, some simple qualitative arguments may appear of illustrative value.

2. TRANSMISSION COEFFICIENTS FOR TUNNELING

The transmission coefficient $T(E)$ of a representative point at the energy E through a potential energy barrier $U(q)$, for an allowed process, has been discussed extensively in the literature (see e. g. refs. 1-4). For a parabolic barrier

$$U(q) = E_o - \frac{1}{2} \mu \omega^{\ddagger 2} q^2 \qquad (1)$$

(reduced mass μ, imaginary frequency $\omega^{\ddagger}/2\pi$ of the barrier, threshold energy E_o), $T(E)$ is given by

$$T(E) = \left\{ 1 + \exp\left[-\frac{2\pi(E - E_o)}{\hbar \omega^{\ddagger}} \right] \right\}^{-1} . \qquad (2)$$

Analytical solutions are also available for rectangular barriers, symmetrical and unsymmetrical Eckart barriers and others (1,4). For sufficiently small transmission coefficients, arbitrary barrier shapes can be treated by the expression (1,2)

$$T(E) = \exp\left[-\frac{2}{\hbar} \int_a^b \sqrt{2\mu(U(q) - E)} \, dq \right] \qquad (3)$$

where a and b denote the q-values with $E = U(a) = U(b)$. Combining eqs. (2) and (3), one may try to approximate $T(E)$ for arbitrary barriers and all values of $T(E)$ by an expression

$$T(E) \approx \left\{ 1 + \exp\left[+\frac{2}{\hbar} \int_a^b \sqrt{2\mu(U(q) - E)} \, dq \right] \right\}^{-1} \qquad (4)$$

which is exact for the parabolic barrier at all $T(E)$ and leads to the correct limit for all barriers at $T(E) \ll 1$.

Different expressions for transmission coefficients are obtained with non-adiabatic, forbidden processes. For this case, the two-state model with "crossing" linear potentials in first approximation leads to the Landau-Zener expression. In second approximation, one obtains (1)

$$T(E) = 4\pi V^2 \left(\frac{2\mu}{\hbar^2 \sqrt{F_1 F_2} \ (F_2 - F_1)} \right)^{2/3} \phi^2 \left[-(E-E_o) \left(\frac{\sqrt{2\mu}}{\hbar} \frac{(F_1 - F_2)}{F_1 F_2} \right)^{2/3} \right]$$

(5)

where $\phi(x)$ is the Airy-function, V denotes the matrix element of the perturbation inducing the transition, the splitting of the linear potentials at the "crossing point" becomes 2 V, F_1 and F_2 are the slopes -dU/dq of the two potentials respectively. Eq. (5) only applies for sufficiently small values of V. In the tunneling range at $E \ll E_o$, the limiting expression (1) of the Airy function for large positive x may be employed giving

$$T(E) \approx \frac{\pi V^2 F_1 F_2}{(F_2 - F_1)^2 \sqrt{E_o - E}} \ \exp \left[-\frac{4}{3} (E_o - E)^{3/2} (\frac{2\mu}{\hbar^2})^{1/2} (\frac{F_1 - F_2}{F_1 F_2}) \right] \quad . \quad (6)$$

It should be noted that the dominating exponential factor in eq. (6) coincides with eq. (3), when the triangular barrier of the linearly crossing potential model used in eqs. (5) and (6) is inserted in eq. (3). Apart from different behaviour close to and above the threshold, the transmission coefficients in allowed and forbidden processes mainly differ by the preexponential factors which, in the case of forbidden, non-adiabatic processes contain the coupling matrix element V.

Generalizations of the two-state linear potential model have been discussed extensively in ref. 2. Multichannel models, non-linear potentials, and, in particular, the extension to a two-dimensional system were treated explicitly. A discussion of the validity of such treatments was given recently in ref. 5.

An alternative, widely used approach to tunneling processes under conditions, where time-dependent perturbation theory applies, are provided by golden rule expressions (6-11). In analogy to eq. (5), the expression for the transmission coefficient is proportional to V^2 and limited to small values of T(E). Multi-dimensional situations have been treated.

The treatment of situations "between allowed and forbidden tunneling" is less straight forward. The character of the expressions for T(E) changes from the one to the other limit. An explicit, accurate quantum-mechanical treatment of multi-dimensional allowed unimolecular reactions is still not available. Golden rule treatments do not apply to situations where the widths of the decaying levels overlap (8). For this reason, in the following we consider various simple ways of implementing one-dimensional expressions of T(E) into conventional unimolecular rate theory.

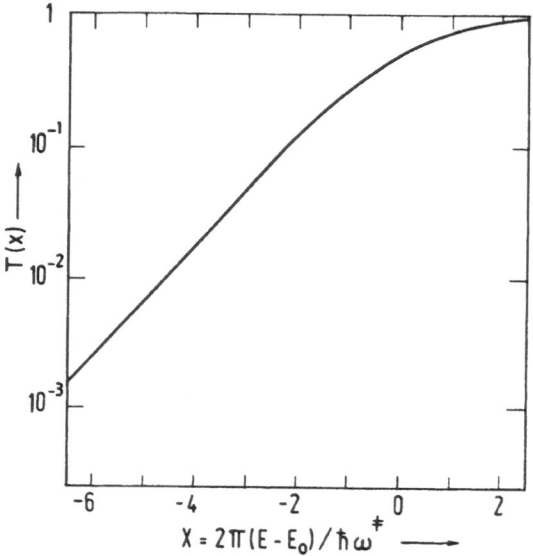

Figure 1.   Transmission coefficient for a parabolic barrier (eq. (2)).

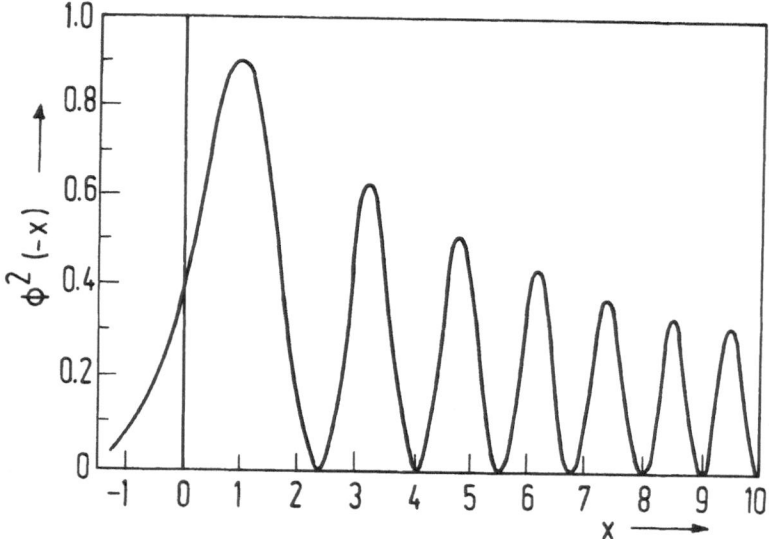

Figure 2.   Transmission coefficient for a non-adiabatic process
            (eq. (5)).

Figs. 1 and 2 demonstrate the different character of the transmission coefficient functions T(E) for "allowed" (eq. (2)) and "forbidden" (eq. (5)) processes. For the parabolic barrier of eq. (2) and
an allowed process, fig. 1 shows a smooth rise from small T(E) values
toward unity; for the crossing linear potential curves of eq. (5) and
a forbidden process, fig. 2 shows the square of the Airy function which
then has to be multiplied with the preexponential factor containing
the coupling matrix element V^2; here, there are interference oscillations above the threshold, the maxima of T(E) (not the maxima of ϕ^2)
are well below unity.

3. TUNNELING IN SPECIFIC RATE CONSTANTS OF UNIMOLECULAR REACTIONS

In the framework of resonance scattering theory, the specific rate
constant for unimolecular decay of a resonance scattering state $|n\rangle$
of the reacting molecule into a product state $|p\rangle$ is given by (8)

$$k_{n \rightarrow p}(E) = \frac{T_{n \rightarrow p}(E)}{h \, \rho_p(E)} \tag{7}$$

where $\rho_p(E)$ is the density of those states $|n\rangle$, which, on the
characteristic time scale of the experiment, are coupled to the pro-
duct state $|p\rangle$, and where T is the transmission coefficient for the
process $|n\rangle \longrightarrow |p\rangle$. In the statistical adiabatic channel model of
ref. 12, single-channel rate constants eq. (7) are combined to a total
specific rate constant

$$k(E,J) = \sum_{a} \frac{T_{n \rightarrow p}(E,J)}{h \, \rho_p(E,J)} \tag{8}$$

where a denotes all product channels accessible from the initial state.
Mode-specificity would reduce $\rho_p(E,J)$ to values below the total ro-
vibrational density of states $\rho(E,J)$.
Some properties of the transmission coefficients in eqs. (7) and
(8) can well be visualized by a look at channel potential curves along
a reaction coordinate q, see fig. 3. In this representation, a "reaction coordinate q" is separated from the rest of the molecule. The
eigenvalues of the s-1 remaining coordinates as a function of q define
the channel potential curves $|a\rangle$. The energy in the reaction coordi-
nate is given by $E - E_a$. If there were a sufficiently strong dynamic
coupling between all channels $|a\rangle$ in a range of q values near q_e,
i. e. if chaotic behaviour and energy randomization would occur, a
molecular trajectory characterized by a total energy E would successi-
vely explore all channels $|a\rangle$ with energy minima below E. An approach
of the barrier on the lowest channel state $|0\rangle$ would be characterized
by the smallest barrier width and, hence, with the largest value of
T(E). Higher channels meet wider adiabatic channel barriers and have
smaller T(E); however, their smaller T(E) may be compensated by a

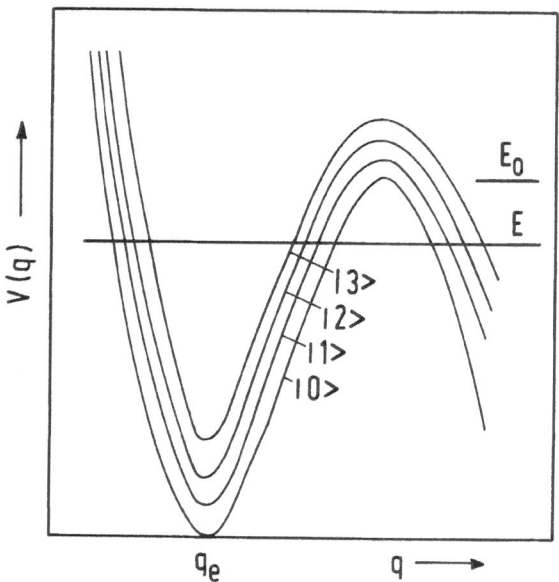

Figure 3. Adiabatic channel potential curves

larger number of such channels. Therefore, statistical factors are of importance in determining the overall transmission coefficient of the reaction.

It has to be emphasized that tunneling processes are intimately related to the problem of regularity or chaos of molecular trajectories and of mode-specificity. We illustrate this in the following by some model considerations based on eqs. (7) and (8). Eq. (7), at first, assumes that there is at least a certain group of zero-order states $| n \rangle$ which couple sufficiently well and can, in the language of eq. (7), be characterized by a density of states $\rho_p(E)$. If one assumes that all states with a given energy E and angular momentum J couple sufficiently well, $\rho_p(E)$ can be replaced by the total rovibrational density (12) of states $\rho(E,J)$. We first consider the evaluation of eq. (8) for this case. Eq. (8) then takes the form (13)

$$k(E,J) = \frac{\sum_a T(E - E_a, J)}{h \, \rho(E,J)} \qquad (9)$$

where E_a denotes the energy fixed in the adiabatic channels $| a \rangle$. Neglecting tunneling, $T(E - E_a, J)$ would be given by step functions with steps at the energy where the channels "open", the sum over T would give (12) the number of open reaction channels $W(E - E_0, J)$. Eq. (9) is evaluated by integrating $T(E - E_a, J)$ over the density $\rho_{s-1}(E_a,J)$ of the adiabatic channel states, for q approaching the barrier,

$$k(E,J) \approx \frac{\int_0^E T(E - E_a, J)\, \wp_{s-1}(\Xi_a, J)\, dE_a}{h\, \wp(E,J)} \qquad (10)$$

or by the corresponding explicit summation when $\wp$ increases too weakly with E_a compared to the decay of $T(E - E_a)$.

Eq. (10) is easily evaluated numerically for specific situations. In the following we consider some special cases. If, for a given energy $E < E_O$, the decay of $T(E - E_a)$ with increasing E_a is much faster than the increase of the statistical weight function $\wp_{s-1}(E_a, J)$, then the channel with $E_a = O$ dominates such that

$$k(E,J) \approx \frac{T(E,J)}{h\, \wp(E,J)} \qquad \qquad . \quad (11)$$

$T(E,J)$ here corresponds to the one-dimensional eq. (4). This situation will be approached for small polyatomic molecules. On the other hand, for large polyatomic molecules, the reverse behaviour may be approached, i. e. the smaller transmission coefficients for channels with E_a approaching E may be compensated by their much larger number. Then, $k(E,J)$ may approach

$$k(E,J) \approx \frac{T(O,J)W(E,J)}{h\, \wp(E,J)} \qquad \qquad . \quad (12)$$

Eq. (12), with the classical approximation for W and $\wp$, gives

$$k(E,J) \approx \gamma\, T(O,J) \qquad \qquad . \quad (13)$$

Of course, eqs. (11) and (12) apply only to the tunneling range; at $E > E_O$, the barrier crossing rate (neglecting tunneling)

$$k(E,J) \approx \frac{W(E - E_O, J)}{h\, \wp(E,J)} \qquad (14)$$

may become dominating. However, one can here imagine situations where the large statistical weight of channels, for which there is only tunneling, may lead to dominating tunneling contributions even at $E > E_O$. An indication for this behaviour may be found, if eq. (12) exceeds eq. (14), i. e. if

$$T(O,J)W(E,J) > W(E - E_O, J) \qquad \qquad . \quad (15)$$

This would be realized either by

$$T(O,J) > 1/W(E_O, J) \qquad (16)$$

or, at higher energies, by

$$T(0,J) \geq \left(\frac{E - E_o + E_z}{E + E_z}\right)^{s-1}$$ (17)

where semiclassical number of states functions are used for $W(E)$ and $W(E - E_o)$.

A practical example for a small molecule case is provided by the molecular elimination reaction

$$H_2CO \longrightarrow H_2 + CO$$ (18)

This case has been treated explicitly with ab initio potential surfaces in refs. 13 and 14. The J-effects were elaborated in refs. 13 and 15. For this case, $\hbar \omega^{\ddagger} \approx 2100 \pm 100$ cm^{-1} was derived from ab initio calculations. $k(E,J=0)$, as calculated in ref. 13, is well approximated with eqs. (2) and (11).

Reaction (18) so far is the only case where calculated $k(E,J)$ can be compared with measurements. Recent determinations (16) of $k(E,J=2)$ for $E = 28309.5$ and 28340.2 cm^{-1} gave 7.14×10^7 and 2.86×10^7 s^{-1} respectively. The apparent decrease of $k(E)$ with increasing energy over a small energy range would point toward non-statistical, mode-selective behaviour. Since $\rho_p(E,J) \leq \rho(E,J)$ in eq. (8), the smaller $k(E)$ value should be closer to a statistical situation. Shifting statistically calculated $k(E,J)$ curves from ref. 15 (see fig. 4)

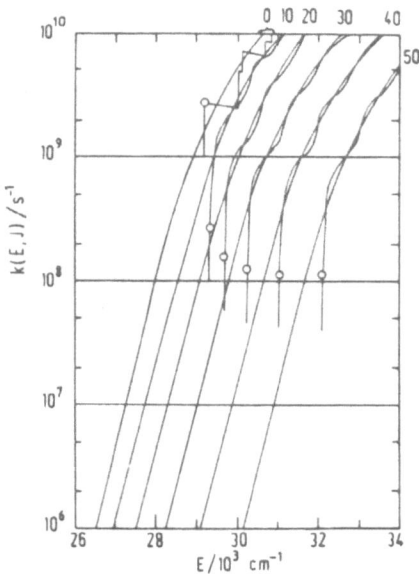

Figure 4.
Specific rate constants $k(E,J)$ for the reaction $H_2CO \longrightarrow H_2 + CO$ (Statistical calculations from refs. 13 - 15, for details see ref. 15. Structured curves: RRKM calculations without tunneling; $E_o(J=0) = 29630$ cm^{-1}; curves shown for $J = 0$, 10, ..., 50).

along the energy scale, the measurement thus would allow to obtain a
new experimental lower limit for the top of the elimination barrier
$E_0 \gtrsim 30000$ cm^{-1} which would be higher than other information (17).
Even if there appears to be some indication of mode-specificity, the
order of magnitude of the measured $k(E,J)$ clearly demonstrates that
there is substantial coupling of the reactant states such that
$\rho_p(E,J)$ is at least $\gtrsim$ 0.1 ρ (E,J). It is also interesting to note
that the width of the investigated tunneling states is about 10 -
100 times smaller than the average distance between the levels. This
means that intramolecular randomization here is not related to over-
lapping widths of the resonance scattering states. In spite of the
indications for some mode-specificity, the H_2CO results nevertheless
support the application of statistical rate theory with tunneling contributions as a useful first approximation.

Eqs. (7) and (8) applies only to situations where at least a
certain number of reactant states couple and realize a "single-channel
barrier-approach rate" constant $1/h$ $\rho_p(E,J)$. Close to the vibrational
ground state of the reactant this condition is not fulfilled. Then,
individual vibrational states will not sample sets of adiabatic channel potentials as shown in fig. 1. Instead, state-specific trajectories may result in completely mode-specific tunneling rate constants
opposite to what we have discussed before. One way of treating this
case would be the explicit calculation of trajectories with a search
for the most favorable tunneling distance at each point along the
trajectory. Similar studies have already been applied for tunneling
in bimolecular reactions (see e. g. ref. 18).

If state-specific tunneling rate coefficients k_i vary from quantum state to quantum state, it is difficult, in a multi-dimensional
situation, to provide a generalized treatment. In the following, we
consider some model cases. At first, we assume that none of the normal
coordinates coincides with the reaction coordinate q, i. e. the most
favorable tunneling direction. In this case, the "projection" of the
normal coordinates (19) Q_i on the reaction coordinate q has to be
determined

$$q = \sum_{j=1}^{s} \alpha_j Q_j \qquad \qquad . \qquad (19)$$

The individual quantum states i then are characterized by transmission
coefficients T_i which depend on the quantum numbers of all individual
harmonic oscillators. A normal mode with negligible projection in the
direction of q does not change T_i; normal modes with large projection
will have the largest effects. As an example, we imagine a model with
two oscillators, the first of which is in the direction of q, the
second has the effect of lowering the barrier for tunneling in the
direction of q to some extent. The specific rate constant in the tunneling range then is given by

$$k(E_1,E_2) \approx \nu \, T(E_1,E_o(E_2)) \qquad (20)$$

where ν denotes the frequency of motion in the direction q. In the simplest case, the barrier reduction may be given by

$$E_o(E_2) \approx E_o - f_2 E_2 \qquad (21)$$

where $0 \leq f_2 \leq 1$ and $T(E_1)$ may be given, e. g., by the parabolic barrier from eq. (2). Tunneling rate constants then are calculated for all states with $E_1 + E_2 < E_o$, see below. Models with oscillating barriers (20) are closely related to this case. Models of this type are of large importance when substitution of the molecule results in marked changes of the tunneling rates. Then, the "projection" of some normal coordinates on the q-direction will have changed, see below.

Specific rate constants in the tunneling range of non-adiabatic processes, although differing in the one-dimensional transmission coefficients from "allowed tunneling", are characterized by an analogous treatment. Eqs. (7) - (13) apply as well, however, above the tunneling range eq. (14) becomes inadequate. Instead the (on the average) decaying transition probability of fig. 3 and eq. (5) and the smaller absolute value (being proportional to V^2) have to be accounted for. A further important aspect is the geometry of the crossing hypersurface of the interacting adiabatic states. Golden rule treatments with suitable Franck-Condon overlap integrals account for this (7,9-11). However, simple geometrical constructions with suitable state counting (21) are much more economic as long as the potential surfaces are not well known.

4. THERMALLY AVERAGED TUNNELING RATE CONSTANTS IN UNIMOLECULAR REACTIONS AT HIGH PRESSURES

Usually not state-specific but thermally averaged rate constants of unimolecular reactions are measured. At high pressures an equilibrium population of molecular states is established such that, with a Boltzmann-distribution f_i or $f(E)$, one has

$$k_\infty (T) = \sum_{i=0}^{\infty} f_i k_i = \int_0^\infty f(E) k(E) dE \qquad (22)$$

(averaging over angular momenta J here has been neglected). In this section, we consider tunneling effects in k_∞ , at first for allowed reactions. Eq. (22) includes the usual contribution from states above the threshold energy E_o and the additional part from tunneling at $E < E_o$. As discussed with eqs. (15) - (17), there may also be an enhancement of k_∞ by tunneling at $E > E_o$. In the small molecule case, eq. (11) combined with an approximate $T(E)$ for a parabolic barrier,

$$T(E) \approx \exp(- \frac{2 \pi (E_o - E)}{\hbar \omega^{\neq}}) \qquad , \qquad (23)$$

via eq. (22) leads to

$$k_\infty \approx \frac{kT}{h}\frac{Q^{\ddagger}}{Q}\exp\left(-\frac{E_o}{kT}\right) + \frac{kT}{hQ}\left(\frac{\hbar\omega^{\ddagger}}{2\pi kT - \hbar\omega^{\ddagger}}\right)\exp\left(-\frac{2\pi E_o}{\hbar\omega^{\ddagger}}\right) *$$

$$\left\{\exp\left(\frac{2\pi E_o}{\hbar\omega^{\ddagger}} - \frac{E_o}{kT}\right) - 1\right\} \qquad . \qquad (24)$$

Q and $Q^{\ddagger}$ denote reactant and activated complex partition functions. At high temperatures, the conventional expression from transition-state theory (first term in eq. (24)) is approached; at low temperatures, eq. (24) leads to

$$k_\infty \approx \frac{kT}{hQ}\exp\left(-\frac{2\pi E_o}{\hbar\omega^{\ddagger}}\right) \approx \nu\, T(0) \qquad . \qquad (25)$$

Because of the classical integration over the energy in the reaction coordinate, the factor kT/hQ in eq. (25) and in the second term of eq. (24) must be put equal to the frequency ν of motion along the reaction coordinate q. By analogy with eqs. (23) - (25) and using eq. (4), transmission coefficients for arbitrary potentials in the reaction coordinate, e. g. two harmonic wells connected by a harmonic barriers (imiginary frequency $\omega^{\ddagger}/2\pi$) are treated in a similar way.

The large molecule case of eq. (13) leads, in analogous way as before, to

$$k_\infty \approx \frac{hT}{h}\frac{Q^{\ddagger}}{Q}\exp\left(-\frac{E_o}{kT}\right) + \nu\, T(0) \qquad . \qquad (26)$$

The same limiting expressions are obtained as from eq. (24); however, the transition from one to the other limit is more abrupt.

It should be emphasized that the proper transition from the low temperature limit of eqs. (25) and (26) to the transition state-high temperature limit requires the summation over the lower vibrational states and their individual mode-specific transmission coefficients. This problem was discussed in connection with eqs. (19) - (21). We illustrate this treatment by a simple two oscillator system with $E_1 = i_1 h\nu$, $E_2 = i_2 h\nu /2$, $E_o = 2h\nu - 0.2\,E_2, \hbar\omega^{\ddagger}/2\pi \approx h\nu$, and $T(E)$ from eqs. (20) and (23). For this model, there are 6 vibrational tunnel states with $E < E_o$. Fig. 5 shows Arrhenius plots for k_∞ from this model (curve 3) and from eqs. (24) (curve 2) and (26) (curve 1). Whereas the low- and high temperature limits of k_∞ are well defined and given by $\nu\, T(0)$ or the transition state expression, the detailed transition curves reflect the number of tunneling states below E_o, their quanta and their individual influences on the transmission coefficient. Obviously, one cannot conclude on these various properties in a unique way, even if a complete transition curve from k_∞ (T→0) to k_∞ (T→∞) has been measured. Instead the multidimensio-

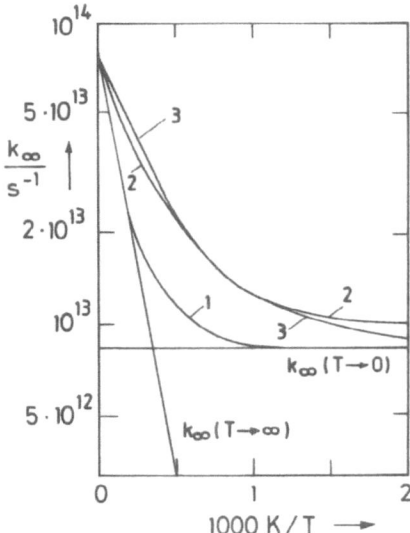

Figure 5.
High pressure rate con-
stants k_∞ including
tunneling (large molecule
model: 1, small molecule
model: 2, discrete level
model: 3, see text).

nal geometry of the barrier and the analysis of normal modes following
eq. (19) with respect to the projections of these modes on the most
favorable tunneling direction q has to be made. For polyatomic systems
one is far from the solution of this problem.

The multi-dimensional character of the problem is quite analogous
for forbidden, non-adiabatic processes. However, in this case the
high temperature limit is not given by the simple transition-state
expression of k_∞ . Instead, the decrease of $T(E)$ at $E > E_o$ (see fig.
2) has to be taken into account. Also, the geometry of the intersection of the crossing potentials has to be taken into account. Finally,
the interaction matrix element V^2 enters k_∞ . In the framework of
golden rule treatments the multi-dimensional tunneling problem was
treated in refs. 7 and 11. In the low temperature range, k_∞ (T) curves
are found which have quite similar shapes as illustrated in fig. 5.
In practice, k_∞ (T) curves at low T are virtually indistinguishable for
adiabatic and non-adiabatic processes. However, at high T, the preexponential factors should reveal the differences. For allowed processes, "normal" preexponential factors are observed; "forbidden processes" would show "anomalously low" preexponential factors.

Fig. 6 demonstrates measurements with pronounced tunneling contributions to k_∞ for a hydrogen shift in the photochemical conversion (22) of an enamine to a hexahydrocarbazole. Similar results were
obtained for the enol-keto tautomerizations in ref. 23. The large
temperature ranges very well show the transition from the low to the
high temperature limits. An interpretation as a non-adiabatic process
was attempted in ref. 7, an interpretation as an allowed process can

be easily made as described in this work. The distinction appears only possible from more measurements at higher temperatures. The present results indicate that a "normal" transition state-high temperature limit is approached or, at least, that the reaction is not very forbidden. The multi-dimensional character of the tunneling has been impressively demonstrated in these examples by H/D isotope effects in dependence on substitution far away from the tunneling area of the molecule (22,23). In practice, weak-coupling golden rule treatments and strong-coupling allowed tunneling models have both to be applied to a given reaction, before the adiabatic or non-adiabatic character of the process is uniquely identified. A similar problem is encountered with the photoisomerization of trans-stilbene which has been interpreted as an allowed (24) or a non-adiabatic (25) process.

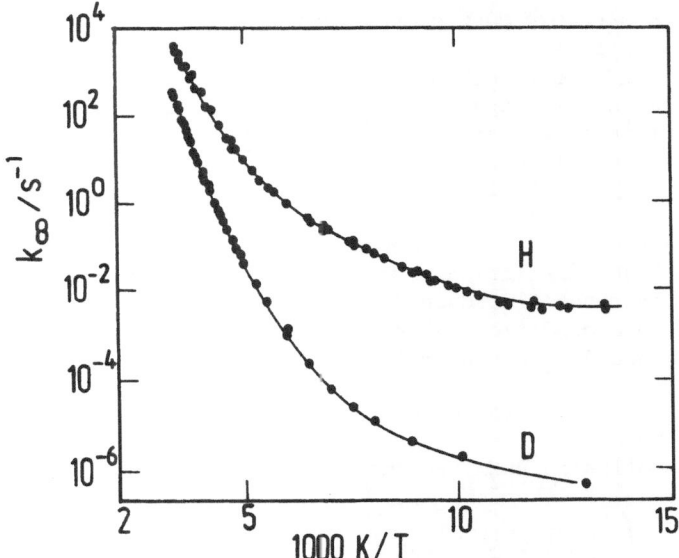

Figure 6. Thermally averaged rate constants k_∞ for intramolecular shift of hydrogen (or deuterium) from an enamine to hexahydrocarbazole (ref. 22).

The question "adiabatic or non-adiabatic" is much easier to answer for spinforbidden processes like $N_2O(^1\Sigma) \rightarrow N_2(^1\Sigma) + O(^3P)$, $CO_2(^1\Sigma) \rightarrow CO(^1\Sigma) + O(^3P)$ and $CS_2(^1\Sigma) \rightarrow CSC(^1\Sigma) + S(^3P)$. Here the preexponential factors of $k_\infty \approx 10^{11.1}$ (ref. 26), $10^{11.4}$, and $10^{12.6}$ s^{-1} (ref. 27), respectively, confirm the more or less spinforbidden character of the reaction. The high pressure rate constant for N_2O dissociation has been analyzed theoretically several times (11,21,26,28). The models have been very elaborate but still highly idealized. Often it was overlooked that the threshold (crossing) energy E_O is not known, the experimental separation of k_∞ into a preexponential and an exponential factor is fairly uncertain, and the measured activation energies $E_{a\,\infty}$ are different from E_O. The statistical problem of counting reaction channels along the intersection

surface was first treated in ref. 21 and extended in ref. 11. This
problem has marked effects on the difference $E_O - E_{a\infty}$. If, e. g., very
many channels open at E_O and a comparably small increase of the number
of open channels follows at $E > E_O$, E_O can be up to skT larger than
$E_{a\infty}$ (s = number of oscillators of the molecule). A decay of $T(E) \propto$
$\sqrt{E - E_O'}$, see fig. 2, can further reduce $E_{a\infty}$ by another 1/2 kT. By
the analysis of the absolute value of k_∞ one can conclude on the
coupling matrix element V. A value close to that for spin-orbit coupling in O atoms was obtained in ref. 26. Other values from other inter-
pretations (28) show the large influence of many not well known molecular parameters such as E_O and the intersection geometry, and of the
measured $E_{a\infty}$.

5. TUNNELING IN FALL-OFF CURVES OF THERMAL UNIMOLECULAR REACTIONS

An interesting additional consequence of tunneling in thermal unimolecular reactions becomes apparent in the pressure or density [M]-depen-
dence of the first order rate coefficient k([M]). The low pressure
limit k_O of the unimolecular reaction is reached only when all activated molecules complete their reaction before collisional deactivation
can take place. It appears trivial to state that this requires a pressure as low as necessary not to quench the slowest possible tunneling
process. With high barriers, however, other process such as wall
collisions or infrared fluorescence will prevent the experimental rea-
lisation of such a condition. Nevertheless, it appears important to
investigate to would extent near low-pressure limiting behaviour can
be realized.

In principle, this problem would require the solution of master
equations in the fall-off range (see ref. 29). We limit ourselves here
to the strong collision RRKM treatment

$$k = \int_0^\infty \left(\frac{k(E) f(E) Z [M]}{k(E) + Z [M]}\right) dE \tag{27}$$

Approximating k(E) by $k(E_O) * \exp(-(E_O - E)/A)$ and f(E) by $f(E_O) *$
$\exp(-(E - E_O)/kT)$, the tunneling part of k is given by

$$k_T \approx Z[M] f(E_O) k(E_O) \int_0^{E_O} \frac{\exp\left\{-(E_O-E)(kT-A)/AkT\right\} dE}{Z[M] + k(E_O) \exp\left\{-(E_O-E)/A\right\}} \tag{28}$$

The integral has analytical solutions for A/kT = 1,2,3 ... and
A/kT $\ll$ 1. E. g., for the latter case,

$$k_T \approx Z[M] f(E_O) A \ln\left\{\frac{1 + Z[M]/k(E_O)}{\exp(-E_O/A) + Z[M]/k(E_O)}\right\} \tag{29}$$

such that a low pressure expression (at $Z[M] \ll k(E_O)$ and
$\exp(-E_O/A) \ll Z[M]/k(E_O)$)

$$k_{T,o} \approx k_{o>} \ln\left\{Z[M]/k(E_o)\right\}^{-A/kT} \tag{30}$$

is approached ($k_{o>}$ denotes the low pressure limit $Z[M]f(E_o)kT$ in the absence of tunneling; the total rate constant k_o is given by $k_{o>} + k_{T,o}$). As long as $\exp(-E_o/A)$ is smaller than $Z[M]/k(E_o)$, one does not reach a situation with $k_o \propto [M]$.

As an example one may consider the thermal decomposition of N_2O. A recent model calculation in ref. 30 has concluded that strong deviations from $k_o \propto [M]$ should have been seen experimentally, in contrast to the actual observations (26). We have evaluated eqs. (28) - (30) again for a tunneling transmission coefficient given by eq. (5). From the slopes of the crossing potentials and with eqs. (5) and (6) one obtains $A/kT \approx 0.04$ at $T = 2000$ K such that eq. (19) applies. $k(E_o)$ can be derived from the measured k_∞. As a result one concludes that tunneling broadening of the fall-off curves does exist, but is probably not observable with actual experimental techniques.

ACKNOWLEDGMENT

Financial support of this work by the Deutsche Forschungsgemeinschaft (Sonderforschungsbereich 93 "Photochemie mit Lasern") and discussions with K. H. Grellmann are gratefully acknowledged.

REFERENCES

1. L. D. Landau and E. M. Lifshitz, "Quantum Mechanics" (2nd edition, Pergamon Press, Oxford, 1965).
2. E. E. Nikitin, "Theory of Elementary Atomic and Molecular Processes in Gases" (Clarendon Press, Oxford, 1974).
3. R. P. Bell, "The Tunnel Effect in Chemistry" (Chapman and Hall, London, 1980).
4. H. S. Johnston, "Gas Phase Reaction Rate Theory" (Ronald Press, New York, 1966).
5. M. R. Spalburg, J. Los, and A. Z. Devdariani, Chem. Phys. 103, 253 (1986).
6. E. Merzbacher, "Quantum Mechanics" (2nd edition, Wiley, New York, 1970).
7. W. Siebrand, T. A. Wildman, and M. Z. Zgierski, J. Am. Chem. Soc. 106, 4083, 4089 (1984).
8. F. H. Mies and M. Krauss, J. Chem. Phys. 45, 4455 (1966).
9. C. E. Caplan and M. S. Child, Mol. Phys. 23, 249 (1972).
10. M. S. Child, "Molecular Collision Theory" (Academic Press, New York, 1974).
11. A. J. Lorquet, J. C. Lorquet, and W. Forst, Chem. Phys. 51, 241, 253 (1980).

12. M. Quack and J. Troe, Ber. Bunsenges. Phys. Chem. 78, 240 (1974).
13. W. H. Miller, J. Am. Chem. Soc. 101, 6810 (1979).
14. S. K. Gray, W. H. Miller, Y. Yamaguchi, and H. F. Schaefer, J. Am. Soc. 103, 1900 (1981).
15. J. Troe, J. Phys. Chem. 88, 4375 (1984).
16. H. L. Dai, R. W. Field, and J. L. Kinsey, J. Chem. Phys. 82, 1606 (1985).
17. H. L. Dai, C. L. Korpa, J. L. Kinsey, and R. W. Field, J. Chem. Phys. 82, 1688 (1985).
18. B. C. Garrett and D. G. Truhlar, J. Chem. Phys. 79, 4931 (1983).
19. N. B. Slater, "Theory of Unimolecular Reactions" (Methuen, London, 1959).
20. W. R. McKinnon and C. M. Hurd, J. Phys. Chem. 87, 1283 (1983).
21. J. Troe, Dr. Thesis, Göttingen, 1965.
22. G. Bartelt, A. Eychmüller, and K. H. Grellmann, Chem. Phys. Lett. 118, 568 (1985).
23. U. Baron, G. Bartelt, A. Eychmüller, K. H. Grellmann, U. Schmitt, E. Tauer, and H. Weller, J. Photochem. 28, 187 (1985).
24. J. Troe, Chem. Phys. Lett. 114, 241 (1985).
25. P. M. Felker and A. H. Zewail, J. Chem. Phys. 89, 5402 (1985).
26. H. A. Olschewski, J. Troe, and H. Gg. Wagner, Ber. Bunsenges. Phys. Chem. 70, 450 (1966).
27. H. A. Olschewski, J. Troe, and H. Gg. Wagner, Ber. Bunsenge. Phys. Chem. 70, 1060 (1966).
28. R. G. Gilbert and I. G. Ross, Aust. J. Chem. 24, 1541 (1971); G. Gebelein and J. Jortner, Theor. Chim. Acta 25, 143 (1972; A. W. Yau and H. D. Pritchard, Can. J. Chem. 57, 1731 (1979); G. E. Zahr, R. K. Preston, and W. H. Miller, J. Chem. Phys. 62, 1127 (1975).
29. J. Troe, Ber. Bunsenges. Phys. Chem. 78, 478 (1974); R. G. Gilbert, K. Luther, and J. Troe, Ber. Bunsenges. Phys. Chem. 87, 169 (1983).
30. W. Forst, J. Phys. Chem. 86, 1776 (1982); H. Loirat, F. Caralp, W. Forst, and C. Schoenenberger, J. Phys. Chem. 89, 4586 (1985).

THE ROLE OF DISSIPATION IN QUANTUM TUNNELING

Hermann Grabert
Institut für Theoretische Physik
Universität Stuttgart
D 7000 Stuttgart 80
Federal Republic of Germany

ABSTRACT. The rate of kinetic processes hindered by a potential barrier is studied in the temperature range where quantum effects are important. Based on a generalized quantum-mechanical version of Kramer's Brownian motion approach, a rate theory accounting for thermally activated and tunneling events is presented. The temperature dependence of the rate is studied and the crossover from the Arrhenius law valid at high temperatures to quantum rate theory valid at low temperatures is discussed. The role of memory friction caused by the dissipative coupling to the environment is emphasized.

1. INTRODUCTION

The Arrhenius law $\Gamma = \omega_a \exp(-V_b/k_B T)$ governs the classical kinetics of many processes in physical and chemical sciences. Here, the preexponential factor ω_a is an attempt frequency and V_b is the height of the potential barrier which must be surmounted in the kinetic process. The Arrhenius law predicts a vanishing rate Γ as the temperature T approaches absolute zero. However, quantum mechanics allows for tunneling through the barrier and leads to a finite rate at zero temperature. The crossover between classical and quantum behavior was observed for phenomena as diverse as diffusion of atoms on surfaces [1], ligand migration in biomolecules [2], decay of the zero-voltage state in current-biased Josephson junctions [3], and domain-wall motion in ferromagnets [4].

Generally, the reaction coordinate q of the kinetic process is coupled to many other degrees of freedom such as internal modes of molecules or solvent excitations in the case of chemical reactions. These environmental influences upon the dynamics of q are of crucial importance since they are the source of frictional and noise forces. Noise causes the over-barrier hopping at high temperatures and friction can strongly disturb the quantum-mechanical barrier penetration prevailing at low temperatures. For an environment in thermal equilibrium the noise strength is intimately connected with the frictional force by virtue of the fluctuation-dissipation theorem so that the environmental influences are characterized by the temperature and a damping coefficient.

165

J. Jortner and B. Pullman (eds.), Tunneling, 165–182.

In Kramers' seminal work [5] on classical reaction rates it was assumed that the frictional influence of the environment can be described by a frequency-independent damping coefficient. However, several recent experiments on classical activation rates have shown a failure of a simple approach based on memoryless damping [6]. This is due to the fact that barrier frequencies are often of the order of 10^{12}-10^{14} Hz, and environmental forces are likely to be correlated on this time scale thereby giving rise to memory effects. Here, I will discuss reaction rates for systems characterized by a classical equation of motion of the form

$$M\ddot{q} + M \int_0^t ds \ \gamma(t-s)\dot{q}(s) + \frac{\partial}{\partial q}V(q) = 0 \qquad (1)$$

where $\gamma(t)$ is a damping kernel describing the time-delayed frictional influence of the environment and $V(q)$ is a potential with a barrier of height V_b [Fig.1].

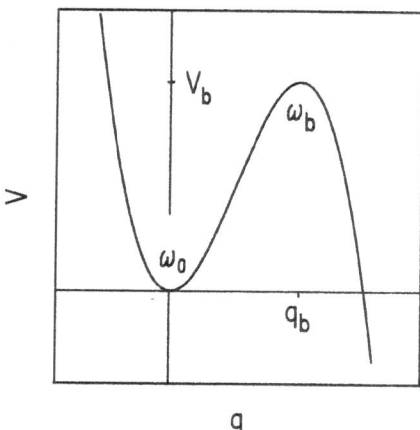

Figure 1.   A potential with a barrier hindering the kinetic process

The form of the potential provides two frequency scales important for the dynamical process. The well frequency

$$\omega_0 = [V''(0)/M]^{1/2} \qquad (2)$$

characterizes the potential in the well which the system occupies initially in a state of quasi-equilibrium. The barrier frequency

$$\omega_b = [-V''(q_b)/M]^{1/2} \qquad (3)$$

characterizes the width of the parabolic top of the barrier separating initial and final configurations. In a simplified description of the reaction by transition state theory, environmental effects are considered only to the extent that they maintain a Boltzmann distribution in the initial well [7]. Then the well frequency determines the attempt frequency $\omega_a = \omega_0/2\pi$ of the classical rate formula and the barrier frequency defines a crossover temperature [8] $T_0 = \hbar\omega_b/2\pi k_B$ below which tunneling events are predominant over thermally activated transitions.

Clearly, dissipation can substantially modify these findings. In the classical region a sluggish reaction coordinate will move across the barrier in a diffusive way with frequent recrossings thereby reducing the preexponential factor ω_a [5]. In the quantum regime, the effect of the environment is even more pronounced since friction supresses tunneling by an exponential factor [9] and thus lowers the crossover temperature T_0 [10]. The size of these effects depends not only on the noise strength but also on the memory correlation time. Increase of the memory correlation time tends to decrease the effective friction [11].

For systems with frequency-dependent damping a theory of classical reaction rates [12] and quantum tunneling rates [13] was developed only recently. Despite the fact that originally the classical region and the quantum region were studied by quite different methods, the entire range of temperatures can now be treated in a unified way by means of a functional integral method [9,14] based on Langer's classical nucleation theory [15]. This imaginary-time functional integral (ITFI) approach to reaction rates not only allows for a straightforward derivation of previous results including quantum corrections to thermally activated events [16] but also provides a complete description of the crossover between noise-activated Arrhenius-type transitions and quantum transitions [14]. In the sequel I will give a brief outline of the ITFI method and present the main results pertaining to the influence of temperature and dissipation on transition rates. In the last section I will discuss limitations and generalizations of the approach.

2. THE FREE ENERGY OF A DAMPED SYSTEM

In the absence of dissipation the coordinate representaion of the canonical operator $\rho_\beta = \exp(-\beta H)$ of a particle of mass M moving in a potential V(q) may be written [17]

$$<q'|\rho_\beta|q> = \int D[q] \, \exp(-\frac{1}{\hbar}S[q]) \tag{4}$$

where the functional integral is over all paths connecting $q(0) = q$ with $q(\hbar\beta) = q'$, and where the path probability is weighted according to the Euclidean action

$$S[q] = \int_0^{\hbar\beta} d\tau \, [\frac{1}{2}M\dot{q}^2 + V(q)] \tag{5}$$

Since the canonical operator may formally be considered as a time evolution operator $\exp(-iHt/\hbar)$ in imaginary time $t = -i\hbar\beta$, the integral (4) is frequently referred to as an imaginary time functional integral (ITFI).

Naturally, environmental influences will modify the stationary density matrix of the particle. In the problem I wish to address the coupling to the environmental degrees of freedom is not of the most general form but such that the damping term in the classical deterministic equation of motion (1) is linear. A linear dissipative mechanism, however, can always be modelled by a heat bath consisting of an infinite set of harmonic oscillators. The system under study is then governed by the Lagrangian

$$L = \frac{1}{2}M\dot{q}^2 - V(q) + \sum_i \frac{1}{2}m_i[\dot{q}_i^2 - \omega_i^2(q_i - \frac{c_i}{m_i\omega_i^2}q)^2]$$ (6)

Here the coupling term is written in a form that does not lead to a coupling-induced renormalization of the potential [9]. However, changes of the effective potential are often important, e.g. when reactions in different solvents are compared. The model characterized by (6) was studied frequently in the last two decades. Within a detailed, realistic model of the environment the model (6) is equivalent to an approximation where the response of the heat bath to the particle's motion is treated linearly. This corresponds to the assumption of linear damping.

Investigating the classical dynamics generated by the Lagrangian (6), one finds that the deterministic equation of motion of the particle is in fact of the form (1) with a damping kernel $\gamma(t)$ the Laplace transform of which is given in terms of the model parameters by [13]

$$\hat{\gamma}(z) = \int_0^\infty dt \; \gamma(t)\exp(-zt) = \frac{1}{M}\sum_i \frac{c_i^2}{m_i\omega_i^2}\frac{z}{z^2+\omega_i^2}$$ (7)

A given phenomenological damping kernel can now easily be modelled by a suitable choice of the parameters in (6). On the other hand, the microscopic model can readily be quantized. In the ITFI representation of the canonical operator of the entire system the integrals over the environmental coordinates are Gaussian and they may be evaluated exactly [17]. Tracing over the heat bath coordinates one then arrives at a functional integral representation of the reduced equilibrium density matrix ρ_β of the damped particle which is again of the form (4), however, with an effective action [9,10]

$$S[q] = \int_0^{\hbar\beta} d\tau \; [\frac{1}{2}M\dot{q}^2 + V(q)] + \frac{1}{2}\int_0^{\hbar\beta}d\tau\int_0^{\hbar\beta}d\tau' \; k(\tau-\tau')q(\tau)q(\tau')$$ (8)

where the last term describes the influence of the environment. The influence kernel $k(\tau)$ is periodic with period $\hbar\beta$ and may be represented as a Fourier series

$$k(\tau) = \frac{1}{\hbar\beta}\sum_{n=-\infty}^{\infty} K(\nu_n)\exp(i\nu_n\tau)$$ (9)

where the $\nu_n = 2\pi n/\hbar\beta$ are the Matsubara frequencies. The Fourier coefficients are given in terms of the model parameters by

$$K(\nu_n) = \sum_i \frac{c_i^2}{m_i \omega_i^2} \frac{\nu_n^2}{\nu_n^2 + \omega_i^2} \tag{10}$$

Comparing (7) and (10) we see that $K(\nu_n) = M|\nu_n|\hat{\gamma}(|\nu_n|)$ which implies

$$k(\tau) = \frac{M}{\hbar\beta} \int_0^\infty ds \, \gamma(s) \frac{\partial}{\partial s} \frac{\sinh(\nu s)}{\cos(\nu\tau) - \cosh(\nu s)} \qquad \text{for } \tau \neq 0 \tag{11}$$

where $\nu = \nu_1 = 2\pi/\hbar\beta$. This equation connects the quantum mechanical influence kernel $k(\tau)$ with the damping kernel $\gamma(t)$ of the classical equation
of motion. Because of this relation further recourse to the microscopic
model is not necessary. Note that the nonlocal form of the last term
in (8) is not due to memory effects since $k(\tau)$ remains finite for finite τ even if $\gamma(t)$ decays infinitely fast (Markov limit).

The partition function Z_β of the damped particle is the trace over
the equilibrium density matrix ρ_β. Using the ITFI representation (4),
we obtain

$$Z_\beta = \int D[q] \, \exp(-\frac{1}{\hbar} S[q]) \tag{12}$$

where the functional integral is over all periodic paths with period $\hbar\beta$.
The free energy is then given by

$$F = -\frac{1}{\beta} \ln Z_\beta \tag{13}$$

By virtue of (8) and (11), the free energy is now specified completely
in terms of quantities appearing in the classical equation of motion
(1). This is the starting point for the further analysis.

3. THE DECAY RATE OF A METASTABLE WELL

3.1 The Crossover Temperature T_0

I shall be concerned here with a damped particle moving in a metastable
well of the form depicted in Fig.1. Thermal and quantum fluctuations
will allow the particle to escape from this well. I will assume that
the barrier height V_b is large compared with other relevant energy
scales, in particular

$$V_b \gg k_B T \quad , \quad V_b \gg \hbar\omega_0 \tag{14}$$

Then, the main contribution to the functional integral (12) comes from
the vicinity of those paths for which the action (8) is stationary. The
extremal action paths satisfy the equation of motion

$$M\ddot{q}(\tau) - \frac{\partial V(\tau)}{\partial q(\tau)} - \int_0^{\hbar\beta} d\tau' \, k(\tau-\tau')q(\tau') = 0 \tag{15}$$

and the boundary condition $q(0) = q(\hbar\beta)$. In the absence of dissipation the evolution in imaginary time corresponds to a real time motion in the potential $-V(q)$. In this inverted potential [Fig.2] there is a trivial periodic solution, $q(\tau)=0$, where the particle just sits on top of the potential barrier corresponding to the well of the original potential $V(q)$, and another solution, $q(\tau)=q_b$, where it sits at the bottom of the well. However, for temperatures below $T_0=\hbar\omega_b/2\pi k_B$, the period $\hbar\beta=\hbar/k_B T$ is extended enough to admit also an oscillation of the particle along a periodic orbit in the classically forbidden region $0<q<q_b$. This latter trajectory is frequently called the bounce.

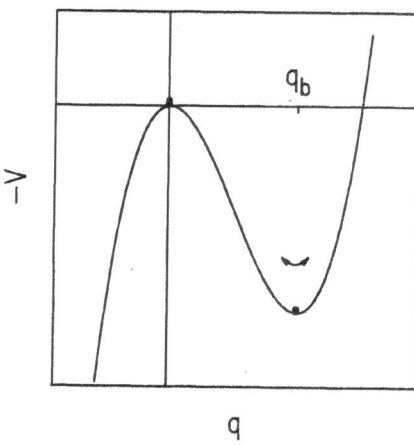

Figure 2.   The inverted potential $-V(q)$

Since the influence kernel satisfies

$$\int_0^{\hbar\beta} d\tau \; k(\tau) = K(0) = 0 \tag{16}$$

the trivial solutions $q(\tau)=0$ and $q(\tau)=q_b$ are not affected by dissipation though the action of paths in the vicinity of these trajectories will be modified. The bounce, however, is changed by dissipation and exists in the damped case for temperatures below the crossover temperature [11,14]

$$T_0 = \hbar\omega_R/2\pi k_B \tag{17}$$

where ω_R is the largest positive root of the equation

$$\omega_R^2 + \omega_R \hat{\gamma}(\omega_R) = \omega_b^2 \tag{18}$$

It will become clear from below that T_0 is the temperature where roughly the transition between thermal hopping and quantum tunneling occurs. This temperature is always reduced by dissipation which affects T_0 through the damping coefficient at frequency ω_R. For weak damping ω_R is of the order of the barrier frequency ω_b. In the particular case of frequency-independent damping, $\hat{\gamma}(z)=\gamma$, we have [10]

$$T_0 = (\hbar\omega_b/2\pi k_B)[(1+\alpha^2)^{1/2}-\alpha]$$ (19)

where $\alpha=\gamma/2\omega_b$ is a dimensionless damping parameter. Often, however, the damping coefficient will be frequency-dependent on the scale ω_b, so that memory effects become important.

3.2 Thermally Activated Transitions Above T_0

For temperatures above T_0 the functional integral (12) may readily be evaluated in the semiclassical approximation. A periodic path near the stationary trajectory $q(\tau)=0$ may be written as

$$x(\tau)= \sum_{n=-\infty}^{\infty} X_n \exp(i\nu_n\tau)$$ (20)

When this is inserted into the action (8) one finds

$$S[x]= \frac{1}{2}M\hbar\beta \sum_{n=-\infty}^{\infty} \lambda_n^0 X_n X_{-n}$$ (21)

where

$$\lambda_n^0 = \nu_n^2 + \omega_0^2 + |\nu_n|\hat{\gamma}(|\nu_n|)$$ (22)

and where terms of third order in the amplitudes X_n were disregarded. The contribution of these paths to the partition function (12) can now be determined by performing the Gaussian integrals over the amplitudes X_n. This gives the partition function Z_0 of a damped particle in a harmonic well.

A periodic path near the other stationary trajectory, $q(\tau)=q_b$, may be written as

$$y(\tau)= q_b + \sum_{n=-\infty}^{\infty} Y_n \exp(i\nu_n\tau)$$ (23)

The second-order action now reads

$$S[y]= \hbar\beta V_b + \frac{1}{2}M\hbar\beta \sum_{n=-\infty}^{\infty} \lambda_n^b Y_n Y_{-n}$$ (24)

where

$$\lambda_n^b = \nu_n^2 - \omega_b^2 + |\nu_n|\hat{\gamma}(|\nu_n|)$$ (25)

When we want to evaluate the contribution Z_b of these paths to the ITFI (12) we encounter a problem. Since the eigenvalue $\lambda_0^b=-\omega_b^2$ is negative, the trajectory $q(\tau)$ is not a minimum of the action but a saddle-point with an unstable direction. Because of this negative mode the integral over the amplitude Y_0 is divergent. This should come as not too big a surprise, after all, we are trying to compute the free energy of an unstable system. Langer [15] has explained that in such a situation the functional integral can still be defined by distorting the integration contour into the upper half of the complex plane along the direction of the steepest descent which is the positive imaginary axis in the present case. This leads to an imaginary part of the partition function. When we write Z_β in the form $Z_\beta=Z_0(1+Z_b/Z_0)$ and note that, as a consequence of the first term in (24), the ratio Z_b/Z_0 contains the exponentially small factor $\exp(-\beta V_b)$, the associated imaginary part of the free energy (13) is found to read

$$\text{Im } F = -(1/\beta Z_0) \text{ Im } Z_b = -(1/2\beta)[D_0/|D_b|]^{1/2}\exp(-\beta V_b) \qquad (26)$$

where D_0 and D_b are the determinants connected with the second-order action functionals (21) and (24)

$$D_0 = \prod_{n=-\infty}^{\infty} \lambda_n^0 \quad , \qquad D_b = \prod_{n=-\infty}^{\infty} \lambda_n^b \qquad (27)$$

Note that the exponentially small contribution Z_b of the paths near $q(\tau)=q_b$ is kept in the semiclassical approximation of the ITFI (12) only because it gives not just a small correction to Z_0 but a contribution which is imaginary and hence of a different type. The imaginary part of the free energy is now interpreted in the same way as the imaginary component of a resonance energy in quantum field theory, namely, as a quantity describing the finite lifetime of the state. By analogy, we would define the decay rate through $\Gamma=-(2/\hbar)\text{Im } F$. This is actually the formula which we shall use in the low temperature region below T_0. However, in the vicinity of T_0 the semiclassical approximation of the ITFI (12) breaks down and a more careful treatment is needed [14] which I will discuss below. This extended theory shows that the low temperature formula matches onto the semiclassical approximation above T_0 where the rate has to be calculated by means of the modified formula [7] $\Gamma=-(2/\hbar)(T_0/T)\text{Im } F$. Hence, above T_0 there is an additional factor (T_0/T) as a remnant of the transition near T_0. Inserting (17) and (26) into the rate formula we obtain

$$\Gamma = \frac{\omega_0}{2\pi} \frac{\omega_R}{\omega_b} f_q \exp(-V_b/k_B T) \qquad (28)$$

where

$$f_q = \prod_{n=1}^{\infty} \frac{\nu_n^2+\omega_0^2+\nu_n\hat{\gamma}(\nu_n)}{\nu_n^2-\omega_b^2+\nu_n\hat{\gamma}(\nu_n)} \qquad (29)$$

As is apparent from the Arrhenius exponential factor, the rate (28) describes thermally activated transitions across the barrier. The quantity f_q arises from quantum corrections [16]. Because of $\nu_n = 2\pi k_B Tn/\hbar$, the factor f_q approaches unity for $T \gg T_0$. The rate (28) then reduces to the classical hopping rate in the presence of memory friction [12] where the attempt frequency depends on the dissipation-renormalized frequency ω_R defined in (18). For frequency-independent damping, (18) can readily be solved and one obtains the familiar Kramers result [5].

On the other hand, as T approaches T_0 the factor f_q grows and it can enhance the rate considerably. The leading quantum corrections are independent of the dissipative mechanism and are given by the simple formula [11]

$$f_q = \exp\left[\frac{\hbar^2}{24}(\omega_0^2 + \omega_b^2)(k_B T)^{-2}\right] \tag{30}$$

where terms of order $(\hbar\omega_0/k_B T)^4$ were disregarded. For frequency-independent damping the product (29) can be evaluated explicitly for all temperatures in terms of gamma functions yielding [16]

$$f_q = \frac{\Gamma(1-\lambda_b^+/\nu)\Gamma(1-\lambda_b^-/\nu)}{\Gamma(1-\lambda_0^+/\nu)\Gamma(1-\lambda_0^-/\nu)} \tag{31}$$

where $\nu = 2\pi k_B T/\hbar$ and where

$$\lambda_b^\pm = -\frac{\gamma}{2} \pm \left(\frac{\gamma^2}{4} + \omega_b^2\right)^{1/2} \quad , \quad \lambda_0^\pm = -\frac{\gamma}{2} \pm \left(\frac{\gamma^2}{4} - \omega_0^2\right)^{1/2} \tag{32}$$

The temperature dependence of the factor f_q is depicted in Fig.3.

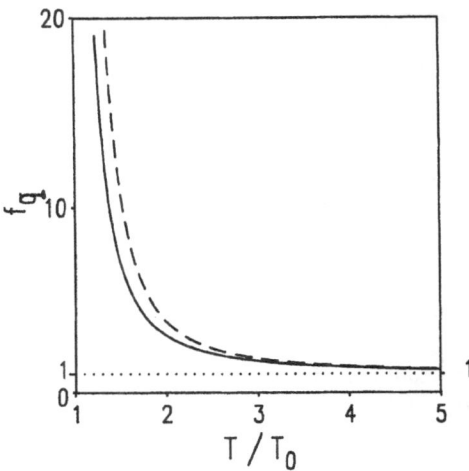

Figure 3. The quantum correction factor f_q is shown as a function of temperature T for a system with $\omega_0 = \omega_b$ and frequency-independent damping $\gamma = \omega_0/2$. The dashed line shows the approximation (30).

3.3 The Crossover Region Near T_0

The semiclassical approximation of the ITFI (12) used so far breaks down in the vicinity of T_0 because the eigenvalue $\lambda_1^b = \lambda_{-1}^b = \nu^2 - \omega_b^2 + \nu\hat{\gamma}(\nu)$ vanishes for $T = T_0$ so that the Gaussian integral over the amplitudes Y_1, Y_{-1} becomes divergent. Using $\nu = 2\pi k_B T/\hbar$ it is readily seen that the definition (17),(18) of T_0 is just derived from the condition $\lambda_1^b(T_0)=0$ [14]. Here I have tacitly assumed that the eigenvalue λ_1^b vanishes first as T is lowered which is the case for most models of the dissipative mechanism of interest. The vanishing eigenvalue points to the fact that below T_0 the evolution equation (15) admits a new oscillatory solution.

To remove the divergence we have to add terms of higher order in the amplitudes Y_1, Y_{-1} to the second order action (24). Expanding the potential $V(q)$ about the barrier top

$$V(q)= V_b - \frac{1}{2}M\omega_b^2(q-q_b)^2 + \sum_{k=3}^{\infty} \frac{1}{k}Mc_k(q-q_b)^k \qquad (33)$$

one readily obtains from (8)

$$S[y]= \hbar\beta V_b + \frac{1}{2}M\hbar\beta[\sum_{n=-\infty}^{\infty} \lambda_n^b Y_n Y_{-n}$$
$$+ 2c_3(Y_{-2}Y_1^2+Y_2Y_{-1}^2+2Y_0Y_1Y_{-1}) + 3c_4Y_1^2Y_{-1}^2] \qquad (34)$$

where terms up to the fourth order in Y_1, Y_{-1} were kept. For temperatures slightly above T_0, the contribution Z_b to the partition function can now be calculated by first integrating over the amplitudes Y_0 and $Y_{\pm n}$, $n\geq 2$ as before. Then, we are left with an integral over the amplitudes Y_1 and Y_{-1}. The integrand is determined by the effective action

$$\Delta S_1= \frac{1}{2}M\hbar\beta(2\lambda_1^b Y_1 Y_{-1} + BY_1^2 Y_{-1}^2) \qquad (35)$$

where

$$B= 4c_3^2/\omega_b^2 - 2c_3^2/\lambda_2^b - 3c_4 \qquad (36)$$

Using $2\pi^{-1/2} \int_z^{\infty}dt \exp(-z^2)=\mathrm{erfc}(z)$, the remaining integration is found to give the factor

$$1/\Lambda_1= (\pi M\beta/2B)^{1/2} \mathrm{erfc}[\lambda_1^b(M\beta/2B)^{1/2}] \exp[(\lambda_1^b)^2(M\beta/2B)] \qquad (37)$$

which replaces the factor $1/\lambda_1^b$ obtained in the semiclassical approximation. Now, λ_1^b may be written

$$\lambda_1^b= -a\varepsilon \qquad (38)$$

where $\varepsilon=(T_0-T)/T_0$ is negative above T_0 which is convenient for later purposes. The coefficient a is positive and reads for $T=T_0$

$$a= \omega_b^2 + \omega_R^2(1+\partial\hat{\gamma}(\omega_R)/\partial\omega_R) \qquad (39)$$

The factor (37) thus takes the form

$$1/\Lambda_1= \sqrt{\pi} (\kappa/a) \mathrm{erfc}(-\kappa\varepsilon) \exp(\kappa^2\varepsilon^2) \qquad (40)$$

where

$$\kappa = a(M\beta/2B)^{1/2} \tag{41}$$

Clearly, $1/\Lambda_1$ remains finite in the limit $\varepsilon \to 0$ and instead of (29) we now obtain for the quantum correction factor

$$f_q = (\lambda_1^0/\Lambda_1) f_R \tag{42}$$

where

$$f_R = \prod_{n=2}^{\infty} (\lambda_n^0/\lambda_n^b) \tag{43}$$

is the regular part of the product (29).

Before I discuss this result let me investigate the behavior of the rate for temperatures slightly below T_0. In this case, there is an additional extremal action path, namely the bounce. Since the bounce is a periodic trajectory, it may be written as a Fourier series

$$q_B(\tau) = q_b + \sum_{n=-\infty}^{\infty} Q_n \exp(i\nu_n \tau) \tag{44}$$

Now, when $q_B(\tau)$ is an extremal action trajectory, $q_B(\tau + \tau_0)$ is also a solution of (15). Hence, there is in fact a whole family of bounces with different phases. We can choose a particular one by requiring $q_B(\tau) = q_B(-\tau)$ or, equivalently, $Q_n = Q_{-n}$. Then, a fluctuation about the bounce which leads to a mere phase shift will not change the action. The amplitudes Q_n are small near T_0 and they can be calculated perturbatively from (15). Using $\varepsilon = (T_0 - T)/T_0 > 0$ as a small parameter one obtains the Q_n as a power series in $\sqrt{\varepsilon}$ [14,18]. When the result is inserted into (8), the bounce action is found to read

$$S_B = \hbar\beta V_b - \frac{1}{2}\hbar\beta(Ma^2/B)\varepsilon^2 \tag{45}$$

where terms of higher order in ε were disregarded. Note that the bounce action is smaller than the action of the trivial saddlepoint $q(\tau) = q_b$.

To study the fluctuation modes, we put $q(\tau) = q_B(\tau) + \xi(\tau)$ and expand $\xi(\tau)$ in a Fourier series

$$\xi(\tau) = \sum_{n=-\infty}^{\infty} \Xi_n \exp(i\nu_n \tau) \tag{46}$$

The fluctuation $\xi(\tau)$ leads to a change of the action (8). Near T_0 the second-order action may be determined explicitly. One finds [14]

$$S[q_B+\xi] = S_B + \frac{1}{2}M\hbar\beta[-\omega_b^2\hat{\Xi}_0^2 + \sum_{n=2}^{\infty} 2\lambda_n^b\hat{\Xi}_n\hat{\Xi}_{-n} + a\varepsilon(\Xi_1+\Xi_{-1})^2] \tag{47}$$

where only terms of leading order in ε were kept. Here, the $\hat{\Xi}_n$ are transformed amplitudes diagonalizing the second-order variation operator. To order $\sqrt{\varepsilon}$ one has

$$\hat{\Xi}_0 = \Xi_0 - (2c_3/\omega_b^2)(a\varepsilon/B)^{1/2}(\Xi_1+\Xi_{-1})$$
$$\hat{\Xi}_{\pm2} = \Xi_{\pm2} + (2c_3/\lambda_2^b)(a\varepsilon/B)^{1/2}\Xi_{\pm1} \qquad (48)$$

while the remaining Fourier coefficients are unchanged. We see that the eigenvalue of the $\hat{\Xi}_0$-fluctuation is negative so that the bounce is also a saddlepoint of the action. The main difference between the second-order action (24) above T_0 and the result (47) valid for temperatures slightly below T_0 is that the two-fold degenerate eigenvalue $\lambda_1^b=\lambda_{-1}^b$ which would become negative below T_0 is now replaced by a small positive eigenvalue $\lambda_1=2a\varepsilon$ and a vanishing eigenvalue $\lambda_{-1}=0$. This latter eigenvalue has the eigenmode $i(\Xi_1-\Xi_{-1})$ which does not contribute to (47).

As a consequence of the small eigenvalues λ_1 and λ_{-1}, the fluctuations Ξ_1 and Ξ_{-1} lead only to a small increase of the action (47) and they can become very large. Therefore, the second-order approximation is again not sufficient but the action of a path $q(\tau)=q_B(\tau)+\xi(\tau)$ must be determined more accurately by taking into account terms of the third and fourth order in the amplitudes Ξ_1 and Ξ_{-1}. These higher-order terms include nonlinear couplings between the fluctuations Ξ_1, Ξ_{-1} and the other modes. Having performed this expansion, we may determine the contribution Z_B of the bounce trajectory to the partition function (12) near T_0. The integrals over the stable modes ($\Xi_{\pm n}$, $n\geq2$) can be performed in semiclassical approximation. The integral over the negative mode (Ξ_0) can likewise be carried out by distorting the integration contour as above. This leads to an imaginary part of Z_B. One is left with an integral over the quasi zero modes (Ξ_1, Ξ_{-1}). These fluctuations have the effective action [14]

$$\Delta S_1 = \tfrac{1}{2}MB\hbar\beta[Q_1^2(\Xi_1+\Xi_{-1})^2 + 2Q_1(\Xi_1+\Xi_{-1})\Xi_1\Xi_{-1} + (\Xi_1\Xi_{-1})^2] \qquad (49)$$

Now, I introduce polar coordinates (ρ,ϕ) by $\rho\cos(\phi)=Q_1+(1/2)(\Xi_1+\Xi_{-1})$, $\rho\sin(\phi)=(1/2i)(\Xi_1-\Xi_{-1})$. Then, ΔS_1 turns out to be independent of ϕ. A change of ϕ just corresponds to a phase fluctuation of the bounce which does not change the action. After a corresponding transformation of the integration measure, the ϕ-integral is trivial. The ρ-integral corresponds to an integration over the amplitude fluctuations of the bounce. These fluctuations can be as large as the bounce amplitude. In particular, trajectories with ρ near zero are in the vicinity of the trivial saddlepoint whose contribution cannot be separated from Z_B for small ε. The ρ-integral can be transformed into an error integral. Using $Z_\beta=Z_0(1+Z_B/Z_0)$ and (13), the imaginary part of the free energy emerges as

$$\text{Im } F = -(1/2\beta)[D_0/|D_B''|]^{1/2} \sqrt{\pi} \ (\kappa/a) \ \text{erfc}(-\kappa\varepsilon) \ \exp(-S_B/h) \qquad (50)$$

Here D_0 is the determinant (27) and

$$D_B'' = \prod_{\substack{n=-\infty \\ n \neq \pm 1}}^{\infty} \lambda_n^b \tag{51}$$

is the determinant connected with the second-order action functional (47) with the zero mode and the quasi zero mode omitted.

The result (50) is valid for temperatures slightly below T_0. Now, inserting (45) and using $\Gamma = -(2/\hbar)\text{Im } F$ one finds for the decay rate [14]

$$\Gamma = \frac{1}{\hbar\beta} \frac{\omega_0}{\omega_b} \frac{\lambda_1^0}{a} f_R \sqrt{\pi} \, \kappa \, \text{erfc}(-\kappa\varepsilon) \, \exp(\kappa^2\varepsilon^2 - V_b/k_BT) \tag{52}$$

Since at the crossover temperature $1/\hbar\beta = \omega_R/2\pi$, the formula (52) coincides in fact with (28) when (42) is inserted there. Hence, (52) describes the behavior of the rate in the crossover region both above and below T_0.

For systems with high barriers and reasonably smooth potentials, the coefficient κ defined in (41) is much larger than 1. This follows from the fact that Ma^2/B is an energy of the order of the barrier height so that κ is of the order of $\sqrt{(V_b/\hbar\omega_R)} \gg 1$. The formula (52), which goes beyond the semiclassical approximation, is only needed in the region $|\kappa\varepsilon| < 1$, or

$$|T - T_0| \leq T_0/\kappa \tag{53}$$

where the argument of the erfc-function is of order 1. Because of $\kappa \gg 1$, the crossover region is narrow on the scale T_0. For temperatures above this region ($\kappa\varepsilon < -1$) we can use

$$\sqrt{\pi} \, \kappa \, \text{erfc}(-\kappa\varepsilon) \, \exp(\kappa^2\varepsilon^2) \simeq -1/\varepsilon \qquad \text{for } \kappa\varepsilon \lesssim -1 \tag{54}$$

Hence, in view of $\lambda_1^b = -a\varepsilon$, the formula (52) then reduces to the semiclassical rate (28), (29). For temperatures below the crossover region ($\kappa\varepsilon > 1$) we can use

$$\text{erfc}(\kappa\varepsilon) \simeq 2 \qquad \text{for } \kappa\varepsilon > 1 \tag{55}$$

and the rate (52) reduces to

$$\Gamma = [D_0/|D_B'|]^{1/2} (S_0/2\pi\hbar)^{1/2} \, \exp(-S_B/\hbar) \tag{56}$$

where D_B' is the determinant connected with the second-order action functional (47) with the zero mode omitted while the quasi zero mode $\lambda_1 = 2a\varepsilon$ is included. The remaining factors are lumped into

$$S_0 = 8\pi^2 (Ma/B\hbar\beta)\varepsilon \tag{57}$$

The result (56) holds for $\kappa^{-1} < \varepsilon \ll 1$ and coincides in this region with the low-temperature semiclassical rate discussed below.

Within the crossover region it is convenient to consider the quantity

$$y = \Gamma \exp(V_b/k_BT) \tag{58}$$

In the classical limit this quantity is independent of T, however, quantum effects lead to an increase of y as T is lowered. Let us study y as a function of $x = T-T_0$. Then, we see from (52) that there is a temperature scale $x_0 = T_0/\kappa$ and a frequency scale

$$y_0 = \frac{1}{2}(\omega_0^2+\omega_b^2)(\frac{M\omega_R}{\hbar B})^{1/2} \frac{\omega_0}{\omega_b} \prod_{n=2}^{\infty} \frac{n^2\omega_R^2+\omega_0^2+n\omega_R\hat{\gamma}(n\omega_R)}{n^2\omega_R^2-\omega_b^2+n\omega_R\hat{\gamma}(n\omega_R)} \tag{59}$$

so that

$$y/y_0 = F(x/x_0) \tag{60}$$

where $F(\xi)=\text{erfc}(\xi)\exp(\xi^2)$ is a universal function which is independent of the form of the metastable potential and also independent of the dissipative mechanism [Fig.4]. Only the scale factors x_0 and y_0 depend on the particular system under consideration. The rate follows the universal law (60) in the crossover region (53).

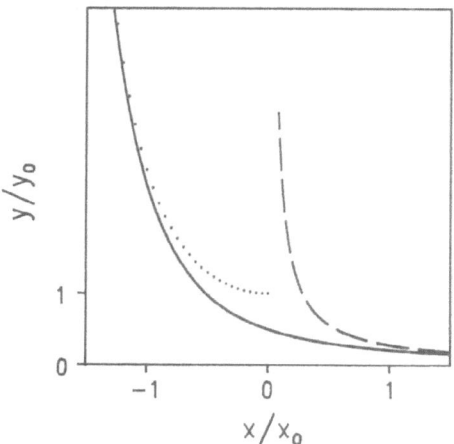

Figure 4. The scaled rate y/y_0 is shown as a function of the scaled temperature x/x_0. The high temperature formula (28),(29) is shown as a dashed line and the low temperature formula (56) as a dotted line. The crossover function smoothly matches onto these formulas valid outside the crossover region (after Ref.14).

3.4 Tunneling Transitions Below T_0

For temperatures below the crossover region the action of the bounce is substantially smaller than $\hbar\beta V_b$ so that the trivial saddlepoint $q(\tau)=q_b$ gives only an exponentially small correction to the decay rate which can be disregarded. However, since the bounce oscillation is no longer small, the equation of motion (15) cannot be solved analytically, in general. On the other hand, the Fourier components of the bounce and the bounce action can easily be calculated numerically [19].

To study the fluctuation modes, one proceeds as above and expands the action (8) about the saddlepoint trajectory. Putting $q(\tau)=q_B(\tau)+\xi(\tau)$ one has

$$S[q]= S_B + \int_0^{\hbar\beta}d\tau[\frac{1}{2}M\dot{\xi}^2 + \frac{1}{2}V''(q_B(\tau))\xi^2] + \frac{1}{2}\int_0^{\hbar\beta}d\tau\int_0^{\hbar\beta}d\tau'\ k(\tau-\tau')\xi(\tau)\xi(\tau') \quad (61)$$

where terms of the third order in $\xi(\tau)$ were omitted. To diagonalize the second-order variation operator one must again resort to numerical methods. This requires some care since the rate involves the product of eigenvalues so that each eigenvalue has to be determined very accurately [19].

One of the eigenvalues is zero. This is due to the fact that the phase of the bounce is arbitrary. To linear order we have $q_B(\tau+\delta)= q_B(\tau)+\dot{q}_B(\tau)\delta$ which shows that the zero mode is proportional to $\dot{q}_B(\tau)$. In fact, using the equation of motion satisfied by the bounce, it is readily seen that a fluctuation proportional to $\dot{q}_B(\tau)$ does not change the second-order action (61). Since the bounce is an oscillation, the zero mode has one node, and there exists a nodeless mode with a smaller, negative eigenvalue. The other eigenvalues are positive. The smallest positive eigenvalue, which merges into the quasi zero mode near T_0, is now sufficiently large in order that all positive modes can be integrated out from the ITFI by steepest descents. The integration contour of the negative mode is distorted as usual. The remaining integral over the zero mode is formally divergent. However, since the mode describes the shift of the bounce, this last integral sums over the family of bounces and it can be transformed to an integral over the bounce phase which varies over a finite interval. From the change of the integration measure one picks up an additional factor which depends on the zero mode normalization factor

$$S_0= M \int_0^{\hbar\beta}d\tau\ \dot{q}_B^2 \quad (62)$$

By virtue of $\Gamma=-(2/\hbar)\mathrm{Im}\ F$, one finally obtains for the rate Γ a formula which is identical with (56) except that S_B and S_0 now have to be calculated using the full nonperturbative bounce solution, and D'_B is the determinant connected with the second-order action (61). Naturally, for small ε one recovers the previous result.

An Arrhenius plot of some numerical results [19] is shown in Fig.5. In this diagram the Arrhenius law is represented by a falling straight line. Because of quantum effects the rate does not decrease continuously as T is lowered but flattens off. We see that for an undamped system ($\alpha=0$) there is a rather sharp transition between the classical regime of thermal hopping and the quantum regime of tunneling. In the presence of damping the classical rate is reduced slightly since the attempt frequency is diminished by the factor ω_R/ω_b. The zero temperature tunneling rate, however, is strongly reduced by an exponential

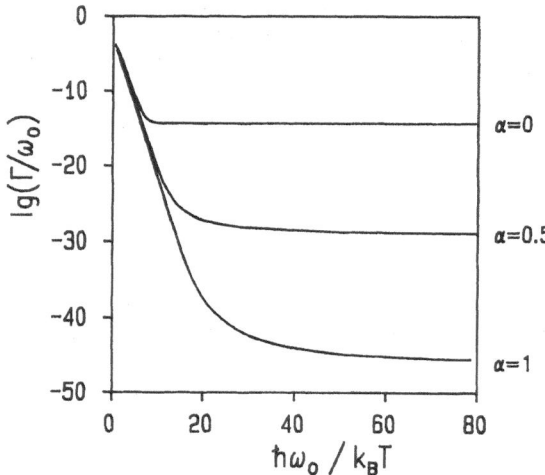

Figure 5.   The logarithm of the transition rate Γ is shown as a function of $1/T$ for a system with a cubic potential ($\hbar\omega_0 = \hbar\omega_b = V_b/5$) and frequency-independent damping $\gamma = 2\alpha\omega_0$ for various values of α (after Ref.19).

suppression factor [9]. For very weak damping thermal fluctuations have little effect on the low temperature behavior of the rate which becomes almost temperature-independent below T_0 for $\alpha=0$. However, for a damped system there is a large regime where thermal and quantum fluctuations interplay. The low temperature behavior of the rate is now governed by a power law [10]

$$\ln[\Gamma(T)/\Gamma(0)] = sT^n \tag{63}$$

where $n=2$ for all systems with finite damping at low frequencies, i.e. $\hat{\gamma}(\omega=0)\neq 0$. It turns out that the exponent of the power law is a characteristic feature of the dissipative mechanism and does not depend on the form of the potential [10]. The slope s increases with damping but depends on details of the model.

4. CONCLUSIONS

The theory of low temperature transition rates which I have presented here is applicable to the phenomenon of macroscopic quantum tunneling in Josephson devices [20]. These systems can be fabricated for a range of parameters so that they allow for a systematic study of the parameter-dependence of the rate. Many of the theoretical predictions have indeed been confirmed by recent experiments [3]. More frequently, tunneling is observed for molecular processes in physical and chemical systems. Often, little a priori information about the potential shape and the environmental coupling is available and one hopes to extract this information from the parameter-dependence of the measured rate. In this context it is important to be aware of the limitations inherent in a given rate formula. Therefore, I will shortly outline the range of validity of the ITFI approach to reaction rates.

Firstly, the reaction must occur under conditions of quasi-equilibrium which means that all relevant variables other than the reaction coordinate are supposed to be in equilibrium. This will be the case whenever Γ is much smaller than other relevant rates. The condition excludes highly underdamped systems where the relaxation within the metastable well is not fast enough to maintain an equilibrium distribution despite the leakage across the barrier [5,21]. For temperatures above T_0 the relevant condition is $\hat{\gamma}(\omega_R)/\omega_b > k_B T/V_b$ while in the quantum regime the theory extends to even weaker damping. This is due to the fact that the tunneling rate remains finite for zero damping while the thermal hopping rate vanishes [5].

Secondly, the barrier height V_b must be large compared with other relevant energy scales. This is necessary because the ITFI was evaluated in the semiclassical approximation. In the quantum regime the relevant condition is $\kappa \gg 1$ which amounts effectively to $V_b > 3\hbar\omega_R$. In the classical regime $\hbar\omega_R$ should be replaced by $2\pi k_B T$. Naturally, high barriers imply small rates so that this condition is not independent of the first one.

Thirdly, I have assumed that the transitions are incoherent. Otherwise the dynamical process cannot be described by a rate. This condition can break down for low temperatures when the potential has another minimum on the other side of the barrier. Then the system may tunnel coherently between these wells. However, coherent oscillations are easily destroyed by environmental influences so that this complication arises only for systems tunneling between two almost degenerate states at temperatures well below the crossover temperature or for extremely underdamped systems [22]. For the most part, the dynamics of the double well can be described by a forward and a backward rate which may both be calculated by the method presented here [23].

Finally, the theory was based on the assumptions of linear dissipation and smooth potential shapes. Naturally, particular phenomena may arise from fancier couplings to the environment. Further, the temperature dependence of the rate was discussed for systems with temperature-independent potential and friction. However, the theory remains valid when additional temperature dependence arising from potential and damping parameters is relevant.

The ITFI method is basically a thermodynamic method avoiding a full dynamical investigation of the kinetic process. However, there are a number of questions, such as the response to applied forces and non-equilibrium effects, that cannot be investigated within such a theory. A more complete real-time description of the kinetic process along similar lines can be based on the Feynman-Vernon theory [24]. This real-time functional integral technique for damped systems has already been applied to particular problems in the theory of dissipative quantum tunneling [22,25] and is expected to allow for further extensions of the theory in the future.

ACKNOWLEDGEMENTS

I wish to thank M.Büttiker, P.Hänggi, A.J.Leggett, S.Linkwitz, P.Olschowski, P.Schramm, S.Washburn, R.A.Webb, and U.Weiss for valuable discussions. This work was supported by the Deutsche Forschungs-gemeinschaft.

REFERENCES

[1] R.DiFoggio and R.Gomer, Phys.Rev.B25,3490(1982)
[2] H.Frauenfelder, in Tunnelling in Biological Systems, edited by B.Chance et al. (Academic, New York, 1979)
[3] S.Washburn and R.A.Webb, in New Techniques and Ideas in Quantum Measurement Theory, Ann.N.Y.Ac.Sci. (in press) and references therein
[4] W.Riehemann and E.Nembach, J.Appl.Phys.55,1081(1984)
[5] H.A.Kramers, Physica (Utrecht) 7,284(1940)
[6] For a recent discussion and a detailed list of references see P.Hanggi, J.Stat.Phys.42,105(1986)
[7] I.K.Affleck, Phys.Rev.Lett.40,388(1981)
[8] V.I.Gol'danskii, Dokl.Akad.Nauk SSSR 124,1261; 127,1037(1959)
[9] A.O.Caldeira and A.J.Leggett, Ann.Phys.(N.Y.)149,374(1983)
[10] H.Grabert, U.Weiss, and P.Hanggi, Phys.Rev.Lett.52,2193(1984)
[11] P.Hanggi, H.Grabert, G.-L.Ingold, and U.Weiss, Phys.Rev.Lett. 55,761(1985)
[12] R.F.Grote and J.T.Hynes, J.Chem.Phys.73,2715(1980)
     P.Hanggi and F.Mojtabai, Phys.Rev.A26,1168(1982)
     B.Carmeli and A.Nitzan, Phys.Rev.A29,1481(1984)
[13] H.Grabert and U.Weiss, Z.Phys.B56,171(1984)
     A.J.Leggett, Phys.Rev.B30,1208(1984)
     D.Esteve, M.H.Devoret, and J.M.Martinis, Phys.Rev.B(in press)
[14] H.Grabert and U.Weiss, Phys.Rev.Lett.53,1787(1984)
[15] J.S.Langer, Ann.Phys.(N.Y.)41,108(1967)
[16] P.G.Wolynes, Phys.Rev.Lett.47,968(1981)
[17] R.P.Feynman, Statistical Mechanics (Benjamin, New York, 1972)
[18] A.I.Larkin and Yu.N.Ovchinnikov, Sov.Phys.JETP 59,420(1984)
[19] H.Grabert, P.Olschowski, and U.Weiss, Phys.Rev.B32,3348(1985)
[20] A.J.Leggett, in Percolation, Localization, and Superconductivity, NATO Advanced Studies Institute, Vol. 109 (eds. A.M.Goldman and S.A.Wolf) (Plenum, New York, 1984)
     H.Grabert in SQUID'85 (eds. H.D.Hahlbohm and H.Lübbig) (de Gruyter, Berlin, 1985)
[21] M.Büttiker, E.P.Harris, and R.Landauer, Phys.Rev.B28,1268(1983)
     H.Risken and K.Voigtländer, J.Stat.Phys.41,825(1985)
[22] A.J.Leggett, S.Chakravarty, A.T.Dorsey, M.P.A.Fisher, A.Garg and W.Zwerger, preprint
[23] U.Weiss and H.Grabert, Phys.Lett.108A,63,(1985)
[24] R.P.Feynman and F.L.Vernon, Ann.Phys.(N.Y.) 24,118(1963)
[25] H.Grabert and U.Weiss, Phys.Rev.Lett.54,1605(1985)

TRANSITION STATE THEORY FOR TUNNELING IN DISSIPATIVE SYSTEMS

Eli Pollak
Chemical Physics Department
Weizmann Institute of Science
Rehovot 76100, Israel

I. INTRODUCTION

The theory of tunneling in dissipative systems has been developed extensively during the past decade. The typical system is that of a particle escaping from a metastable potential well which is separated by a barrier of height $V^{\#}$ from a different well or a continuum. The particle is immersed in a medium which exerts on it a frictional force. Escape from the well may occur either by thermal activation or quantal tunneling. The object of the theory is to predict the temperature dependence of the rate of escape of the particle.

If tunneling is ignored one has the generalized Kramers problem[1] whose solution is well known. The problem including tunneling was solved for 'high temperatures' by Wolynes.[2] His solution was obtained by means of a path integral evaluation of the reactive flux correlation function. The low temperature limit for a two state system was studied by Harris and Stodolsky[3] who predicted that dissipation will enhance the stability of optical isomers. A general formulation of the problem as well as an elegant solution, using instanton methods, has been given by Caldeira and Leggett.[4] The main result of their analysis is that at zero temperature the bath will exponentially dampen the tunneling rate. Grabert, Weiss and Hänggi[4,5] have extended the Caldeira-Leggett methodology to estimate the effect of temperature on the tunneling rate. Their main result is that for ohmic dissipation, increasing the temperature of the bath will (in the low-temperature limit) increase the tunneling rate by a factor of $\exp(AT^2)$.

In all these studies, the quantal theory is based on modeling the Generalized Langevin Equation (GLE) that governs the tunneling by an infinite set of harmonic oscillators coupled linearly to the tunneling degree of freedom. In a recent series of papers,[7-9] we have shown that one can derive many of the results obtained by using instanton methods via harmonic quantal transition state theory (TST).[10] Here we review the TST approach, providing a more unified treatment for the whole temperature range. In Sec. II we describe the quantum mechanical TST formulation and derive the Wolynes rate expression. The extension of

J. Jortner and B. Pullman (eds.), Tunneling, 183–192.

the theory to the low temperature region is given in Sec. III. In Sec. IV we discuss the differences between the TST approach and the instanton method.

II. Quantum TST for Dissipative Systems - the High Temperature Limit

The motion of the particle is governed by the GLE

$$M\ddot{q} + \int_0^t d\tau \eta(\tau)\dot{q}(t-\tau) + \frac{dV}{dq} = F_{ext}(t) \tag{1}$$

$V(q)$ is the system potential, $\eta(t)$ is the time dependent friction, related to the zero centered Gaussian random force $F_{ext}(t)$ by the usual fluctuation dissipation relation. As stressed by Caldeira and Leggett,[4] the GLE may be derived from the following Hamiltonian

$$H = \frac{1}{2M} p_q^2 + V(q) + \sum_{j=1}^N (\frac{p_j^2}{2m_j} + \frac{m_j}{2} [\omega_j x_j + \frac{c_j}{m_j \omega_j} q]^2) \tag{2}$$

(p_j, x_j) are the momentum and coordinate of the j-th bath oscillator. c_j couples the j-th oscillator to the system. The system dynamics in Eq. (2) are precisely of the GLE form with the identification of the time dependent friction as

$$\eta(t) = \sum_{j=1}^N \frac{c_j^2}{m_j \omega_j^2} \cos\omega_j t \tag{3}$$

In the limit $N \to \infty$, Eq. (3) is just the Fourier cosine series of $\eta(t)$. So for arbitrary $\eta(t)$ one finds the Fourier expansion and so determines the coupling coefficients and frequencies needed to construct the equivalent Hamiltonian. For future use we note from Eq. (3) that the Laplace transform of $\eta(t)$ is

$$\hat{\eta}(\epsilon) \equiv \int_0^\infty e^{-\epsilon t} \eta(t) dt = \sum_{j=1}^N \frac{c_j^2}{m_j \omega_j^2} \cdot \frac{\epsilon}{\epsilon^2 + \omega_j^2} \tag{4}$$

To derive the quantal rate we will use the following methodology. The Hamiltonian given in Eq. (2) may be treated as a quantal Hamiltonian. For a finite discrete set of oscillators one may evaluate the quantal thermal decay rate using harmonic quantum transition state theory. After obtaining the TST expression one may take the continuum limit, to obtain an estimate for the quantal decay rate of a particle governed by the GLE.

The classical TST rate expression is well known[10]

$$\Gamma_{cl}(T) = \frac{k_B T}{h} \, e^{-\beta V^{\#}} \, \frac{Z^{\#}}{Z_0} \quad (\beta \equiv \frac{1}{k_B T})$$ (5)

Here $Z^{\#}$ is the partition function of the bath at the barrier while Z_0 is the partition function of the combined system and bath at the well. $k_B T/h$ is the classical flux of particles crossing a one dimensional barrier whose height is $V^{\#}$. The quantal rate is obtained by evaluating quantal partition functions. These partition functions are evaluated via a normal mode analysis at the barrier and the well. Specifically, at the well ($q=0$) we assume that the potential $V(q)$ may be approximated as $\sim M\omega_0^2 q^2/2$ and at the barrier ($q=q_{\#}$) as $V^{\#}-M\omega_{\#}^2(q-q_{\#}^2)/2$. This harmonic approximation implies that the Hamiltonian in the vicinity of the well and barrier may be written in separable form as a sum of N+1 harmonic oscillators. This is achieved in standard fashion[11] by transforming to mass weighted coordinates ($q'=\sqrt{M}q$, $x_j'=\sqrt{m_j}x_j$) and then diagonalizing the (N+1)x(N+1) force constant matrix $\underset{\sim}{K}$ defined by the second derivatives of the potential at the well and the barrier.

The N+1 eigenvalues of $\underset{\sim}{K}$ at the well are denoted as λ_i^2; i=0,1,...,N. λ_i are the normal mode frequencies at the well. Because of the relatively simple structure of the Hamiltonian it is not difficult to prove the following identity[9]

$$\det(\underset{\sim}{K}+\epsilon^2 \underset{\sim}{I}) = (\lambda_0^2+\epsilon^2) \prod_{j=1}^{N} (\lambda_j^2+\epsilon^2) = [\omega_0^2+\epsilon^2 + \frac{\epsilon}{M} \hat{\eta}(\epsilon)] \prod_{j=1}^{N} (\omega_j^2+\epsilon^2)$$ (6)

Here $\underset{\sim}{I}$ is the (N+1)x(N+1) identity matrix and ϵ an arbitrary number. Similarly at the saddle point, the eigenvalues of the second derivative matrix $\underset{\sim}{K}^{\#}$ are denoted as $-\lambda_0^{\#2}$, $\lambda_j^{\#2}$; j=1, N. The lowest eigenvalue is negative and so associated with the unstable mode. In Ref. 8 we have shown that $\lambda_0^{\#}$ is identical to the Grote-Hynes[12] reactive frequency λ_r which is determined by the equation

$$\lambda_r = \frac{\omega_{\#}^2}{\lambda_r + \frac{\hat{\eta}(\lambda_r)}{M}}$$ (7)

Again it is not difficult to prove an identity which is analogous to Eq. (6):

$$\det(\underset{\sim}{K}^{\#}+\epsilon^2 \underset{\sim}{I}) = (-\lambda_0^{\#2}+\epsilon^2) \prod_{j=1}^{N} (\lambda_j^{\#2}+\epsilon^2) =$$

$$= [-\omega_\#^2 + \varepsilon^2 + \frac{\varepsilon}{M} \hat{\eta}(\varepsilon)] \prod_{j=1}^{N} (\omega_j^2 + \varepsilon^2) \tag{8}$$

We are now able to evaluate quantal partition functions. For a harmonic oscillator

$$Z_{qu.} = Z_{cl.} \frac{(\hbar\beta\omega/2)}{\sinh(\hbar\beta\omega/2)} \quad , \quad Z_{cl} = (\hbar\beta\omega)^{-1} \tag{9}$$

The quantum tunneling correction for the classical flux of particles over the harmonic barrier is also well known:[13]

$$\frac{1}{\hbar\beta} \to \frac{1}{\hbar\beta} \frac{(\hbar\beta\lambda_o^\#/2)}{\sin(\hbar\beta\lambda_o^\#/2)} \tag{10}$$

Inserting Eqs. (9,10) into Eq. (5), using the normal mode frequencies we find that the harmonic quantal TST expression for the rate is

$$\Gamma_{Qu}(T) = \Gamma_{Cl}(T) \frac{\lambda_o^\#}{\lambda_o} \frac{\sinh(\hbar\beta\lambda_o/2)}{\sin(\hbar\beta\lambda_o^\#/2)} \prod_{j=1}^{N} \frac{\lambda_j^\#}{\lambda_j} \frac{\sinh(\hbar\beta\lambda_j/2}{\sinh(\hbar\beta\lambda_j^\#/2)} \tag{11}$$

$$\Gamma_{Cl}(T) = \frac{\lambda_o}{2\pi} (\prod_{j=1}^{N} \frac{\lambda_j}{\lambda_j^\#}) e^{-\beta V^\#} \tag{12}$$

Equations (11,12) aren't very useful since it would be very tedious to compute all the eigenvalues needed to get the converged infinite bath limit. However, using Eqs. (6,8) with $\varepsilon=0$ it is easy to show that

$$\Gamma_{Cl}(T) = \frac{\omega_o}{2\pi} \frac{\lambda_o^\#}{\omega^\#} e^{-\beta V^\#} \tag{13}$$

but, $\lambda_o^\#$ is determined exclusively by the time dependent friction via Eq. (7) so that the classical rate is readily determined directly from the GLE. It should be stressed that Eq. (13) is not new, it has been derived previously by Kramers,[1] Grote and Hynes.[12] However, the present method[8] of derivation is new, it proves that Eq. (13) is really the true TST expression in the continuum limit.

Bringing the quantal expression into a tractable form is just slightly more tedious. One must use the infinite product expansions

$$\frac{\sinh x}{x} = \prod_{k=1}^{\infty} (1 + \frac{x^2}{k^2 \pi^2}) \qquad \frac{\sin x}{x} = \prod_{k=1}^{\infty} (1 - \frac{x^2}{k^2 \pi^2}) \tag{14}$$

and the identities given in Eqs. (6,8) to show that

$$\Gamma_{Qu}(T) = \Gamma_{Cl}(T) \prod_{k=1}^{\infty} \frac{\omega_0^2 + \nu^2 k^2 + \nu k \frac{1}{M} \hat{\eta}(\nu k)}{-\omega_{\#}^2 + \nu^2 k^2 + \nu k \frac{1}{N} \hat{\eta}(\nu k)} \tag{15}$$

Here $\nu \equiv 2\pi/\hbar\beta$. This result is identical to Wolynes' result[14] (Eq. 7 of Ref. 2).

The main problem with Eq. (15) is that it diverges at the critical temperature $T_c = \hbar\omega_{\#}/2\pi k_B$. The source of this divergence is the harmonic quantal tunneling correction (Eq. 10). To extend the TST treatment to lower temperatures one must evaluate the tunneling correction with more care, this is done in the next section.

III. THE LOW TEMPERATURE LIMIT

a. The System Dynamics

To clarify the divergence at the so called critical temperature we review the semiclassical theory of tunneling for the system when it is decoupled from the bath. We assume that the oscillator is characterized by 'bound' states with energy (in the harmonic limit) $E_n = (n+1/2)\hbar\omega$. The semiclassical decay rate from the n-th oscillator state is[15]

$$\Gamma_n = \frac{\omega_0}{2\pi} [1 + \exp \frac{2\pi(V^{\#} - E_n)}{\hbar\omega^{\#}}]^{-1} \tag{16}$$

The thermal rate is

$$\Gamma(T) = \sum_{n=0}^{\infty} e^{-\beta E_n} \Gamma_n / \sum_{n=0}^{\infty} e^{-\beta E_n} \tag{17}$$

Here we stress that $\Gamma(T)$ is finite at any temperature – there isn't any divergence. In other words, the so called critical temperature is an artifact of various approximations used to evaluate the sum found in Eq. (17). To stress this point further we analyze two common approximations. At very low temperatures, it is obvious that only the lowest oscillator states will contribute to the rate - the dominant pathway for decay is via tunneling. So in Eq. (16) one can substitute $1+\exp(B_n)$ with $\exp(B_n)$. The sum in Eq. (17) is then trivial and one finds

$$\Gamma(T) = \Gamma_o \cdot \frac{1-\exp(-\hbar\beta\omega_o)}{1-\exp\left[-(\hbar\beta\omega_o - 2\pi\omega_o/\omega^{\#})\right]} \tag{18}$$

This expression diverges at the critical temperature

$$T_c = \frac{\hbar\omega^{\#}}{2\pi k_B}$$

but of course, at this temperature the assumption that tunneling is the dominant pathway breaks down so that the approximate relation in Eq. (18) is no longer valid.

For high temperatures one evaluates Eq. (17) using two approximations.[13] First, one changes the discrete summation to a continuous integration, at this point

$$\Gamma(T) \simeq \frac{e^{-\beta V^{\#}}}{2\pi\hbar Z_{qu}} \int_{-V^{\#}}^{\infty} dx \, \frac{e^{-\beta x}}{1+\exp(-2\pi x/\hbar\omega^{\#})} \tag{19}$$

Note that $\Gamma(T)$ is still finite, however one now introduces the high temperature limit by noting that at high temperatures thermal activation is the dominant pathway so one can extend the lower limit from $-V^{\#}$ to $-\infty$. The integral is now analytical and one finds

$$\Gamma(T) = \frac{e^{-\beta V^{\#}}}{Z_{qu}} \frac{1}{\hbar\beta} \frac{\hbar\omega^{\#}\beta/2}{\sin(\hbar\omega^{\#}\beta/2)} \tag{20}$$

This is of course the expression used in the previous section (cf. Eq. 10) to evaluate the high temperature limit of the rate. The expression diverges at T_c but is clearly inapplicable since at T_c tunneling starts to become dominant and the extension of the lower limit is inapplicable. It is then clear that to apply TST to dissipative systems in the low temperature regime one must start from the dissipative system analog of Eqs. (16,17). This is done in the next subsection.

b. Theory of Dissipative Tunneling at Low Temperature

As already noted in the previous section, the coupling to the bath modifies both the system and the bath frequencies. Furthermore, the bath frequencies at the barrier are different from the bath frequencies at the well. Since the bath is no longer separable from the system it is seemingly difficult to evaluate the decay rate from an initial system-bath state. To proceed,[7] we will make an adiabatic approximation for the bath modes. This implies that if initially the j-th bath oscillator is in its n_j-th vibrational state, it will stay in this state forever. If at the well its frequency is λ_j and at the barrier $\lambda_j^{\#}$ then as the system moves from the well to the barrier it will have to overcome (or gain) the energy difference $(n_j+1/2)\hbar(\lambda_j^{\#}-\lambda_j)$. In the adiabatic approximation, the bath modes provide an effective potential for the tunneling mode. If the system and the bath are in the ground state then within the harmonic approximation the tunneling rate is simply

$$\Gamma_0 = \frac{\lambda_0}{2\pi} \exp(-\frac{2\pi E_0}{\hbar\lambda_0^{\#}}) \tag{21}$$

$$E_0 = V^{\#} - \frac{1}{2}\hbar\lambda_0 + \sum_{j=1}^{N} \frac{\hbar}{2}(\lambda_j^{\#}-\lambda_j) \tag{22}$$

As shown in Ref. 7 usually one will find that the main effect of the bath is to change the imaginary barrier frequency. Both for ohmic dissipation as well as an exponential friction one finds that the difference in ground state zero point energy is small and that $\lambda_0 \sim \omega_0$. However $\lambda_0^{\#}/\omega^{\#}$ may be substantially smaller then one. Hence the observation that dissipation decreases the tunneling rate. For ohmic dissipation $(\eta(t)=\eta\cdot\delta(t))$ and large η one can show[7] that the decrease in the rate goes as $\exp(-B\eta\cdot\Delta q^2)$ where Δq is the tunneling path. Comparison with the corresponding instanton expression shows quantitative agreement in the prediction of the exponential decrease of the rate.

To obtain the temperature effect one must consider thermal excitations of the bath. Within the adiabatic approximation, if the system is in the ground state and the bath is in state $\underline{n}$, that is n_1 quanta in mode 1, n_2 in mode 2 etc., then the tunneling rate from this state will be

$$\Gamma_0(\underline{n}) = \frac{\lambda_0}{2\pi} \exp\{-\frac{2\pi}{\hbar\lambda_0^{\#}}[E_0 + \hbar\sum_{j=1}^{N} n_j(\lambda_j^{\#}-\lambda_j)]\} \tag{23}$$

The thermal rate, averaged over the bath is

$$\Gamma_o(\beta) = \frac{\sum\limits_{\underline{n}} e^{-\beta \hbar \underline{n} \cdot \underline{\lambda}} \Gamma_o(\underline{n})}{\sum\limits_{\underline{n}} e^{-\beta \hbar \underline{n} \cdot \underline{\lambda}}} \tag{24}$$

Here the summation for the j-th oscillator is from $n_j=0,\infty$ and the zero point energy contribution appears in both numerator and denominator and so is already cancelled. One may now manipulate Eq. (24), this is given in detail in Ref. 7. The important point though is the result that the adiabatic approximation predicts that an increase in temperature will cause a <u>decrease</u> in the rate. This prediction is exactly the opposite of the instanton method prediction and is in disagreement with experiment and must therefore be wrong!

An analysis of the adiabatic result points out the source of the error. The coupling of the bath modes to the system causes a shift of frequencies. At the well one will always find that the coupling will lower the bath frequencies that are less than ω_o and increase those that are above. The same holds true at the barrier, except that here the lowest eigenvalue is the system eigenvalue which is negative so that all the bath frequencies are increased. The net result is that the adiabatic barrier height becomes larger for the low frequency modes. However at low enough temperature one predominantly excites the low frequency modes and the net result is an exponential decrease of the rate.

The only fallacy in the adiabatic theory is the adiabatic assumption. We just saw that the main effect at low temperatures comes from the low frequency bath modes which are moving slowly relative to the system. For these modes the adiabatic assumption is of course wrong[16,17] and one must use a sudden approximation.[17]

The sudden approximation implies that the dissociation is practically instantaneous. The initial bath coordinate x_j does not have time to change. As a result, at the transition state the bath will no longer be in an eigenstate of the normal mode Hamiltonian at the barrier. If at the well, the i-th oscillator is in the n_i-th state, there will be a finite transition probability from this state to the $n_i \pm 1$, $n_i \pm 2$ etc. corresponding normal mode states at the barrier. Such transitions can effectively reduce the tunneling barrier and so increase the decay rate. An estimate of this increase is given in detail in Ref. 7. Here it suffices to note that in fact the sudden approximation gives the same $\exp(AT^2)$ increase as obtained via instanton methods.

IV. DISCUSSION

The transition state theory approach to tunneling in dissipative systems is conceptually very simple and clear. Although, in the present context, we have used a harmonic approximation, one may use numerical methods to extend this approach to anharmonic systems. However, at present, the theory has one main defect. In order to obtain the rate for all temperatures one must provide a theory which goes in a consistent manner from the sudden limit at low temperatures to the adiabatic limit at high temperatures. The instanton method does not suffer from this shortcoming since it deals with the bath modes in an exact manner. However, within the instanton formalism one has to worry about divergences around T_c, these, don't exist, at least in principle in the TST method.

An additional advantage of the TST approach is that this theory will provide decay rate estimates for individual excited system oscillator states. These rates are today experimentally accessible.[18] The instanton method, is valid only for thermally averaged rates and so cannot be used for more detailed quantities.

At present, both approaches ignore completely the problem of whether the bath truly stays in thermal equilibrium. It is well known from the classical problem that at low friction the rate determining step is the energy flow from the bath to the system. Rips[19] has provided a theoretical framework for including this effect in the low friction limit. However a more general theory, is at present not available.

In summary, the TST approach outlined in this chapter although not yet complete does provide a good intuitive approach for understanding the phenomenon of tunneling in dissipative systems. Future work will hopefully show that this approach can provide a general quantitative theory.

ACKNOWLEDGMENT

This work has been supported by grants of the U.S.-Israel Binational Science Foundation and the Minerva Foundation.

REFERENCES

1. Kramers, H.A.: 1940, Physica 7, 284.

2. Wolynes, P.G.: 1981, Phys.Rev.Lett. 47, 968.

3. Harris, R.A., and Stodolsky, L.: 1981, J.Chem.Phys. 74, 2145.

4. Caldeira, A.O., and Leggett, A.J.: 1983, Ann.Phys. 149, 374.

5. Grabert, H., Weiss, U., and Hänggi, P.: 1984, Phys.Rev.Lett. 52, 2193.

6. Grabert, H., and Weiss, U.: 1984, Z.Phys.B 56, 171.

7. Pollak, E.: 1986, Phys.Rev.A. 33, xxx.

8. Pollak, E.: 1986, J.Chem.Phys. 85, xxx.

9. Pollak, E.: 1986, Chem.Phys.Lett., in press.

10. Pechukas, P.: 1981, Ann.Rev.Phys.Chem. 32, 159.

11. Wilson, E.B. Jr., Decius, J.C., and Cross, P.C.: 1955, Molecular Vibrations (McGraw-Hill, N.Y.).

12. Grote, R.F., and Hynes, J.T.: 1980, J.Chem.Phys. 73, 2715.

13. Pechukas, P.: 1976, in Dynamics of Molecular Collisions, part B, edited by Miller, W.H. (Plenum Press, N.Y.) Ch. 6.

14. Hänggi, P.: 1986, J.Stat.Phys. 42, 105.

15. Connor, J.N.L., and Smith, A.D.: 1982, Mol.Phys. 45, 149.

16. Büttiker, M., and Landauer, R.: 1982, Phys.Rev.Lett. 49, 1739.

17. Pollak, E.: 1985, J.Chem.Phys. 83, 1111.

18. Martinis, J.M., Devoret, M.H., and Clarke, J.: 1985, Phys.Rev.Lett. 55, 1543.

19. Rips, I.: 1986, paper in this volume.

QUANTUM DYNAMICAL SIMULATIONS OF TUNNELING SYSTEMS

Z. Kotler and A. Nitzan
School of Chemistry
The Sackler Faculty of Sciences
Tel Aviv University
Tel Aviv 69978, Israel
        and
R. Kosloff
Department of Physical Chemistry
and the Fritz Haber Institute
The Hebrew University
Jerusalem 91904, Israel

ABSTRACT. We describe several applications of the Fast Fourier transform algorithm for numerical solutions of the few body Schrodinger equation. Simple models for quantum diffusion of interacting particles are studied in detail. Thermal relaxation and friction effects in non-adiabatic tunneling systems is studied within the semiclassical time dependent Self Consistent Field approximations.

With the increasing availability of large computing facilities, simulations of quantum mechanical systems which have already become focus of much research effort, are expected to gain in importance as tools for theoretical research. Most of the methods developed so far involve calculations of stationary properties, e.g. ground state energies or the equilibrium density matrix, although some advances in dynamical simulations have been recently reported [1].

It has been recognized that a useful way to meet the large demand on computer time that such simulations require is to separate, when possible, a relatively small part of the system for which quantum mechanics is essential, from the rest of the system which is treated classically. A convenient way for simulating such mixed quantum-classical system is offered by the discretized path integral representation of the cannonical partition function of the system. This partition function can be evaluated by Monte Carlo techniques [2] or alternatively by molecular dynamics simulation in which each quantum degree of freedom is represented by a large number of fictitious classical particles, in addition to the genuine classical subsystem [3-5].

As noted above, these methods have been so far limited to equilibrium problems. In the present work we describe an approach to quantum dynamical simulations of systems of similar complexity. Our method is

J. Jortner and B. Pullman (eds.), Tunneling, 193–201
© *1986 by D. Reidel Publishing Company.*

based on the Fast Fourier Transform (FFT) algorithm for the numerical
solution of the time dependent Schrodinger equation [6,7], in conjunction with the Time Dependent Self Consistent Field (TDSCF) approximation for the coupling between different parts of the system [8]. For a
system described by the Hamiltonian

$$H = H_1 + H_2 + V(x_1 x_2) \tag{1}$$

(1 and 2 denote 2 particles or generally two subsystems) the TDSCF
equations of motion are, in the fully quantum mechanical theory
$$\Psi(x_1 x_2; t) = \Psi_1(x_1; t) \Psi_2(x_2; t) \exp(-iS(t))$$

$$\frac{\partial \Psi_1}{\partial t} = -\frac{i}{\hbar} (H_1 + <\Psi_2(t)|V|\Psi_2(t)>)\Psi_1 \tag{2}$$

$$\frac{\partial \Psi_2}{\partial t} = -\frac{i}{\hbar} (H_2 + <\Psi_1(t)|V|\Psi_1(t)>)\Psi_2 \tag{3}$$

$$\delta(t) = -\frac{1}{\hbar} \int_o^t <\Psi_1(\tau)\Psi_2(\tau)|V|\Psi_1(\tau)\Psi_2(\tau)>d\tau \tag{4}$$

and in the mixed quantum-classical theory

$$\frac{\partial \Psi_1}{\partial t} = -\frac{i}{\hbar} (H_1 + V(x_1,x_2))\Psi_1 \tag{5}$$

$$\dot{x}_2 = \partial H_2/\partial p_2 \tag{6a}$$

$$\dot{p}_2 = -\frac{\partial}{\partial x_2} (H_2 + <\Psi_1|V|\Psi_1>) \tag{6b}$$

Generalization of these equations to many particle systems is straightforward.

The success of a simulation based on Eqs.(2)-(4) or (5)-(6)
depends on the validity of the TDSCF approximation. While this approximation is successful in many applications, explicit conditions for its
validity are not available. For this reason it is important to have,
for simple systems, exact numerical or analytical results to compare
to. The FFT algorithm without the TDSCF approximation can be used for
this purpose for systems of a few (2-6) degrees of freedom, depending
on the availability of computer resources.

In the mixed quantum (q)-classical (c) situation an additional obvious requirement is that the c subsystem by itself could be described using classical mechanics. It is also required that the state of the c-system does not strongly depend on the particular quantum state of the q system, at least not on quantum states accessible during the simulation. For many applications this is expected to be valid, a simple example is the relaxation of a quantum oscillator coupled to a bath of classical particles. On the other hand there are important cases where this approximation is expected to fail: In a charge transfer process involving an impurity molecule in a condensed polar medium, the (otherwise classical) medium will respond to the charge transfer process in a discrete, quantum like way. TDSCF approximation cannot describe such a situation which is better handled by the surface hopping procedure [9] when the latter is valid.

It should be noted that in the molecular dynamics evaluation of the mixed quantum-classical partition function [3-5] the genuine classical particles move in the average field of all the fictitious classical particles which represent the quantum subsystem. This is in essence the same semiclassical SCF approximation implied by Eqs.(5)-(6). In this sense the mixed quantum-classical simulation described here is the dynamical equivalent of the method of refs. [3-5].

The mixed quantum mechanical approach may also be used to simulate systems interacting with their thermal environment. This can be done by using a large classical subsystem composed of many particles. Alternatively Eq.(6) may be supplemented by noise and damping terms in the same way that this is done in classical Langevin dynamics [10]. Eq.(6) is then replaced by

$$\ddot{x} = - \frac{1}{M} \frac{\partial}{\partial x} <\Psi|V|\Psi> - \int_0^\tau d\tau Z(t-\tau)\dot{x}(\tau) + \frac{1}{M} R(\tau) \qquad (7)$$

where the friction kernel $Z(t)$ and the random noise $R(t)$ are related by the classical fluctuation dissipation theorem

$$<R(t)R(t')> = Mk_B T Z(t-t') \qquad (8)$$

where k_B is the Boltzmann constant and T - the temperature [11]. Again, extension of (7) to many particle systems is straightforward.

An alternative coupling to a heat bath has been recently described by Cerjan and Kosloff [12] and applied to atom-surface scattering. In this approach the classical subspace is enlarged to include also variables bilinear in the x_c's (i.e. x_c^2 in the one classical variable case) and the corresponding equations of motion are derived from the classical limit of the Heizenberg equations for these variables.

In what follows we present results of simulations done on some very simple prototype systems. First we use the FFT algorithm to investigate tunneling behavior of two interacting particles in a 1-dimensional periodic potential. Secondly we use the mixed quantum classical TDSCF approach to simulate thermal relaxation of quantum systems.

Tunneling in a system of interacting particles is an important and generally unsolved problem in quantum mechanics. While screening makes the almost free quasi-particle picture a useful model for electronic properties of metals, electron motion in narrow band semiconductors are strongly affected by electron-electron interactions. A strong influence of H-H interaction on tunneling diffusion of hydrogen on metal surfaces is suggested by the strong coverage dependence of the diffusion rate [13].

The following questions concerning tunneling of interacting particles are of general importance and can be investigated in simple model systems:

(a). In what way does the interparticle interaction (its sign, strength, and range) affect the tunneling rate?

(b). How is the tunneling behavior affected by the symmetry (Bose-Einstein or Fermi-Dirac) of the particles?

(c). Can the (fully quantum) TDSCF approximation describe the correct dynamical behavior of the system?

The following results were obtained for a system of two identical particles in a periodic potential $V(x)=0.5A \cos Bx$. The length L of the system is $L=8\pi/B$ (so there are 4 wells) and periodic boundary conditions are imposed. The inter-particle interaction is taken to be repulsive Gaussian: $V(x_1-x_2)=a \exp(-b|x_1-x_2|^2)$. The initial wavefunction is $\Psi = \sqrt{2}/(\pi d^2) \exp(-(x_1-x_{10})^2/d^2)\exp(-(x_2-x_{20})^2/d^2)$ or its symmetrized or antisymmetrized version. The values of the parameters reported below are dimensionless in which $\hbar=1$, m(mass)=1 and ω(frequency at the well bottom)=1.24. The simulations yield the total wavefunction $\Psi(x_1,x_2;t)$ at each desired time t. This can be used for: (1). Obtaining the eigenvalue spectrum of the system by Fourier transforming in time the function $f(t) = <\Psi(t=o)|\Psi(t)>$ (Figs.(1a-e)) (2). Following the population of a particle in its initial well (Figs.(2a-b)) and (3). Evaluating the density-density autocorrelation function (this can be done for each well separately or as average over all wells; the resulting correlation functions were all almost identical if enough time-averaging was taken; Fig.(3a-c). It should be noted that to obtain the results of Fig.3 the first transient decay of the trajectory is discarded. Also, the last third of the lines displayed in Fig.3 are not accurate due to insufficient statistics incurred in the evaluation of the long time correlation function.

The spectra displayed in Figures 1 are rich, and in addition to giving a detailed picture of the low energy levels of the system, also demonstrate the usefulness of the method for evaluating eigenvalues. The relative intensities of the peaks are not significant as they merely reflect the structure of the initial state. The following observations are significant:

(a) The repulsive interparticle interaction causes a blue shift in the spectral structure. The slight blue shift seen in Fig. 1 increases with increasing the strength or the range of this interaction. This is the expected behavior in a system where the coverage is commensurate with the number of sites.

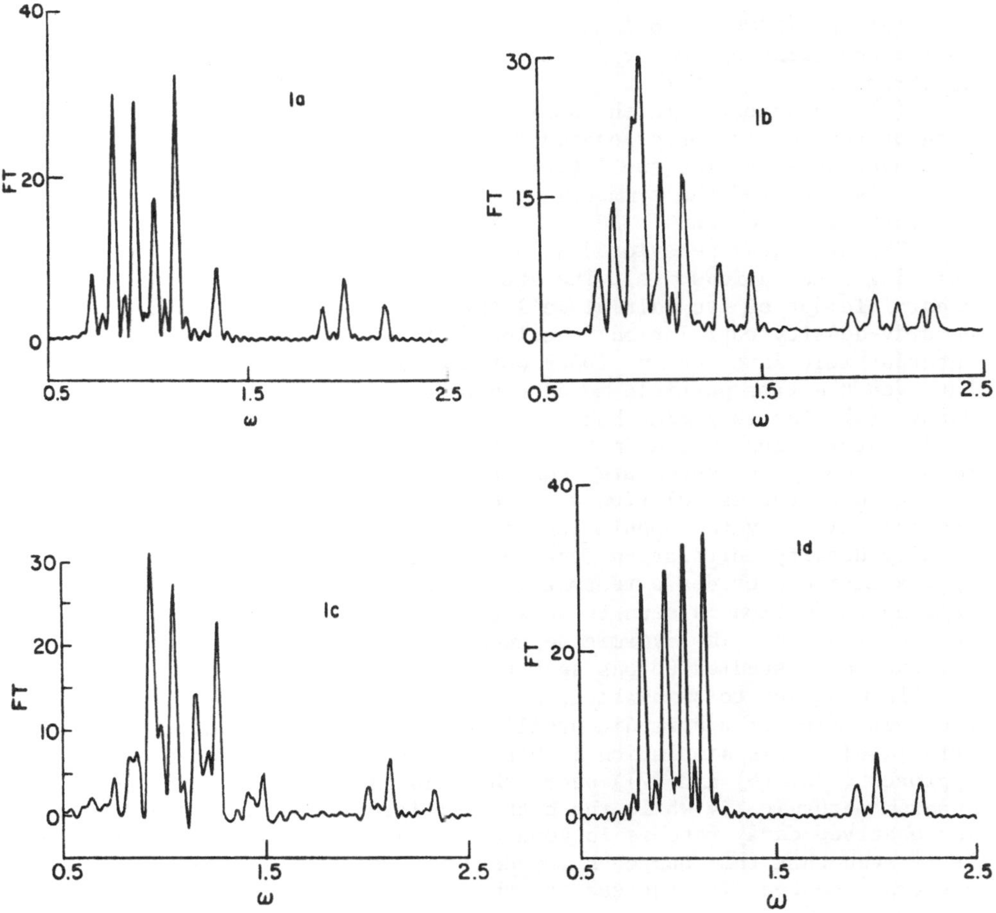

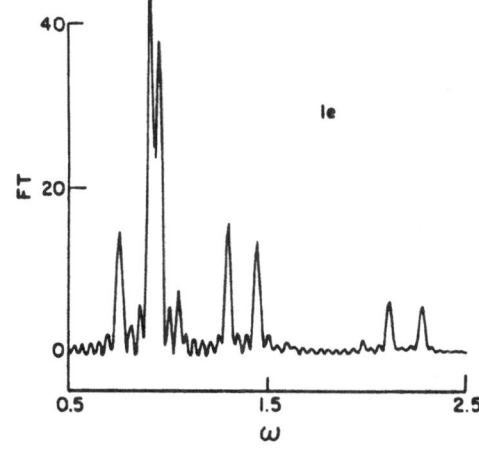

Figure 1. Fourier transform of the function $f(t)=\langle\Psi(0)|\Psi(t)\rangle$. The parameters used in these simulations are $m_1=m_2=1$, A=0.8, B=1.963 (corresponding to 4 wells within the system length L=12.8, a=0.5, b=1, $x_{10}=4.8$, $x_{20}=8.0$, d=1. In Fig.1a a=0.0 (no interparticle interaction). 1b) Exact calculation for a simple product initial wavefunction. 1c) TDSCF approximation. 1d) Antisymmetrized initial wavefunction. 1e) Symmetrized initial wavefunction.

(b) The TDSCF approximation gives a reasonably good approximation to the positions of energy levels, even though the shape of the spectrum is very different.

(c) The spectrum in the non-symmetric case (Fig.1a) is a combination of the levels corresponding to the antisymmetric (1c) and symmetric wavefunctions. The two latter spectra are quite different from each other. As expected the symmetric ground state is lower in energy than the antisymmetric one.

The dynamical behavior displayed in Figs.(2) and (3) yields three striking observations: (a). The short time behavior - the tunneling of a particle out of its initial well (Fig.2) or the initial decay of the density-density correlation function (Fig.3) is not very sensitive to the interparticle interaction. TDSCF works well on this short time scale but even the free particle model can account for this first initial decay. (b). The long time behavior is very sensitive to the interparticle interaction. While in the TDSCF case (as well as for free particles) strong recurrences are seen, the exact solution shows on the timescale of the calculation an ergodic-like behavior where following the initial decay the population fluctuate about its mean. (c). The density-density correlation function is well behaved and the TDSCF approximation provides a reasonable description of its time evolution. This approximation is seen to work better for the case of longer range interaction. (d). The dynamic depends on symmetry and, surprisingly, for the model studied bosons seem to tunnel more slowly than fermions.

Turning now to simulations of thermal system we show in Fig. 4 the relaxation of a harmonic oscillator coupled to an ohmic thermal bath [14a]. This simulation is based on the mixed quantum-classical approach (Eqs. (5) and (7)) where the relaxing oscillator is described quantum mechanically while the bath oscillators are taken classical. The observed decay rate is in good agreement with the imposed friction. It is seen that this approach can describe relaxation in quantum systems, however its success for the present system should not be too surprising as it has been shown that even a fully classical simulation can reproduce well the thermal relaxation of a harmonic oscillator [11].

Finally, in Fig. 5 we describe the results of simulations of activated tunneling in a non-adiabatic double well system, in the presence of ohmic friction. The two adiabatic surfaces are identical shifted harmonic oscillator potentials of frequency $\omega = 1$, made to cross at height $E_B = 3.0$ above their bottom. The non adiabatic coupling between these surfaces is taken to be a constant $V_I = 0.1$. Shown is the population in the initial state as a function of time for two values of the ohmic friction $\eta = 0.01$ and $\eta = 0.1$. For the large friction tunneling is seen to be completely blocked.

While these results are in qualitative agreement with recent theoretical work [14], no quantitative agreement can be achieved with the present simplified model. The source of the problem has been traced to the use, within the mixed quantum-classical TDSCF approximation, of a single ohmic bath which is coupled to the two quantum states. That (initially small) part of the wavefunction which crossed to the other

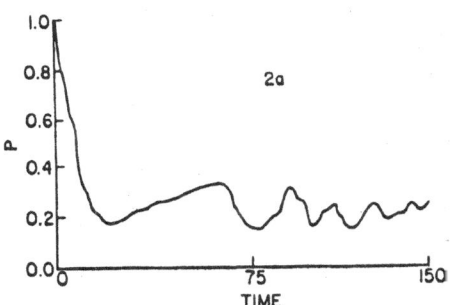 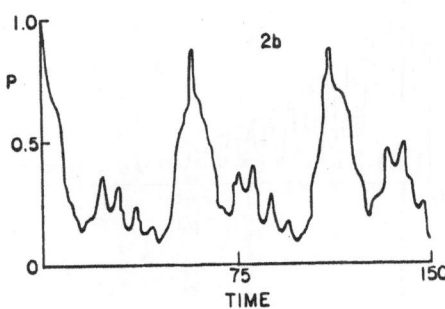

Figure 2. Time evolution of the initial state defined in the text with the parameters of Fig.1. Shown is the population of particle 1 in its initial well. 2a) Exact calculation. 2b) TDSCF approximation. Parameters are the same as for Fig.1.

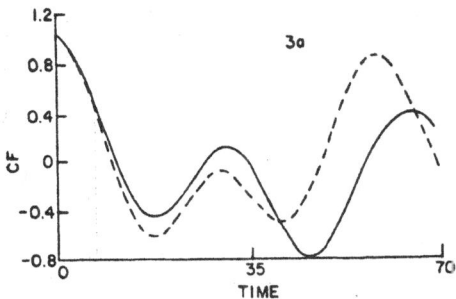 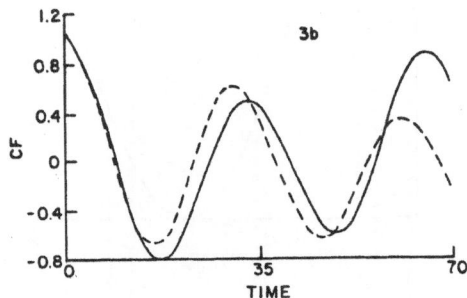

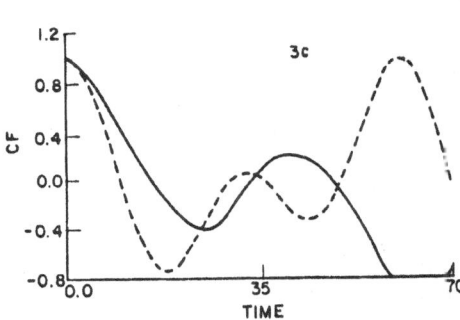

Figure 3. Density site correlation function

$$\frac{1}{4} \sum_i \lim(T\to\infty)\frac{1}{T} \int_0^T dt'\rho_i(t')\rho_i(t'+t)$$

a) Simple product initial wavefunction parameters as in Fig.1b-d. b) Longer range interaction, b=0.1 in Figs.a,b full line is an exact calculation, dashed line is TDSCF approximation. c) Full line – symmetrized initial wavefunction; dashed line – antisymmetrized initial wavefunction. Parameters as in (a).

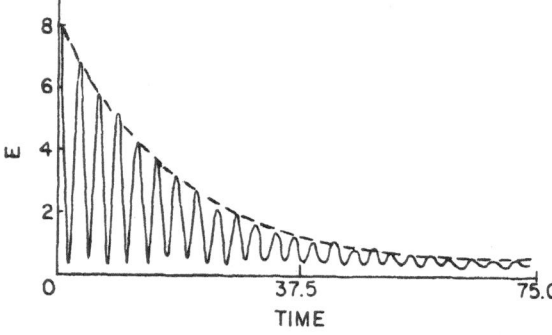

Figure 4.   Thermal relaxation
of an harmonic oscillator
coupled to an ohmic thermal
bath. Oscillator frequency is
1 and the thermal coupling
correspond to friction =
0.0628. The oscillating full
line is the kinetic energy
while the dashed line is the
total energy.

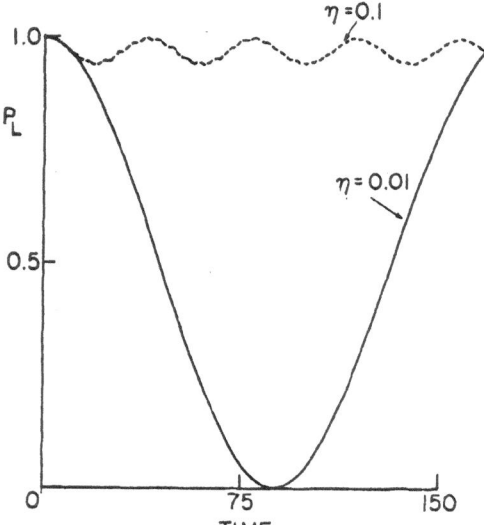

Figure 5.   Tunneling in a
system of two crossing degenerate harmonic wells. P is
the population in the initial
well. Parameters are given in
the text.

potential surface still feels a thermal bath which is in turn strongly
affected by the large quantum particle amplitude still left on the
original surface. This interaction is strong because the bath modes as
defined in ref.14a do not switch off with increasing distance. This
unphysical situation leads to considerable deviations of the numerical
time evolution from exact analytical results. This problem can be
overcome by taking the thermal bath to be (as it is in reality) spatially distributed so that a quantum particle interacts strongly only with
bath particles relatively close to it. Work along these lines is
currently in progress.

    In conclusion, the fast Fourier transform algorithm is a powerful
tool in analyzing few body quantum mechanical problems, and in conjunction with the TDSCF approximation can be used to simulate thermally
relaxing quantum systems. More work is needed to assess the limitations
associated with the TDSCF approximation.

Acknowledgement. This research was partially supported by grants from the Israel Academy of Science, the U.S.-Israel Binational Science Foundation and by a PACER fellowship from Control Data Corporation.

REFERENCES

1. D. Thirumalai and B.J. Berne, J. Chem. Phys. 79, 5029 (1983); 81, 2512 (1984); W.H. Miller, S.D. Schwartz and J.W. Tromp, J. Chem. Phys. 79, 4889 (1983).
2. J.A. Barker, J. Chem. Phys. 70, 2914 (1979); M.F. Herman, E.J. Brushin and B.J. Berne, J. Chem. Phys. 78, 5150 (1982).
3. M. Parrinello and A. Rahman, J. Chem. Phys. 80, 860 (1984).
4. B. deRaedt, L.M. Sprik and M.L. Klein, J. Chem. Phys. 80, 5719 (1984).
5. U. Landman, D. Scharf and J. Jortner, Phys. Rev. Letters 54, 1860 (1985).
6. D. Kosloff and R. Kosloff, J. Comput. Phys. 52, 35 (1983); R. Kosloff and D. Kosloff, J. Chem. Phys. 79, 1823 (1983); R. Kosloff and C. Cerjan, J. Chem. Phys. 81, 3722 (1984).
7. M.D. Feit and J.A. Fleck, Jr., J. Chem. Phys. 78, 301 (1983); 80, 2578 (1984); M.D. Feit, J.A. Fleck, Jr. and A. Steiger, J. Comput. Phys. 47, 412 (1982).
8. P.A.M. Dirac, Proc. Cambridge Phil. Soc. 26, 376 (1930); A.D. McLachlan, Mol. Phys. 7, 139 (1964); D. Kumamoto and R. Silbey, J. Chem. Phys. 75, 5164 (1981) and references therein. For recent applications of the semiclassical method see e.g. G.C. Schatz, V. Buch, M.A. Ratner and R.B. Gerber, J. Chem. Phys. 79, 1808 (1983); Z. Kirson, R.B. Gerber and A. Nitzan, Surf. Sci 137, 527 (1984).
9. J.C. Tully and R.K. Preston, J. Chem. Phys. 55, 562 (1971).
10. See, e.g. J.C. Tully, Acc. Chem. Res. 14, 188 (1981) and references therein.
11. In some situations it may be better to use instead of T an effective "quasiclassical temperature" to account for zero point motion. See A. Nitzan and J.C. Tully, J. Chem. Phys. 78, 3959 (1983).
12. C. Cerjan and R. Kosloff, 'Atom Phonon Interaction Using a Consistent Quantum Treatment', to be published.
13. R. Diffogio and R. Gomer, Phys. Rev. B25, 3490 (1982); S.C. Wang and R. Gomer, J. Chem. Phys. 83, 4193 (1985).
14. a). A.O. Caldeira and A.J. Leggett, Ann. of Phys. 149, 374 (1983); D. Waxman and A.J. Leggett, Phys. Rev. B32, 4450 (1985) and references therein.
    b). R.A. Harris and R. Silbey, J. Chem. Phys. 78, 7330 (1983); 80, 2615 (1984); 83, 1069 (1985).
    c). A. Garg, J.N. Onuchic and V. Ambegaokar, J. Chem. Phys. 83, 4491 (1985), and references therein.

ACTIVATED HYDROGEN ATOM TUNNELING REACTIONS IN SOLUTION

Domenic P. Ali and James T. Hynes
Department of Chemistry and Biochemistry
University of Colorado
Boulder, CO 80309 USA

ABSTRACT. A theory of the tunneling reaction of an H atom transfer between radicals in solution is presented. The quantum motion of the H atom is treated in the adiabatic approximation while the heavy flanking groups are treated classically. The solvent-reaction system interaction is described via a Generalized Langevin equation. The theory is applied to the methyl radical—methane reaction in a compressed inert gas solvent. The tunneling rate constant is found to be influenced by solvent dynamics in only a minor way, and H/D isotope effects are predicted to be nearly the same as the gas phase values. The physical origin of these features is discussed.

I.   INTRODUCTION

Hydrogen (H) atom transfer reactions play an important role in chemistry, with examples ranging from bimolecular H abstractions in free radical reactions [1] to intramolecular H transfers in keto-enol [2] and other [3] isomerizations. Here we present a semiclassical theory for symmetric intermolecular H atom transfer reactions

$$R'H + \cdot R \rightarrow R'\cdot + RH \tag{1}$$

in solution $(R = R')$. We focus both on the quantum tunneling aspects of the reaction and the solvent influence on that tunneling. The present description is a partial account; a more detailed discussion will be presented elsewhere [4].

   Our theory has several ingredients. First we take advantage of and extend the "heavy-light-heavy" Delves polar coordinate representation discussed by several authors [5-8] for gas phase H transfers. In particular, we recast the formulation of Babamov and Marcus [8] slightly into a time dependent form convenient for discussion of the solution phase problem. In this perspective, the reaction is viewed in terms of a double well for the H atom transfer from reactants to products, which evolves dynamically as the heavy flanking units R' and R approach and recede along a steeply repulsive potential. Tunneling from reactant to

J. Jortner and B. Pullman (eds.), Tunneling, 203–212.

product wells is possible for sufficiently small R-R' separation.
Second, the reaction rate constant is given by a correlation function
formula involving fluxes into and out of a high energy interaction zone
within which tunneling can occur. Third, the dynamical influence of the
solvent is accounted for by a Generalized Langevin Equation (GLE)
description of the fluctuating solvent forces on the heavy flanking
groups R' and R. This allows us to estimate the solvent influence on
the tunneling rate constant and on H/D isotope effects. The theory is
illustrated by model calculations for the $CH_3\cdot$ + CH_4 reaction in a
compressed inert gas solvent.

II. FORMULATION

As is now well known from gas phase studies [5-8], Delves polar
coordinates (Fig. 1) provide a convenient description for H transfers of
the heavy-light-heavy variety. The radial coordinate ρ governs the
relatively slow separation of the outer massive species R and R', while
the angle θ describes the rapid H atom motion between them.

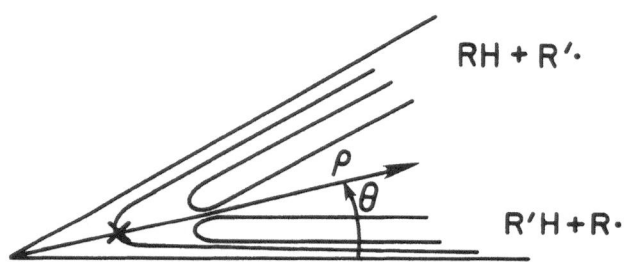

Figure 1.   Schematic illustration of Delves polar coordinates in the
skewed and scaled potential energy diagram for a symmetric H atom
transfer reaction.   The cross denotes the classical transition state.

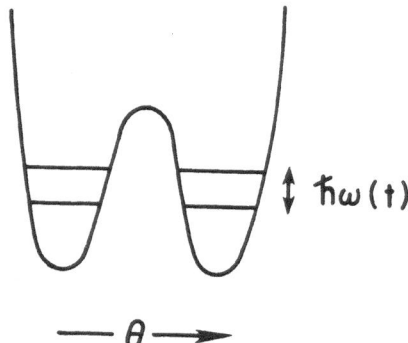

Figure 2.   Schematic illustration of the double well for H atom quantum
tunneling at fixed ρ. The tunnel splitting is indicated.

This perspective clearly reveals the essential features of the tunneling reaction (Fig. 2). At large R-R' separation, a cut in θ displays a reactant-product double well with a high barrier. At such large ρ values, an R'H bond must be nearly completely broken before the new HR bond is formed. Thus tunneling here is negligible and the R'H and RH vibrational levels in the appropriate wells are degenerate. But as R' and R approach, two things occur. First the "floor" of the H atom double well rises as the RHR' system is compressed – energy is required to access smaller ρ values. Second, the double well barrier both drops and narrows – now the RH bond begins to form before there is extensive breakage of the R'H bond. Tunneling is now possible, leading to split symmetric and asymmetric H vibrational levels, with the splitting given by

$$\varepsilon^a - \varepsilon^s = \hbar\omega(\rho) . \qquad (2)$$

An adiabatic, semiclassical treatment along the lines given by Babamov and Marcus [8] allows a simple treatment of the tunneling reaction. First, a Born-Oppenheimer approximation for the light fast H atom and the slow heavy radicals gives the effective potential energies ε^a and ε^s for the ρ motion as an average of the θ Hamiltonian, including the total potential energy $V(\rho,\theta)$, over the fast θ coordinate eigenfunctions for fixed ρ. The energies ε^a and ε^s give the tunnel splitting at each ρ by eq. (2). Second, a semiclassical treatment for the symmetric reaction of the heavy particle ρ coordinate introduces

$$\varepsilon(\rho) = \frac{1}{2}\left(\varepsilon^s(\rho) + \varepsilon^a(\rho)\right) \qquad (3)$$

as the effective potential energy for the classical ρ motion.

The tunneling reaction now appears as a classical approach of the reactants along the steeply rising $\varepsilon(\rho)$ potential, followed by tunneling as the splitting ω becomes significant, and terminating with a recession back out along ρ with a reaction probability determined by the net successful tunneling probability over to the product well. This is a quantum motion of the H atom in a dynamically evolving double well which in turn is governed by the classical ρ motion along the repulsive effective potential $\varepsilon(\rho)$. The semiclassical reaction probability for given incident total energy

$$E = K_\rho + \varepsilon(\rho), \qquad (4)$$

with K_ρ the ρ coordinate kinetic energy, is [4]

$$P_{rxn}(E) = \sin^2\left\{\frac{1}{2} \int_0^\infty dt\ \omega\ [\rho(t)]\right\} . \qquad (5)$$

Here at time zero, ρ is large enough so that tunneling is negligible, while at "infinite" time ρ is again large enough such that the reaction

fate is settled.   In practice, we will find that this latter time is
quite short (much less than 1 ps).

      Two features here are worth special comment.   First, and as
stressed by others [5-8], significant tunneling can occur before the
classical transition state is reached.   Since reaching the latter along
ρ costs additional energy, tunneling can often be the dominant reaction
mechanism.   Second, since tunneling only becomes significant for
sufficiently small ρ, we can henceforth focus on events in a short range
<u>interaction zone</u>, outside of which tunneling is negligible.

      Indeed, this last feature allows us to exploit the general Stable
States formalism of Northrup and Hynes [9] and write the reaction rate
constant k as [Fig. 3]

$$k = \int_0^\infty dt \; \langle j_i(0) j_o(t) \rangle_R \; . \tag{6}$$

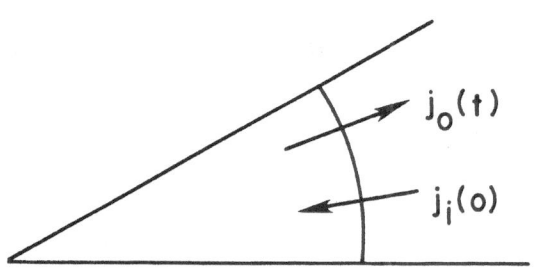

Figure 3.   Schematic illustration of the fluxes across the interaction
zone surface for the rate constant expression eq. (6).

This is the time integral of the correlation function of (a) the
incoming initial flux $j_i(0)$ along ρ from the reactants into the
interaction zone and (b) the outgoing flux $j_o(t)$ at time t from that
zone out into the product well.   An equilibrium average $\langle (\cdots) \rangle_R$ is taken
over both the RHR' system and any solvent, with a normalization by the
reactants' partition function.

      On the steep ε(ρ) curve, there will be negligible solvent-induced
recrossing of the interaction zone surfaces [10].   Then eq. (6) can be
integrated directly to give

$$k = \langle j_i(0) P_{rxn}(\infty) \rangle_R , \tag{7}$$

with $P_{rxn}(\infty)$ the reaction probability subject to initial entrance at the
interaction zone surface as reactant with a given initial momentum in ρ,
and where "∞" simply means a (very short) time after which the reaction
fate is sealed.   The initial flux can be explicitly written out [4,11]
and the integral over ρ variables performed formally.   In terms of the
total energy $E = K_\rho + \varepsilon(\rho)$, this gives

$$k = Q_R^{-1} \int_{\varepsilon(\rho_i)}^{\infty} dE \ e^{-E/k_BT} \ \overline{P_{rxn}(E)} \ , \tag{8}$$

where $\varepsilon(\rho_i)$ is the reactants' energy at the zone entrance, and the overbar denotes an average over the solvent.

To evaluate the solvent averaged reaction probability $\overline{P_{rxn}(E)}$ after entrance into the interaction zone with energy E, we must consider the interactions between the R'HR system and the solvent. Within the zone, the R'-R separation will be markedly smaller than the sum of the van der Waals radii of R and R' [see Sec. III]. Then the H atom will be shielded from the solvent and we accordingly neglect any direct H atom-solvent interaction. On the other hand, the R and R' flanking groups will be buffeted by the solvent, thus modulating the ρ motion and concomitantly the tunnel splitting. To emphasize this, we write k as

$$k = Q_R^{-1} \int_{\varepsilon(\rho_i)}^{\infty} dE e^{-E/k_BT} \overline{\sin^2\{\tfrac{1}{2}\int_0^{\infty} dt \omega[\rho(t); \ E]\}} \ . \tag{9}$$

In the general case, eq. (9) could be used for a Molecular Dynamics simulation to determine k. Instead we follow an approximate approach here.

Since the interaction zone is not large, we develop the splitting up to terms linear in the solvent fluctuation $\Delta\rho(t) = \rho(t) - \bar{\rho}(t)$ of the trajectory, so that

$$\omega[\rho(t)] = \bar{\omega}(t) + \Delta\omega(t)$$
$$\tilde{=} \ \omega[\bar{\rho}(t)] + \frac{\partial\omega}{\partial\rho}[\bar{\rho}(t)] \ \Delta\rho(t) \ . \tag{10}$$

A cumulant expansion [12] up to the leading second order in the splitting fluctuations then gives

$$k = Q_R^{-1} \int_{\varepsilon(\rho_i)}^{\infty} dE \ e^{-E/k_BT} \ \tfrac{1}{2}\{1 - [\cos\int_0^{\infty} dt \bar{\omega}(t)]e^{-I}\} \ , \tag{11}$$

with I determined by the solvent averaged correlation function of the fluctuating splitting,

$$I = \frac{1}{2} \int_0^{\infty} dt \int_0^{\infty} dt' \overline{\Delta\omega(t)\Delta\omega(t')} \ , \tag{12}$$

with eq. (10) relating $\Delta\omega$ to $\Delta\rho$. The time integrals in eqs. (11) and (12) are all short-lived, with a lifetime on the order of the residence time in the interaction zone in the presence of the strong repulsive forces $-\partial\varepsilon(\rho)/\partial\rho$.

A satisfactory description of the solvent-modulated ρ motion for these short times of interest is provided by a Generalized Langevin Equation (GLE): [4,10]

$$\ddot{\rho}(t) = F_\rho(t) - \int_0^t d\tau \zeta(t-\tau)\dot{\rho}(\tau) + R(t) , \qquad (13)$$

where $F_\rho = -\partial\varepsilon(\rho)/\partial\rho$ is the potential force, and

$$\zeta(t) = \left(k_B T\right)^{-1} \overline{RR(t)} , \qquad (14)$$

is the time dependent friction, i.e., the correlation function of the fluctuating solvent force $R(t)$ on the ρ coordinate.

[In what follows we will explicitly ignore solvent mean potential effects [13] arising from static solvent caging of the reaction system. These can influence the absolute value of the reactant rate constant in solution compared to the gas phase. Discussion of these in the present forum would take us too far afield from the focus here, which is the dynamic influence of the solvent on the tunneling. For a more detailed discussion including static solvent caging, see ref. [4]].

What is required for the rate constant eqs. (11) and (12) is the correlation function of the fluctuation $\Delta\rho(t)$, which from eq. (13), is

$$\overline{\Delta\rho(t)\Delta\rho(t')} = \int_0^t d\tau \int_0^{t'} d\tau' \gamma(\tau)\gamma(\tau') \overline{F_+(\tau)F_+(\tau')} \qquad (15)$$

with $\gamma(t) = \int_0^t d\tau \pi(\tau)$, where $\pi(t)$ is the normalized ρ velocity tcf, and

$$F_+(t) = R(t) + \bar{F}_\rho(t) - F_\rho(t) \qquad (16)$$

includes the deviation of the instantaneous ρ force from its average. Three reasonable approximations motivated by the short residence time in the interaction zone allow progress: (a) the neglect of the difference between F_ρ and $\bar{F}_\rho$, (b) the short time approximation $\pi(t) = 1$ so that $\gamma(t) = t$, and (c) the replacement of the average trajectory $\bar{\rho}(t)$ by the gas phase trajectory $\rho_E(t)$ at fixed incident energy. These then yield our final expression for k as

$$k = Q_R^{-1}\int_{\varepsilon(\rho_i)}^\infty dE e^{-E/k_B T} \frac{1}{2}\{1-e^{-I_E} \cos\int_0^\infty dt\omega[\rho_E(t)]\} ;$$

$$I_E = \left(\zeta(0)/8\right)\int_0^\infty dt\int_0^\infty dt' \frac{\partial\omega}{\partial\rho}[\rho_E(t)]\frac{\partial\omega}{\partial\rho}[\rho_E(t')] t^2 t'^2, \qquad (17)$$

in which $\zeta(0)$ is the initial value of the time dependent friction.

III. MODEL CALCULATION

We apply the preceding theory to the transfer of an H atom between
methyl radical $CH_3 \cdot$ and the methane molecule CH_4 in a model rare gas
solvent to explore the numerical consequences of the solvent dynamic
influence on the tunneling reaction.

We assume a collinear reaction and adopt the potential parameters
of Babamov and Marcus [8], who have also determined the asymmetric and
symmetric energies $\varepsilon^a(\rho)$ and $\varepsilon^s(\rho)$ for this system. The resulting
semiclassical energy $\varepsilon(\rho)$ for the ρ motion has a value $\varepsilon^{\neq} = 13.75$
kcal/mol at the classical transition state $\rho^{\neq} = 2.66$ Å and a splitting ω
which can reach up to ≈ 1 kcal/mol at this short separation. We select
the entrance ρ_i to the interaction zone at 3.06 Å where $\varepsilon(\rho_i) = 8$
kcal/mol, at which point the splitting is $\approx 10^{-3}$ kcal/mol and tunneling
is negligible. We take the initial time dependent friction value $\zeta(0)$
to be 260 ps^{-2} which is that appropriate for liquid Ar at 200 K,
appropriately mass-scaled for the CH_3-CH_4 reaction system [4].

Table I displays the results of the calculated rate constants eq.
(17), scaled by the corresponding standard semiclassical result [4,7]

$$k^{sc} = Q_R^{-1} \int_{\varepsilon(\infty)}^{\infty} dE \ e^{-E/k_B T} \theta(E - \varepsilon^{\neq}) \ , \qquad (18)$$

in which $\varepsilon(\infty)$ is the zero point energy of the reactants and $\theta(E - \varepsilon^{\neq})$ is
a step function presuming certain reaction for energies above the
activation energy $\varepsilon^{\neq}$ (referenced to zero energy and including the
orthogonal nonreactive symmetric stretch zero point energy).

Table I. Calculated ratios k/k^{sc} of the $CH_3 + CH_4$ rate
constant to its standard semiclassical
approximation

T, °K	GAS	SOLUTION
200	120	140
300	10	12
400	4	5

The first point of interest is the increase of k over k^{sc} in the gas
phase due to tunneling. Indeed, upon examination of the gas phase
reaction probability $P_{rxn}(E)$ in eq. (8), we find that even at 300 K,
approximately 90% of the rate constant integral

$$\int_{\varepsilon(\rho_i)}^{\infty} dE \ e^{-E/k_B T} P_{rxn}(E) \qquad (19)$$

comes from classically nonreactive energies below $\varepsilon^{\neq}$. This finding of
tunneling dominance is in line with the previous work of others [5-8].
The second striking feature in Table I is the very small influence of
the solvent on the tunneling. The source of this is simple. We
estimate that the characteristic time that the reaction system spends in
the interaction zone where tunneling can occur is ≈ 0.05 ps, due to the
strongly repulsive forces encountered there. Thus, the solvent has very
little time in which to influence the tunneling event since solvent
forces on this time scale are negligible. The modest enhancement of $\approx$
20% in k that does occur in solution may be understood in semiclassical
terms. For the $\cdot CH_3 + CH_4$ reaction, we find that $P_{rxn}(E)$ exceeds the
gas phase value $P_{rxn}(E)$ for the tunneling region $E < \varepsilon^{\neq}$ due to a solvent
induced spread in the splitting ω and thus the semiclassical phase
$\int_0^\infty dt\omega(t)$ accumulated during the passage through the interaction zone.
This increased phase heightens the probability that the H atom will be
found, upon exit, on the product side, i.e., has tunneled.

[The solution phase ratios in Table I will be influenced by mean
potential, static caging effects [13] - even when k^{sc} in solution
includes mean potential variations throughout the tunneling region.
Nonetheless the present conclusion as to the importance of dynamic
solvent effects still holds; for a more detailed discussion, see ref.
[4]].

We have also performed calculations of the model D atom transfer
$\cdot CH_3 + CH_3D$ reaction, again based on input potentials from collinear gas
phase quantum calculations by Babamov and Marcus [8]. Here the
classical activation energy (referenced to zero energy) is $\varepsilon^{\neq} = 13.95$
kcal/mol at $\rho^{\neq} = 2.63$ Å, where the splitting has increased to ≈ 0.5
kcal/mol. We take the interaction zone entrance at $\rho_i = 2.93$ Å where
$\varepsilon = 8.5$ kcal/mol and $\omega = 0.001$ kcal/mol. Table II displays the
resulting calculated H/D isotope effects for the reduced tunneling
contribution for D transfer.

Table II.   Calculated isotope effects k_H/k_D for the H
            and D transfer reactions $CH_3 + CH_3H(D)$ in gas
            and solution

T, °K	GAS	SOLUTION
200	60	64
300	22	23
400	13	13

The central point for our purposes is the small differences between the
gas and solution phase values. This feature arises from the small
influence of the solvent dynamics on the tunneling itself as discussed
above. Our theory then predicts that the isotope effect for the methyl
radical-methane reaction should remain approximately the same in gas
phase and in inert solvents at the same temperature.

These predictions may be testable via suitable ESR experiments involving $C^{13}H_3\cdot$ and C^{12} methane and deutero methane molecules [14], in which the reaction may be followed by changing ESR signals.

IV. CONCLUDING REMARKS

Here we have briefly described a theory for activated tunneling H atom transfer reactions in inert solution and applied it to the example of the methyl radical-methane reaction. We find that the influence of the solvent dynamics on the dynamic tunneling is small and predict that the H/D isotope effect should be essentially the same in gas phase and in an inert solvent. The source of these features is the very short time spent by the reaction system in the region where tunneling is possible.

In contrast, one might expect a more pronounced solvent effect for lower barrier H transfers such that the tunneling region is more extended and there is more time for the solvent to exert an influence. Similarly, the solvent might play more of a role when there is a significant dipole moment change, as in F + HF → FH + F, in a polar solvent. We hope to return to these topics.

In concluding, we would like to stress the sharp contrast between the activated molecular H transfer reaction class studied here and some other tunneling studies. First, for H atom transfers in low temperature matrices, it is likely that matrix caging effects lead to a reasonable description in terms of a nearly quadratic bound potential in ρ for the radicals [3]. Tunneling would then occur from larger ρ values than the shorter high energy separations proposed here to be of paramount importance in the fluid solution phase. Second, much activity (for a review, see Ref. [15]) has been centered on model intramolecular tunneling reactions in which (a) the surroundings induce fluctuating asymmetry in the two wells for the tunneling particle and (b) no other strongly coupled internal coordinate of the reaction system is involved. For the reaction class discussed in the present work, fluctuating level effects are negligible compared to the ρ modulation of the coupling [4], and the high potential energy ρ internal coordinate is all-important.

A more extensive version of this study, including attention to mean potential effects [13], the influence of fluctuation-induced asymmetry of the quantum H vibrational levels, and a detailed examination of the approximations introduced, will be presented elsewhere [4].

ACKNOWLEDGMENTS

This work was supported in part by NSF grants CHE 81-13240 and CHE 84-19830. We thank the organizers of the Jerusalem Tunneling Symposium for their kind invitation enabling us to present this work.

REFERENCES

1. P. Gray, A. A. Herod and A. Jones, Chem. Rev. **71**, 246(1971).
2. J. R. de la Vega, Acc. Chem. Res. **15**, 185(1982).
3. W. Siebrand, T. A. Wildman and M. Z. Zgierski, J. Am. Chem. Soc. **106**, 4089(1984).
4. D. P. Ali and J. T. Hynes, J. Chem. Phys. (submitted).
5. A. Kupperman, J. A. Kaye and J. P. Dwyer, Chem. Phys. Letters **74**, 257(1980).
6. C. Hiller, J. Manz, W. H. Miller and J. Römelt, J. Chem. Phys. **78**, 3850(1983).
7. D. K. Bondi, J. N. L. Connor, B. C. Garrett and D. G. Truhlar, J. Chem. Phys. **78**, 5981(1983).
8. V. K. Babamov and R. A. Marcus, J. Chem. Phys. **74**, 1790(1981).
9. S. H. Northrup and J. T. Hynes, J. Chem. Phys. **72**, 2700(1980).
10. J. T. Hynes, in The Theory of Classical Reactions, M. Baer, ed. (CRC Press, Boca Raton, FL, 1985). Vol. **4**, p. 171.
11. See, e.g., P. Pechukas, in Dynamics of Molecular Collisions, Part B, W. H. Miller, ed. (Plenum, New York, 1976).
12. R. Kubo, J. Math. Phys. **4**, 174(1963).
13. B. M. Ladanyi and J. T. Hynes, J. Am. Chem. Soc. **108**, 585(1986).
14. K. Peters, private communication.
15. P. Hänggi, J. Stat. Phys. **42**, 105(1986).

THE QUANTUM KRAMERS' PROBLEM IN THE UNDERDAMPED LIMIT

Ilya Rips and Joshua Jortner
School of Chemistry
Tel Aviv University
69978 Tel Aviv
Israel

ABSTRACT. The problem of particle escape from the quantum potential well in the case of weak coupling to thermal bath is considered. Quantum-mechanical transparency of the barrier is taken into account. Explicit analytic solutions are presented for the distribution function and for the escape rate in the extremely underdamped limit. A model, which rests on an integral self-consistent equation for the distribution function, is utilized for the description of particle escape in the underdamped domain at high temperatures. An equivalent form of the Mel'nikov expression for the escape rate is derived. The resulting expression is analysed for the single- and double-well systems.

1. INTRODUCTION

The ubiquity of experimental data in tunneling effects in various systems [1-3] has stimulated extensive theoretical studies of the quantum Kramers' model [4], with the quantum-mechanical transparency of the barrier being taken into account [5-16]. Three different approaches have been employed for the evaluation of the escape rate of the particle:

a) Thermal averaging over the Boltzmann distribution [6].
b) The use of the Kubo formula [7].
c) The semiclassical imaginary-time path-integral technique (the instanton technique) [5,8-15].

The validity condition for the applicability of these approaches is the existence of the thermal equilibrium within the well (or, at least, the time necessary for the system to reach thermal equilibrium is short compared with the inverse escape rate). This condition is satisfied provided that:

1) The metastability condition $V_0/kT \gg 1$ (V_0 is the depth of the well) is satisfied.
2) The coupling to thermal bath is sufficiently strong (high- or intermediate-friction domain).

Due to the absence of thermal equilibrium the underdamped (low friction) limit [4,17-25] necessitates a separate study, which has already

213

J. Jortner and B. Pullman (eds.), Tunneling, 213–226

been undertaken for the classical case [4,17-22]. Two aspects make the tunneling particularly important in the underdamped limit (UL). Firstly, the energy distribution function (EDF) in this domain decreases faster with increasing energy than is expected for equilibrium distribution. Accordingly, the number of particles with sufficiently high energy to escape through the classically allowed region is very small. Secondly, the suppression of tunneling by dissipation is weaker than in the intermediate- or high-friction limits.

The present work is devoted to a study of the quantum Kramers' problem in the UL. We were able to obtain [25] an exact solution for the distribution function and for the escape rate of the particle in the extremely underdamped limit (EUL). We also derive an expression for the rate, which is valid in the entire UL and which continuously matches the intermediate-friction result. Our expression is equivalent to that obtained by Mel'nikov [23] but more convenient for both analytical and numerical analysis.

2. THE EXTREMELY UNDERDAMPED LIMIT

The extremely underdamped limit (EUL) is characterized by the following features:
1) In view of the weak interaction with the bath, the majority of the particles possess very low total energy.
2) The escape of particles from the well occurs in an extremely narrow energy domain (i.e., in the classical over-barrier region and in the tunneling region below the barrier top).
3) The particle motion is nearly conservative. Averaging out the "fast" motion within the constant energy surface allows a description of the "slow" motion in terms of diffusion in the energy space.

The system can be described in this limit by the diffusion equation

$$\delta \left(kT \frac{d^2\rho}{dE^2} + \frac{d\rho}{dE} \right) = T(E) \; \rho(E) \quad , \tag{1}$$

where E is the energy of the particle (E=0 at the top of the barrier), $\rho(E)$ is the energy distribution function (EDF) and $T(E)$ is the transmission coefficient of the barrier, which for the parabolic barrier is

$$T(E) = \left[1 + \exp \{- 2\pi E/\hbar\omega\} \right]^{-1} \quad . \tag{2}$$

Finally, δ is the average energy dissipated per period of the conservative motion at the top of the barrier. It is related to the well frequency, Ω, to the barrier height, V_0, and to the friction coefficient, γ, by

$$\delta = c\gamma V_0/\Omega \tag{3}$$

where the constant factor c depends on the detailed form of the potential.

The diffusion equation, Eq. (1), has to be solved with the boundary condition:

$$\rho(E) = \frac{\Omega}{2\pi kT} \exp\{- \frac{V_o}{k\Xi} - \frac{E}{kT}\} \quad (E \rightarrow - V_o) , \tag{4}$$

which implies the existence of thermal equilibrium in the vicinity of the bottom of the well. Our diffusion equation differs from that studied by Büttiker, Harris and Landauer [19] in that their treatment was totally classical. It also differs from the model solved by Mel'nikov [23], who used an approximate WKB expression for the transmission coefficient.

The introduction of the dimensionless energy variable $\varepsilon = E/kT$, as well as the parameters

$$\Delta \equiv \delta/kT \equiv \mu^{-1} \tag{5a}$$

and

$$\alpha \equiv 2\pi kT/\hbar\omega \equiv \lambda^{-1} \tag{5b}$$

allows us to recast Eq. (1) in the form

$$\frac{d^2\rho}{d\varepsilon^2} + \frac{d\rho}{d\varepsilon} = \frac{1}{\Delta[1 + \exp(- d\alpha)]} \rho . \tag{6}$$

The solution of Eq. (6) can be represented in two equivalent forms:

$$\rho(\varepsilon) = D\exp\{- \frac{1}{2} [1 + (1 + 4\mu)^{\frac{1}{2}}]\varepsilon\} F(a,b,a+b+1; - e^{-\alpha\varepsilon}) \tag{7}$$

or

$$\rho(\varepsilon) = D[A_1(\varepsilon) + A_2(\varepsilon) \exp(-\varepsilon)] , \tag{8}$$

where

$$A_1(\varepsilon) = \frac{\Gamma(a+b+1) \ \Gamma(b-a)}{\Gamma(b) \ \Gamma(b+1)} \quad F(a,-b, 1-b+a; - e^{\alpha\varepsilon}) \tag{9a}$$

$$A_2(\varepsilon) = \frac{\Gamma(a+b+1) \ \Gamma(a-b)}{\Gamma(a) \ \Gamma(a+1)} \quad F(b,-a, 1-a+b; - e^{\alpha\varepsilon}) . \tag{9b}$$

$\Gamma(x)$ is the gamma function, and $F(a,b,c,;z)$ is the hypergeometric function. Finally, the parameters a and b are defined as

$$a = [(1 + 4\mu)^{\frac{1}{2}} + 1]/2\alpha \tag{10a}$$

and

$$b = [(1 + 4\mu)^{\frac{1}{2}} - 1]/2\alpha \tag{10b}$$

$$W = \frac{\Omega}{2\pi} \frac{\Gamma(a)\ \Gamma(a+1)\ \Gamma(b-a)}{\Gamma(b)\ \Gamma(b+1)\ \Gamma(a-b)} \left[\Delta\ F(a,-b,\ 1+a-b;-z_0)\ \exp(-V_0/kT)+ \right.$$

$$+ \frac{1}{1+\alpha}\ F(a+1,\ 1-b,\ 2+a-b;-z_0)\ \exp\{-\frac{(\alpha+1)V_0}{kT}\} \Big] -$$

$$\left. - \frac{\Omega}{2\pi} \cdot \frac{1}{1-\alpha} \cdot F(1+b,\ 1-a,\ 2-a+b;-z_0)\ \exp\{-\ \alpha V_0/kT\} \right., \quad (17)$$

where $z_0 = \exp(-\alpha V_0/kT)$. If $\alpha \gg 1$, the first term in Eq. (17) is the dominating one, and the rate can be written in the form akin to the transition-state theory (TST),

$$W \simeq \frac{\Omega}{2\pi}\ A \cdot \exp\{-\ V_0/kT\}, \quad (18)$$

where the TST correction factor, A, is given by

$$A = \Delta \cdot \frac{\Gamma(a)\ \Gamma(a+1)}{\Gamma(b)\ \Gamma(b+1)} \cdot \frac{(\pi/\alpha)}{\sin(\pi/\alpha)\ [\Gamma(1+1/\alpha)]^2} \cdot \quad (19)$$

Divergence of the rate, Eq. (17), for $\alpha = 1$ is of no physical significance since our results are valid only at sufficiently high temperatures (cf. Eq. (14)). This divergence has an analogue in the intermediate-friction limit. In this case, the rate in the high-temperature domain is given by [7,8]

$$W_I = \frac{\Omega}{2\pi} \cdot \frac{\omega^{\neq}}{\omega} \frac{\Gamma(1-\sigma_-/\nu)\ \Gamma(1-\sigma_+/\nu)}{\Gamma(1-\theta_-/\nu)\ \Gamma(1-\sigma_+/\nu)}\ \exp(-\ V_0/kT)\ , \quad (20)$$

where $\omega^{\neq} = \sigma_+$ is the frequency of the unstable mode, $\nu \equiv 2\pi kT/\hbar$, and

$$\sigma_\pm = -\frac{\gamma}{2} \pm (\gamma^2/4 + \omega^2)^{1/2} \quad (21a)$$

$$\theta_\pm = -\frac{\gamma}{2} \pm (\gamma^2/4 - \Omega^2)^{1/2} \quad . \quad (21b)$$

As was first noted by Wolynes [7], this expression for the rate diverges when $\sigma_+ = \nu$, i.e., at the temperature

$$T = T_c = \hbar\omega^{\neq}/2\pi k \quad . \quad (22)$$

Larkin and Ovchinnikov have shown [10] that T_c is the transition temperature which marks the crossover from the tunneling to the activated Arrhenius regime. T_c is the monotonically decreasing function

Equation (8), together with $F(a,b,c;0) = 1$, gives

$$\rho(\varepsilon)\Big|_{\substack{\varepsilon<0 \\ |\varepsilon|>>1}} = D\left[\frac{\Gamma(a+b+1)\ \Gamma(b-a)}{\Gamma(b)\ \Gamma(b+1)} + \frac{\Gamma(a+b+1)\ \Gamma(a-b)}{\Gamma(a)\ \Gamma(a+1)}\ \exp\ (-\varepsilon)\right]$$
(11)

Equation (11), together with the equilibrium boundary condition,
$\rho(\varepsilon) \simeq \frac{\Omega}{2\pi} \exp\{-\frac{V_o}{kT} - \varepsilon\}$, allows for the determination of the coefficient

$$D = \frac{\Omega}{2\pi}\ \frac{\Gamma(a)\ \Gamma(1+a)}{\Gamma(a+b+1)\ \Gamma(a-b)}\ \exp(-\ V_o/kT) \quad .$$
(12)

The equilibrium boundary condition near the bottom of the well can be satisfied only if the first term in Eq. (11) is negligibly small as compared with the second one, which implies that

$$\left|\frac{\Gamma(a)\ \Gamma(1+a)\ \Gamma(-1/\alpha)}{\Gamma(b)\ \Gamma(1+b)\ \Gamma(1/\alpha)}\right| << \exp(V_o/kT) \quad .$$
(13)

An approximate condition under which this inequality holds can be written as

$$\frac{\hbar\omega}{2\pi kT} \lesssim \frac{1}{\pi}\ \arcsin\ \{\pi\ e^{-V_0/kT}\} \simeq 1 - e^{-V_0/kT} \simeq 1 \quad .$$
(14)

Thus, the results have physical meaning only for sufficiently high temperatures.

Equation (7) implies the following asymptotic behaviour of the EDF

$$\rho(\varepsilon) \sim \exp\{-\frac{1}{2}[1 + (1 + 4kT/\delta)^{\frac{1}{2}}]\varepsilon\} \sim \exp\{-\left(\frac{kT}{\delta}\right)^{\frac{1}{2}} \varepsilon\} \quad .$$
(15)

Our knowledge of EDF enables us to derive an expression for the escape rate. The normalization of EDF has been chosen so that the escape rate is equal to the total flux out of the well, i.e.,

$$W = \int_{-V_o/kT}^{\infty} d\varepsilon\ T(\varepsilon)\ \rho(\varepsilon) \quad .$$
(16)

Straightforward but somewhat tedious manipulations lead to the final result:

of friction with

$$T_c^{max} = T_c(\gamma \to 0) = \hbar\omega/2\pi k \quad . \tag{23}$$

This value of the transition temperature was predicted long ago by Goldanskii [26] on the basis of much simpler arguments.

Since T_c^{max} corresponds to $\alpha = 1$, we note that the divergence of our expression for the rate is related to the divergence of the inter-mediate- and the high-friction results.

3.   CLASSICAL ESCAPE VERSUS TUNNELING

We inquire what is the relative contribution of the tunneling to the total flux out of the potential well. To answer this question we obtain the expression for the flux due to the classically allowed region, J_+, and compare it with the total flux. Our final result for J_+ is

$$J_+ = \frac{1}{2} \Delta \cdot [(1 + 4\mu)^{\frac{1}{2}} - 1] \cdot D \quad x$$

$$x \quad [F(a,b,a+b+1; -1) - \frac{a}{(a+b+1)} F(a+1,b+1,a+b+2; -1)] \tag{24}$$

Calculation of J_+/W, where W is given by Eq. (17), gives the relative contribution of the escape through the classically allowed region. This can be done only numerically and the results are presented in Fig. 1. Detailed analysis of the results has been given elsewhere [25]. Here, we only wish to point out that:

a) The character of the dependence of J_+/W on the friction is differ-ent in the classical and in the quantum case.
b) The physical parameter, which determines the relative contribution of the tunneling is

$$d = \frac{\hbar\omega}{\pi k T} \left(\frac{kT}{\delta}\right)^{\frac{1}{2}} \quad .$$

For d>1 a larger portion of the flux is due to tunneling. In the case d<1 particle escape occurs mainly through the classically allowed region.

4.   EXTENSIONS

The results of the previous analysis can be extended in two direc-tions.

A.   Non-Markoffian Dissipation.

The assumption that the dissipation is Markoffian can be

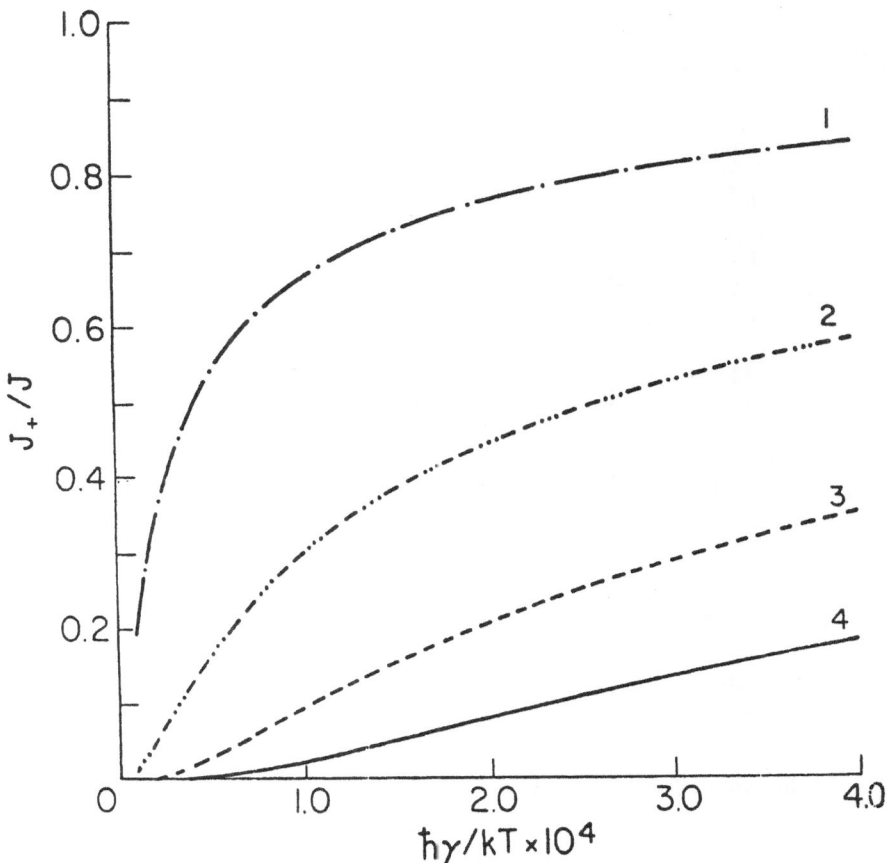

Fig. 1: The fraction of the particle flux J_+/J from the classically allowed region as a function of the friction coefficient.
Curve 1: $\hbar\Omega/kT = 0.2$, $\hbar\omega/kT = 0.1$; Curve 2: $\hbar\Omega/kT = 0.3$, $\hbar\omega/kT = 0.2$;
Curve 3: $\hbar\Omega/kT = 0.4$, $\hbar\omega/kT = 0.3$; Curve 4: $\hbar\Omega/kT = 0.5$ $\hbar\omega/kT = 0.4$.
$V_0/kT = 3$ and $c = 1.0$ for all the curves.

abandoned. Extension of the diffusion equation in the classical case
has been considered by Hänggi and Weiss [21]. Our expression for the
escape rate, Eq. (17), is valid also in the case of non-Markoffian
dissipation provided that the friction coefficient γ is replaced by
$\hat{\gamma}(0) = {}_0\!\int^{\infty}dt\gamma(t)$. $\hat{\gamma}(0)$, in its turn, is related to the autocorrelation
function of the stochastic force, $F(t)$, by the fluctuation-dissipation
theorem:

$$\hat{\gamma}(0) = \frac{1}{MkT} \int\limits_0^{\infty} dt \ <F(t) \ F(0)> \qquad (25)$$

B.   The Double Well Problem

This problem reduces to a solution of the system of differential equations,

$$\Delta_1 \left\{ \frac{d^2\rho_1}{d\epsilon^2} + \frac{d\rho_1}{d\epsilon} \right\} = T(\epsilon)[\rho_1(\epsilon) - \rho_2(\epsilon)]$$

$$\Delta_2 \left\{ \frac{d^2\rho_2}{d\epsilon^2} + \frac{d\rho_2}{d\epsilon} \right\} = T(\epsilon)[\rho_2(\epsilon) - \rho_1(\epsilon)]$$

(26)

where $\rho_1(\epsilon)$ and $\rho_2(\epsilon)$ are the EDF for the first and second potential well, respectively, with the boundary conditions:

$$\begin{cases} \rho_1(\epsilon) = (\Omega/2\pi) \exp\{- V_1/kT - \epsilon\} & (\epsilon \to - V_1/kT) \\ \rho_2(\epsilon) = 0 & (\epsilon \to - V_1/kT) \end{cases}$$

(27)

Introducing $f(\epsilon) = \rho_1(\epsilon) - \rho_2(\epsilon)$, we get from Eq. (27)

$$\frac{d^2f}{d\epsilon^2} + \frac{df}{d\epsilon} = \left(\frac{1}{\Delta_1} + \frac{1}{\Delta_2} \right) T(\epsilon) \, f(\epsilon) \quad ,$$

(28)

which has to be solved with the boundary condition,

$$f(\epsilon) = \frac{\Omega}{2\pi} \exp\{- \frac{V_1}{kT} - \epsilon\} \quad (\epsilon \to - V_1/kT)$$

(29)

Equation (28) is equivalent to that for the single-well system (Eq. (6)), so that we can use our previous results with the trivial replacement $1/\Delta \to 1/\Delta_1 + 1/\Delta_2$. Thus, for example, the transition probability, W_{1-2}, is given by

$$W_{1 \to 2} \simeq \frac{\Omega}{2\pi} \cdot \frac{\Delta_1 \Delta_2}{(\Delta_1 + \Delta_2)} \cdot \frac{\Gamma(\tilde{a}) \, \Gamma(1 + \tilde{a})}{\Gamma(\tilde{b}) \, \Gamma(1 + \tilde{b})} \cdot \frac{(\pi/\alpha)}{\sin(\pi/\alpha)[\Gamma(1+1/\alpha)]^2}$$

(30)

with

$$\tilde{a} = \left\{ \left[1 + \frac{4(\Delta_1 + \Delta_2)}{\Delta_1 \Delta_2} \right]^{\frac{1}{2}} + 1 \right\}/2\alpha$$

(31a)

$$\tilde{b} = \left\{ \left[1 + \frac{4(\Delta_1 + \Delta_2)}{\Delta_1 \Delta_2} \right]^{\frac{1}{2}} - 1 \right\}/2\alpha$$

(31b)

5. LIMITATIONS

The model considered above is applicable only in the EUL. The actual criterion for the validity of the diffusion equation is that the energy dissipated per period of conservative motion, δ, is much smaller than the energy scale ($\sim kT$) on which the EDF varies and is larger than the energy spacing between the levels ($\sim \hbar\Omega$), so that $\gamma/\Omega \ll kT/V_0 \ll 1$. On the other hand, the range of validity of the thermal equilibrium assumption is $\gamma/\Omega \gtrsim 1$. Thus it is necessary to match the results for the EUL with the intermediate-friction region results. A similar problem is also encountered in the classical Kramers' model [17-19,22]. However, there is an important difference between the classical and the quantum problems. The classical approach to the matching of the EUL and intermediate-friction limits rests on first-passage time arguments, which are inapplicable in the quantum case. This problem has been solved by Mel'nikov [23], who has shown that in the UL the Fokker-Planck equation can be reduced to the integral equation for the EDF. Mel'nikov's model is the subject matter of the next section.

6. AN INTEGRAL EQUATION FOR THE ENERGY DISTRIBUTION FUNCTION

The general picture advanced by Mel'nikov [23] is based upon multiple scattering of the particle from the energy barrier. In the time intervals between the subsequent collisions with the barrier the particle moves in the main part of the well under the action of the viscous friction and thermal fluctuations. Furthermore, the following assumptions are invoked:
1) The particle motion in the main part of the well can be described as diffusion in the energy space accompanied by the energy relaxation.
2) Neither friction nor thermal fluctuations influence the transmission coefficient of the barrier.
Mel'nikov's analysis rests on the integral self-consistency equation for EDF [23]

$$\rho(\epsilon) = \int_{-\infty}^{\infty} d\epsilon' \, k(\epsilon - \epsilon') \, R(\epsilon') \, \rho(\epsilon') . \qquad (32)$$

$R(\epsilon')$ is the reflection coefficient of the barrier, while $k(\epsilon - \epsilon')$ is the dissipation kernal

$$k(\epsilon - \epsilon') = (4\pi\Delta)^{-\frac{1}{2}} \exp\{-(\epsilon - \epsilon' + \Delta)^2/4\Delta\} \qquad (33)$$

and describes diffusion (with a drift) in the energy space. The self-consistency equation (32) is solved with the equilibrium boundary condition close to the bottom of the well. The observable calculated is the escape rate

$$W = \int_{-\infty}^{\infty} d\varepsilon \; T(\varepsilon) \; \rho(\varepsilon).$$

Mel'nikov has solved the integral equation using a modification of the Wiener-Hopf method and has obtained the expression for the escape rate of the form [18] with the TST correction factor

$$A = \frac{\pi\lambda}{\sin(\pi\lambda)} \; \exp\{\lambda\sin(\pi\lambda) \int_{-\infty}^{\infty} dx \; \frac{\ln[1 - \exp[-\Delta(x^2 + 1/4)]]}{\cosh(2\pi\lambda x) - \cos(\pi\lambda)} \} \tag{34}$$

where $\lambda = \hbar\omega/2\pi kT$. The shortcoming of Mel'nikov's final expression, Eq. (34), is that it is difficult to analyse. However, it can be simplified considerably. The application of the well-known Euler-Maclaurin formula to the integral in Eq. (34) results in

$$\int_{-\infty}^{\infty} dx \; \frac{\ln[1 - \exp[-\Delta(x^2 + 1/4)]]}{\cosh(2\pi x\lambda) - \cos(\pi\lambda)} = \frac{\ln[1 - e^{-\Delta/4}]}{2\sin^2(\pi\lambda/2)} +$$

$$+ 2 \sum_{n=1}^{\infty} \; \frac{\ln[1 - \exp\{-\Delta(n^2 + 1/4)\}]}{\cosh(2\pi n\lambda) - \cos(\pi\lambda)} \tag{35}$$

Equation (35) is exact. The function in the integral is even, so that all the odd order derivatives at x=0 vanish. Detailed examination shows that also derivatives at x=∞ vanish. Consequently, it is possible to write A in the form of an infinite product,

$$A = [\pi\lambda/\sin(\pi\lambda)] \; \prod_{n=0}^{\infty} \; [1 - \exp\{-\Delta(n^2 + 1/4)\}]^{a_n} \tag{36}$$

where we have defined

$$\begin{cases} a_0 = \lambda \, \text{ctg}(\pi\lambda/2) \\ a_n = 2\lambda\sin(\pi\lambda)/[\cosh(2\pi n\lambda) - \cos(\pi\lambda)] \end{cases} \tag{37}$$

Expression (36) can easily be analysed, both analytically and numerically. As far as "numerology" is concerned, the infinite product reduces for all practical purposes to a product of few terms. It is apparent from Eq. (37) that the $\{a_n\}$ coefficients converge to zero rapidly. Convergence to unity of the separate factors in Eq. (36) is also extremely fast. Even in the worst case, when $\Delta \ll 1$ and $\lambda \ll 1$, it is sufficient to retain a few terms from the infinite product in order to get reasonable accuracy.

The expression for the TST correction factor can further be simplified in the classical limit when $\lambda \to 0$, resulting in

$$A_{c\ell} = \prod_{n=0}^{\infty} [1 - \exp\{- \Delta(n^2 + 1/4)\}]^{\overline{a}_n} \tag{38}$$

where

$$\begin{cases} \overline{a}_0 = 2/\pi \\ \overline{a}_n = 1/\pi(n^2 + 1/4) \quad (n = 1,2,\ldots.) \end{cases} \tag{39}$$

If $\Delta/4 \gtrsim 1$, the expression for the TST correction factor reduces to $\pi\lambda/\sin(\pi\lambda)$ in the quantum case and to unity in the classical case. Thus, the result matches that for the intermediate-friction domain.

The infinite product form of the TST correction factor is particularly convenient for dealing with the double-well system. Mel'nikov shows that the rate of escape from the first well can be represented in the form

$$W_{12} = (\Omega/2\pi) \; \tilde{A}(\Delta_1;\Delta_2) \; \exp\{- V_1/kT\} \quad , \tag{40}$$

where it is assumed that both potential wells are harmonic with the same frequency Ω, and V_1 is the activation energy for the first well. The TST correction factor, $\tilde{A}(\Delta_1;\Delta_2)$, is given by [23]

$$\tilde{A}(\Delta_1;\Delta_2) = \tilde{A}(\Delta_1) \; \tilde{A}(\Delta_2)/\tilde{A}(\Delta_1 + \Delta_2) \tag{41}$$

The use of Eq. (41), together with our result for $\tilde{A}(\Delta)$, Eq. (36), gives

$$\tilde{A}(\Delta_1;\Delta_2) = \frac{\pi\lambda}{\sin(\pi\lambda)} \prod_{n=0}^{\infty} \left[\frac{2}{\coth\left[\frac{\Delta_1(4n^2+1)}{8}\right] + \coth\left[\frac{\Delta_1(4n^2+1)}{8}\right]} \right]^{a_n} \tag{42}$$

The results of the numerical evaluation of the TST correction factor, A, for the single well and the double well are presented in Figs. 2 and 3. The average dissipated energy has been taken in the form appropriate for the EUL, i.e., $\Delta = c\gamma V_0/\Omega$ for the single well and $\Delta_{1(2)} = c\gamma V_{1(2)}/\Omega$ for the double well; c has been taken as unity.

The following features of the single-well problem (Fig. 2) should be noted:
a) A linear dependence of the TST correction factor on the friction in the EUL.
b) The approach to the TST result (A→**1**) on the upper boundary of the low friction domain.
c) The TST result is reached for the lower friction values with increasing barrier height.

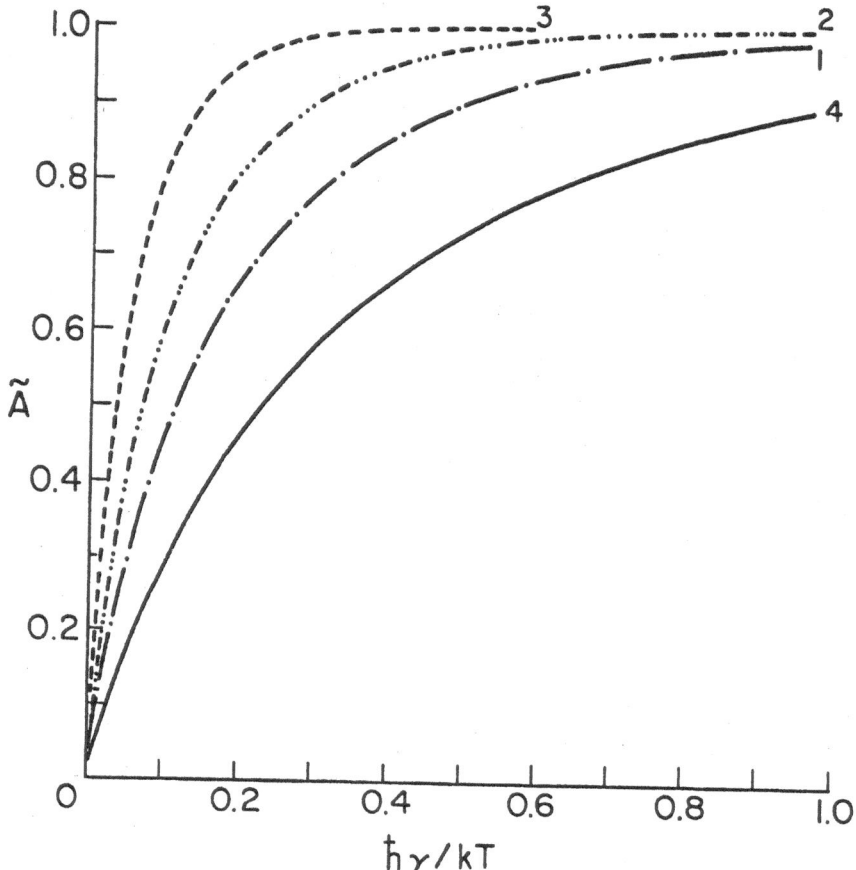

<u>Fig. 2</u>: The TST correction factor, A, for the single-well system.
Different curves on the figure correspond to the following values
of the physical parameters: 1) $V_0/kT = 3$, $\hbar\Omega/kT = 0.2$, $\hbar\omega/kT = 0.1$;
2) $V_0/kT = 5$, $\hbar\Omega/kT = 0.2$, $\hbar\omega/kT = 0.1$; 3) $V_0/kT = 10$, $\hbar\Omega/kT = 0.2$,
$\hbar\omega/kT = 0.1$; 4) $V_0/kT = 3$, $\hbar\Omega/kT = 0.4$, $\hbar\omega/kT = 0.3$; c = 1 for all
the curves.

In Fig. 3, we show the same dependence of the A factor on the
friction for the double-well system. Different curves in Fig. 3 correspond to the varying depth of the second well, while the depth of
the first well was kept fixed. These data reveal how the double-well
system behaviour approaches that of the single well with the increase
of the depth of the second well. The curve which corresponds to
$V_2/kT = 20$ is already indistinguishable from that corresponding to the
single potential well.

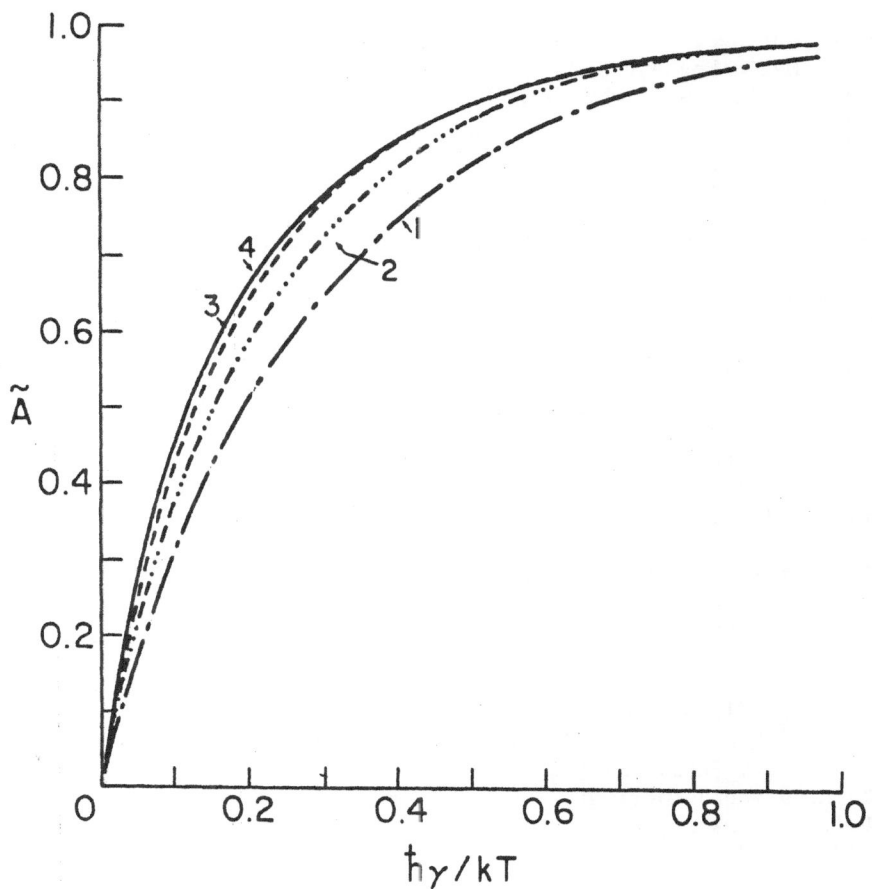

Fig. 3: The TST correction factor, A, for the double-well system. The curves correspond to the following values of the depth of the second potential well: 1) $V_2/kT = 3$; 2) $V_2/kT = 5$; 3) $V_2/kT = 10$; 4) $V_2/kT = 20$. The rest of the parameters are $V_1/kT = 3$, $\hbar\Omega/kT = 0.2$, $\hbar\omega/kT = 0.1$ and $c = 1.0$.

REFERENCES

[1] K.K. Likharev, Usp. Fiz. Nauk. 139 (1983) 169; Sov. Phys.-Usp 26 (1983) 87.

[2] R.P. Bell, in: The Tunnel Effect in Chemistry, Chapman & Hall, London, (180).

[3] H. Frauenfelder, in: Tunneling in Biological Systems, eds. B. Chance et al., Academic Press, New York, (1979), p. 627.

[4] H.A. Kramers, Physica 7 (1940) 284.

[5] A.O. Caldeira and A.J. Leggett, Phys. Rev. Lett. 46, (1981) 211; Ann. Phys. 149 (1983) 374.

[6] I. Affleck, Phys. Rev. Lett. 46 (1981) 388.

[7] P.G. Wolynes, Phys. Rev. Lett. 47 (1981) 968.

[8] V.I. Mel'nikov and S.V. Meshkov, Pis'ma Zh. Eksp. Teor. Fiz. 38 (1983) 111; Sov. Phys.-JETP Lett. 38 (1983) 130.

[9] A.I. Larkin and Yu. N. Ovchinnikov, Pis'ma Zh. Eksp. Teor. Fiz. 37 (1983) 322; Sov. Phys.-JETP Lett. 37 (1983) 382.

[10] A.I. Larkin, and Yu.N. Ovchinnikov, Zh. Eksp. Teor. Fiz. 86 (1984) 719; Sov. Phys.-JETP 59 (1984) 420.

[11] H. Grabert, U. Weiss and P. Hänggi, Phys. Rev. Lett. 52 (1984) 2193.

[12] H. Grabert and U. Weiss, Z. Physik B56 (1984) 171.

[13] H. Grabert and U. Weiss, Phys. Rev. Lett. 53 (1984) 1787.

[14] U. Weiss, P. Riseborough, P. Hänggi and H. Grabert, Phys. Lett. 104A (1984) 10; ibid. 104A (1984) 492E.

[15] P. Riseborough, P. Hänggi and E. Freidkin, Phys. Rev. A32 (1985)

[16] P. Hänggi, J. Stat. Phys. 42 (1986) 105; and references therein.

[17] P.B. Visscher, Phys. Rev. B. 13 (1976) 3272.

[18] B. Carmeli and A. Nitzan, Phys. Rev. Lett. 51 (1983) 233.

[19] M. Büttiker, E.P. Harris and R. Landauer, Phys. Rev. B. 28 (1983) 1208.

[20] M. Büttiker and R. Landauer, Phys. Rev. B.30 (1984) 1551.

[21] P. Hänggi and U. Weiss, Phys. Rev. A. 29 (1984) 2265.

[22] B.J. Matkovsky, Z. Schuss and C. Tier, J. Stat. Phys. 35 (1984) 443.

[23] V.I. Mel'nikov, Zh. Eksp. Teor. Fiz. 87 (1984) 663; Sov. Phys.-JETP 60 (1984) 380.

[24] A.I. Larkin and Yu.N. Ovchinnikov, Zh. Eksp. Teor. Fiz. 87 (1984) 1842; Sov.Phys.-JETP, 60 (1984) 1060.

[25] I. Rips and J. Jortner, Phys. Rev. B. (to be published).

[26] V.I. Goldanskii, Dokl. Akad. Nauk. SSSR 124 (1959) 1261; ibid. 127 (1959) 1037.

RELAXATION PROCESSES IN GLASSES AT LOW TEMPERATURE

H.P.H. Thijssen[1], R. van den Berg and S. Völker
Center for the Study of the Excited States of Molecules
Gorlaeus and Huygens Laboratories
University of Leiden, P.O. Box 9504, 2300 RA Leiden
The Netherlands

ABSTRACT. Optical relaxation processes in organic molecules as guests in amorphous hosts (glasses and polymers) have been studied by means of hole-burning at temperatures between 0.3 and 20 K. It is shown that the homogeneous linewidth Γ_{hom} of 0-0 $S_1 \leftarrow S_0$ transitions obeys a $T^{1.3}$ temperature dependence, and extrapolates or actually reaches the fluorescence lifetime-limited value of the guest molecules for $T \rightarrow 0$, independent of whether hole-burning occurs by a photochemical or non-photochemical mechanism. An analysis of the holewidth dependence on various parameters provides the conditions required for the determination of Γ_{hom}, and proves that the excessive holewidths reported in the literature are the unfortunate results of power broadening, sample heating and poor resolution of the detection system. A study of semi-crystalline materials shows that the temperature dependence of Γ_{hom} is much steeper than $T^{1.3}$, and varies with the degree of crystallinity of the polymer. The results are interpreted in terms of various theoretical models for dephasing in glasses.

1. INTRODUCTION

Physical properties of amorphous materials at low temperature are very different from those of crystals. For example, the specific heat in glasses below 1 K is much larger than in crystals and varies linearly with temperature T, while the thermal conductivity is much smaller and varies at T^2. Further, for many amorphous solids these physical properties are rather insensitive to the chemical composition of the glass [1a]. Of the many models proposed to explain these "anomalies", the most successful has been the "tunneling" model by Phillips [1b] and by Anderson et al. [1c]. The central hypothesis of this model is that very low frequency excitations are present in glasses, the so called "two-level-systems", TLS. It is assumed that certain atoms or

---

[1] Present adress: Medical Systems Division, Nederlandse Philips
Bedrijven B.V., Heerlen, The Netherlands.

J. Jortner and B. Pullman (eds.), Tunneling, 227–243.

groups of atoms sit in two energetically non-equivalent equilibrium positions separated by an energy barrier. At very low temperature the atoms cannot thermally jump over the barrier, but they have a certain probability to tunnel through it. Owing to the random distribution of atoms or molecules in glasses, there will be a wide distribution of such TLS which can be represented by a nearly constant density of states. Detailed predictions of this model depend on assumptions made about the distribution of tunneling parameters. However, a microscopic picture of these TLS and their physical nature is still unknown.

Not only thermal, acoustic and dielectric properties, but also *optical* properties of glasses differ from those of crystals. In the last ten years several studies have been devoted to impurity linewidths in amorphous materials. Absolute values of these linewidths at low temperatures were reported to be orders of magnitude larger than in crystals, and their temperature dependences very different: the linewidths followed a weak T^n power law, with n = 2 for inorganic glasses [2-5], and n = 1 or 2 for organic glasses, n depending on the system investigated [6-8]. Furthermore, in none of these experiments the lifetime-limited value of the transition, $\Gamma_0 = (2\pi T_1)^{-1}$, was reached on extrapolation to zero temperature. This is to be expected from the relation

$$\Gamma_{hom} = (\pi T_2)^{-1} = (2\pi T_1)^{-1} + (\pi T_2^*)^{-1}, \qquad (1)$$

where the first term on the right represents the uncertainty principle, with T_1 the lifetime of the electronically excited state, and the second term T_2^* is the "pure dephasing time" of the transition, which accounts for thermally induced scattering processes. Until 1982 two techniques in the frequency domain had been used to study optical dephasing in glasses: fluorescence line-narrowing (FLN) for inorganic glasses and optical hole-burning for organic glasses. Theories based on the "tunneling model" [1a,b] were developed to interpret the observed optical linewidths, but they were not able to explain the aggregate of experimental data in a satisfactory way [2,6,9-11].

Because the results for organic amorphous systems seemed inconsistent to us, we started in 1982 a series of hole-burning experiments with the aim to resolve these contradictions and to test the validity of the theoretical models for optical dephasing. Sections 2.1 and 2.2 will give a brief account of our findings in organic glasses at temperatures between 0.3 and 20 K. The systems investigated were a variety of organic guest molecules (see fig. 1) in a large number of organic glasses (mostly alcohols) and polymers.
Our results proved to be in contrast with the data reported in the literature up to 1982: the homogeneous linewidths of the 0-0 $S_1 \leftarrow S_0$ transitions were found to be about two orders of magnitude smaller than those previously reported, they invariably follow a $T^{1.3}$-dependence, and for $T \rightarrow 0$ they extrapolate or even reach the lifetime-limited value [12-15]. A comment about the contradictions in the literature will be given in section 2.3. In order to find the source of these contradictions we have carried out a systematic analysis of the influence on the holewidth of sample preparation and optical density,

sample heating, laser power and burning time [16-18]. We will show that a properly designed hole-burning experiment gives unambiguous information on the optical dynamics occurring in glasses.

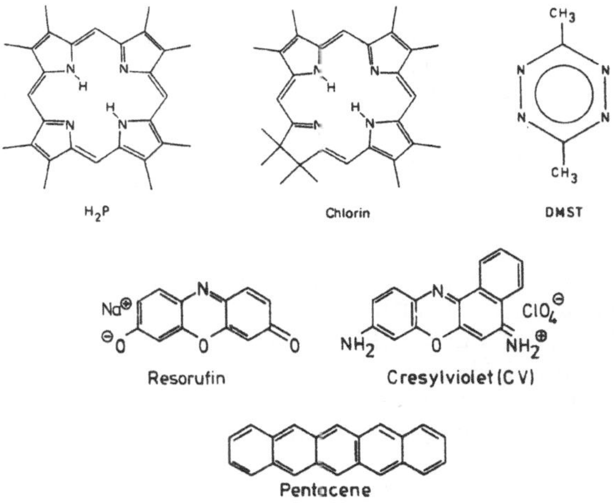

Figure 1. Organic molecules studied as guests in amorphous and semi-crystalline hosts.
Top: free-base porphin (H_2P), free-base chlorin (H_2Ch) and dimethyl-s-tetrazine (DMST) undergo intramolecular PHB (see section 2.1) [12-16].
Middle: ionic dyes resorufin (Reso) and cresylviolet (CV) undergo intermolecular PHB (see section 2.1) [17,35].
Bottom: pentacene undergoes NPHB (see section 2.2) [18].

As a consequence of our results, new models [19-21] were developed to explain the non-integral power law for the temperature dependence. More recently the same $T^{1.3}$ dependence of Γ_{hom} found by us has been observed by means of photon echoes in an inorganic glass between 0.1 and 1 K [22], and in an organic amorphous system between 1.5 and 12 K [23]. Also hole-burning [24,25] and phosphorescence line-narrowing [26] experiments by other groups idicate that a nearly equal power law is obeyed in a variety of systems.

In section 3, theories for dephasing in glasses which appeared in the literature in the last few years will be mentioned. We have tried to fit some of these models to our results on amorphous and semi-crystalline materials, the physical implications of which are discussed in section 3.1. The principal conclusions are summarized in the last section.

2. OPTICAL DEPHASING IN GLASSES STUDIED BY HOLE-BURNING

Amorphous organic solids at low temperature have inhomogeneously broadened linewidhts of a few hundred cm^{-1}, that is about two orders of magnitude larger than in organic crystals. This is due to the intrinsic disorder of the glassy host, which gives rise to a large spread in the interaction of the guest molecules with their amorphous environment.

The homogeneous or natural linewidth Γ_{hom} of an electronic transition of a guest molecule in its amorphous host is hidden under the broad inhomogeneous spectral band. However, information on Γ_{hom} can be obtained with several laser techniques. One of them is optical *hole-burning*, which we have used in the work described here. When an inhomogeneously broadened line is irradiated with a laser whose bandwidth is small compared to Γ_{hom}, a homogeneous package of molecules may be selectively removed from the sample. The "hole" produced in this way in the original absorption band has a width which is proportional to Γ_{hom}, and can be probed with a scanning laser at low intensity. Three mechanisms are known to be responsible for "hole-burning":
1. Photochemical hole-burning (PHB) [27-29],
2. Non-photochemical hole-burning (NPHB) [6,30,31] and
3. Transient hole-burning [32-34].
We shall discuss here only PHB and NPHB, because these are the mechanisms that appear most frequently in organic glasses.

2.1 Photochemical hole-burning (PHB)

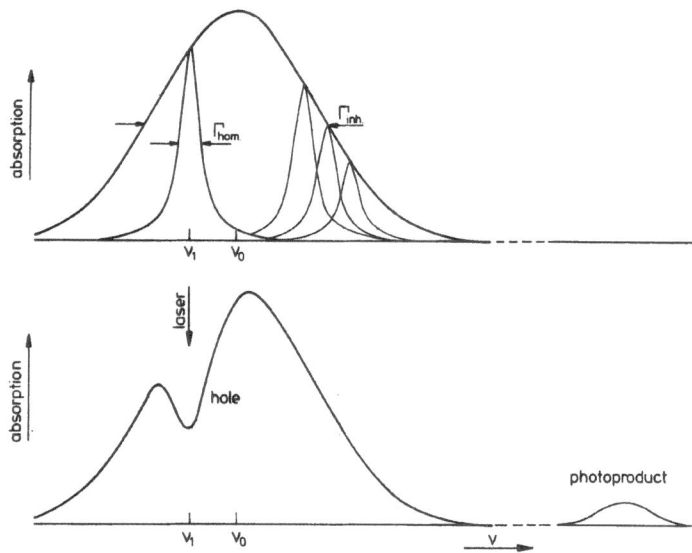

Figure 2. Top: Diagram of an inhomogeneously broadened absorption line of width Γ_{inh}, which is the envelope of individual electronic transitions of homogeneous width Γ_{hom}. Bottom: Illustration of a laser induced photochemical hole burnt at low temperature. The photoproduct absorbs in a different part of the spectrum.

The irradiated molecules undergo selective photochemistry and transform into a photoproduct which absorbs at a different wavelength. Within PHB one can further distinguish between *intramolecular* photochemical reactions, taking place inside the guest molecule [12-15,24], or *intermolecular* occurring between the guest and the host [8,17,35,36]. An example of the first type is free-base porphin (H_2P) as a guest in a large variety of glasses and polymers. These amorphous systems undergo a reversible one-photon intramolecular photochemical reaction involving proton transfer in the H_2P molecule [29]. We have performed hole-burning experiments on the 0-0 $S_1 \leftarrow S_0$ transitions of these systems between 0.3 and 20 K and found holes of about ten MHz to a few GHz [12-13]. In addition, the homogeneous linewidths Γ_{hom} extrapolate to the lifetime-limited value of ~ 10 MHz of H_2P for $T \rightarrow 0$, and follow a $T^{1.3}$ temperature law, independent of the organic amorphous host used [12-15]. This is illustrated in figures 3 and 4.

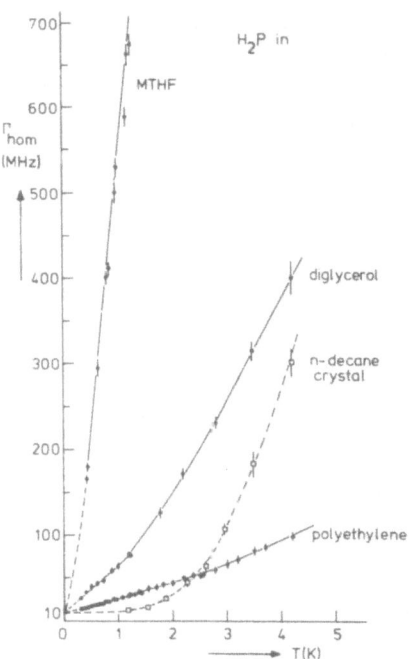

Figure 3. Temperature dependence of the homogeneous linewidth Γ_{hom} of the 0-0 $S_1 \leftarrow S_0$ transition of H_2P in various organic amorphous hosts: MTHF [13], diglycerol and polyethylene (PE) [15]. For comparison data for an n-decane crystal are also plotted [37].

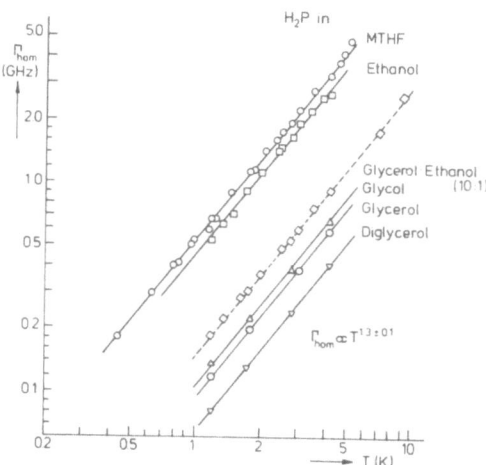

Figure 4. Log-log plot of Γ_{hom} versus T for H_2P in various glasses. Notice that for a given temperature Γ_{hom} depends on the number of OH groups in the glass [14]. Furthermore, $\Gamma_{hom} \propto T^{1.3\pm0.1}$ independent of the amorphous host [12-15].

On the other hand, hole-burning experiments performed by other groups in different organic amorphous systems gave results strongly in contrast to ours: broad holes (of the order of cm^{-1}) at low T, large residual linewidths for $T \to 0$, and a different T-dependence [6-8]. It was argued that the contradictory results from different laboratories were due to the different guest molecules used, and in particular might vary with the mechanism of the hole burning process.

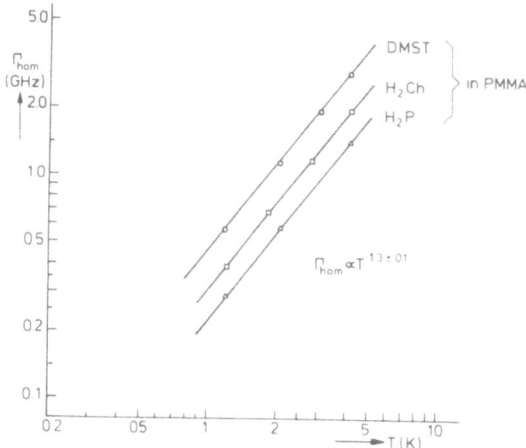

Figure 5. Log-log plot of Γ_{hom} versus T for the $S_1 \leftarrow S_0$ 0-0 transition of DMST, H_2Ch and H_2P in polymethylmethacrylate (PMMA). These systems undergo intramolecular PHB. Notice that $\Gamma_{hom} \propto T^{1.3\pm0.1}$ independent of the guest molecule [14].

In order to clear up this point we investigated other amorphous systems with the following guests: free-bases chlorin (H_2Ch), a derivative of H_2P which undergoes a reversible one-photon *intramolecular* reaction similar to that in H_2P [38], and dimethyl-s-tetrazine (DMST) in which the reaction is an irreversible two photon *intramolecular* photodissociation [28,39]. Again we found very narrow holes [14], that extrapolate to or even reach the lifetime-limited values of the guest [15,40], and a proportionality of Γ_{hom} to $T^{1.3}$, as previously found for H_2P in glasses (see fig. 5). From these results we concluded that different organic amorphous systems undergoing *intramolecular PHB* have very similar optical dephasing properties at low temperature.

The same conclusion was reached by us for amorphous systems undergoing *intermolecular PHB* . Examples of this category are the ionic dyes resorufin (Reso) and cresylviolet (CV) in which a charge redistribution seems to take place on laser excitation [17]. The photoproduct of these ionic dyes lies at ~ 50 to 400 cm^{-1} to the high energy side of the original molecule, and is very dependent on the polarity of the host [35]. In fig. 6 Γ_{hom} versus T is plotted for resorufin in ethanol, PMMA and PE. We see that Γ_{hom} varies from a few tens of MHz to a few GHz, depending on the temperature and the host [17].

Figure 6. Γ_{hom} versus T for resorufin in ethanol, PMMA and polyethylene (PE). Insert: for $T \to 0$ the extrapolation to the lifetime-limited value of resorufin,
$$\Gamma_0 = (2\pi T_1)^{-1} \approx$$
20 MHz, is shown in detail for PMMA and PE [17].

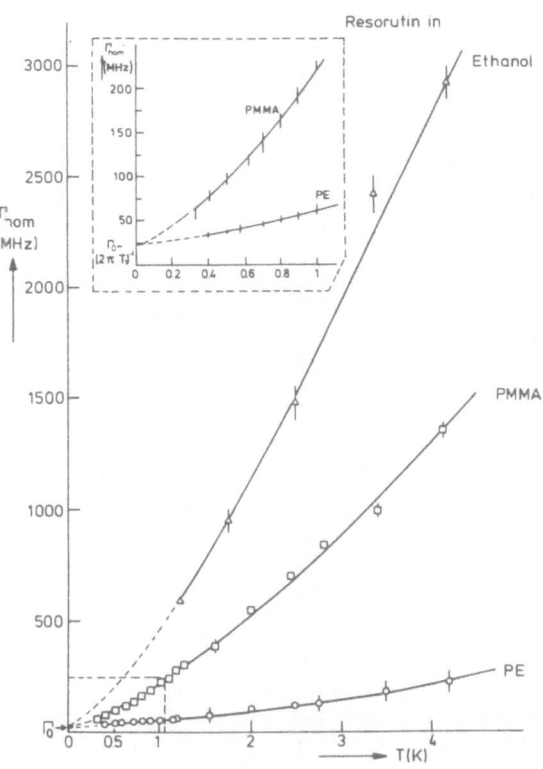

These values are about one to two orders of magnitude smaller than those previously reported for the same guest in PMMA [41,42]. This large difference is most likely due to the hole-detection technique and saturation effects [16,17] (see section 2.3). For $T \to 0$ the curves in fig. 6 either extrapolate to or even reach the lifetime-limited value of resorufin, $\Gamma_0 = (2\pi T_1)^{-1} \approx 20$ ns [17,43]. This value corresponds to $T_1 \approx 8$ ns, which is in agreement with fluorescence decay measurements performed by us on this guest in different glasses at 1.2 K.

In fig. 7 we have plotted log $(\Gamma_{hom} - \Gamma_0)$ versus log T between 0.3 and 4.2 K for both ionic dyes cresylviolet (CV) and resorufin (Reso) as guests in ethanol and PMMA [17,43].

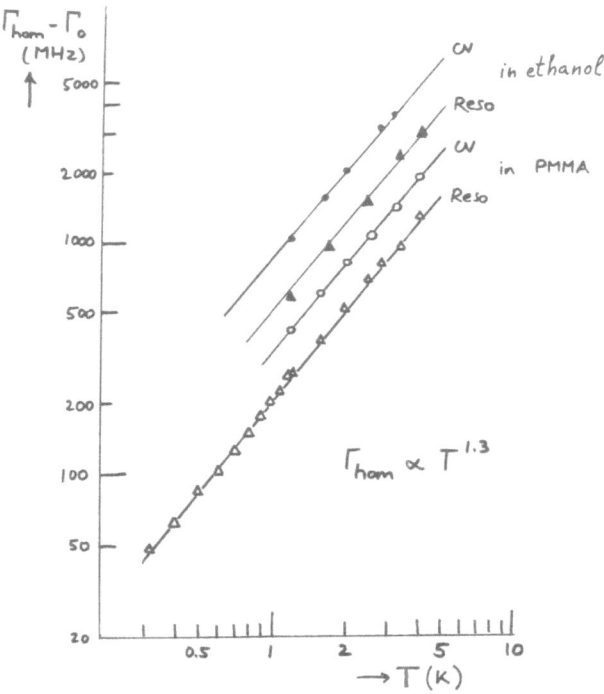

Figure 7. Log-log plot of $\Gamma_{hom} - \Gamma_0$ versus T for cresylviolet (CV) and resorufin (Reso) as guests in ethanol and PMMA. These systems undergo intermolecular PHB. Notice that $\Gamma_{hom} \propto T^{1.3}$ independent of the PHB mechanism.

We see here that for *intermolecular* PHB-systems, as well as for *intramolecular* PHB-systems, Γ_{hom} is proportional to $T^{1.3}$ between 0.3 and 4.2 K, which indicates that this T-dependence is independent of the PHB process involved.

2.2 Non-photochemical hole-burning (NPHB).

This process is characteristic for amorphous systems. On excitation, a reorientation of the guest molecules with respect to its environment takes place [6,30,31]. A tunneling mechanism has been proposed by Small et al. [6,31], which is schematically represented in fig. 8: after selective excitation in the left hand conformation, tunneling takes place in the excited state, and reorientation occurs. In this way, a *non*-photochemical "hole" is burnt in the absorption band.

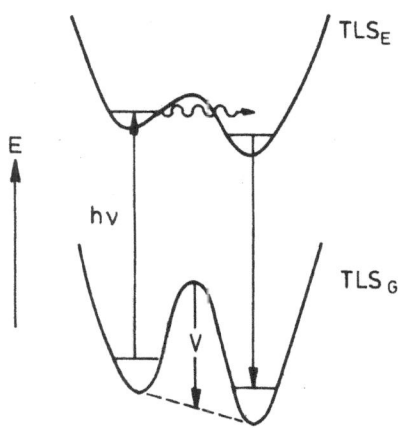

Figure 8. Diagram of the NPHB mechanism according to ref. [31]. TLS_G and TLS_E represent potential energy curves of a TLS in a glass interacting with the ground and electronically excited states of a guest molecule. After laser irradiation, NPHB occurs through tunneling in the excited state.

The absorption frequency of the reoriented molecule is assumed in this model to be very close (< 2 cm^{-1}) to that of the original molecule. It should be noted, that holes resulting from NPHB are permanent at temperatures much lower than that corresponding to the barrier V of TLS_G.

It has been claimed that systems undergoing NPHB have homogeneous linewidths of a few cm^{-1} and large residual linewidths for T → 0 [6,7,25,44]. Furthermore, no consensus has been reached with respect to the temperature dependence of the holewidths, allegedly linear in some

cases and quadratic in others. The contrast between the results obtained
with PHB and NPHB has very recently even been emphasized [23] with the
statement that NPHB does not reflect the optical dynamics of glasses,
but only measures slow host relaxation processes. The argument was based
on a comparison between picosecond photon echo results and NPHB applied
to the system pentacene in PMMA [23]. Whereas the former technique
yielded $\Gamma_{hom} \propto T^{1.3}$ between 1.5 and 20 K, as previously obtained from
PHB results [12-15], the NPHB results yielded much larger values
of Γ_{hom} following a T^2-dependence.

Intrigued by this unclear situation, we have re-investigated the
system pentacene in PMMA using NPHB. Our results (see fig. 9, bottom
curve) show much narrower holes than those of ref. [23] and they follow
a $T^{1.3}$-dependence, supporting our previous conclusions for PHB [18].

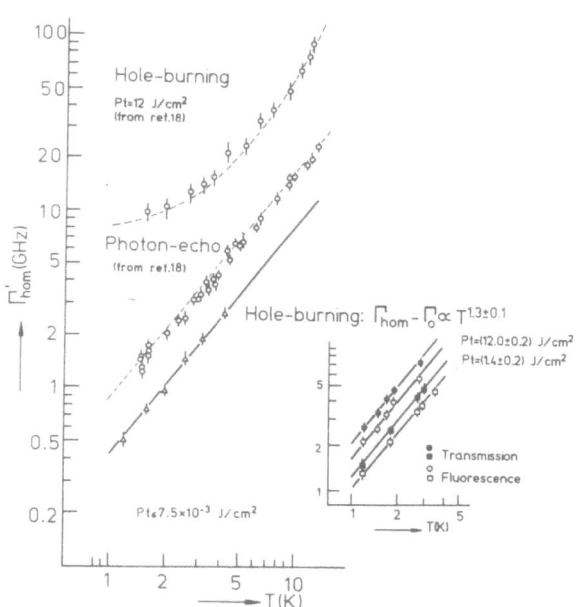

Figure 9. Log-log plot of Γ'_{hom} versus T for pentacene in PMMA (we have
defined $\Gamma'_{hom} = \frac{1}{2}\Gamma_{hole}$ for hole-burning data, and $\Gamma'_{hom} = (\pi T_2)^{-1}$ for
photon echo data). The two upper curves (dashed lines) are reproduced
from ref. [23]. The lowest curve represents our hole-burning data for
fluences Pt ≤ 7.5 x 10^{-3} J cm^{-2} [18]. Notice that the hole-burning
results of ref. [23] follow $\Gamma'_{hom} \propto T^2$ (Pt = 12 J cm^{-2}), whereas both the
photon echo data [23] and our hole-burning data show $\Gamma'_{hom} \propto T^{1.3}$.
Insert: similar plot for our hole-burning data obtained with 200 and
1600 times higher fluences than those used for the lowest curve on the
left part of the figure. All data follow $\Gamma'_{hom} \propto T^{1.3}$.

It should be noticed that the curves in the insert of fig. 9, were measured with 2 to 3 orders of magnitude higher laser fluences Pt (with P: laser power, t: burning time). They are not representative of the true homogeneous linewidths, as will be shown in section 2.3.

2.3. Pitfalls in the determination of homogeneous linewidths

Although hole-burning is a rather straightforward technique, power broadening and local heating may cause serious problems and should be taken into account. Our PHB and NPHB studies of the influence of laser power and burning time [16-18] suggest that many of the holewidths reported in the literature [6-8,23,25,36,41,42,44,45] may suffer from saturation effects. Thus, if one wants to determine the value of the homogeneous linewidth Γ_{hom}, it is not sufficient to verify that the hole shape is lorentzian, but, in a series of experiments one has to measure the width as a function of laser power and burning time. These effects are illustrated in figures 10 and 11. In fig. 10 the influence of laser power on the holewidth for H_2P in ethanol with a fixed burning time of 20s at 4.2 K is shown [16]. Notice that the holewidth in this amorphous intramolecular PHB system becomes already "saturated" at laser powers $\geq 20 \ \mu W/cm^2$.

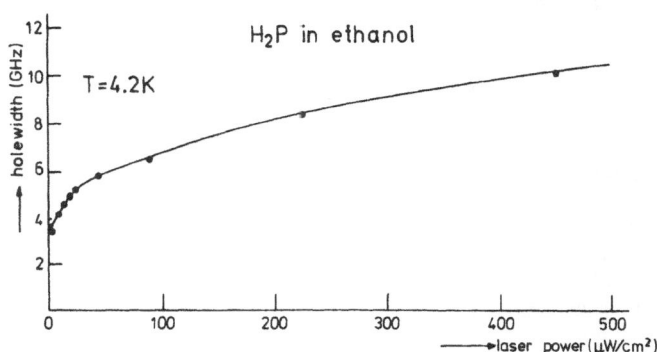

Figure 10. Holewidth as a function of laser burning power for H_2P in ethanol at T = 4.2 K, burnt for 20 seconds [16].

Fig. 11 represents the holewidth ($\frac{1}{2}\Gamma_{hole}$) as a function of burning time t, for the NPHB-system pentacene in PMMA. The results for two different burning times at T = 1.2 K are plotted. Notice that the values of $\frac{1}{2}\Gamma_{hole}$ on the lowest curve extrapolate to Γ_{hom} = 0.5 GHz when t → 0, whereas the two upper curves reach the same value of Γ_{hom}, but with a much steeper slope. From these results it follows that the true value of Γ_{hom} at a given temperature is only obtained at the lowest possible laser fluences [18].

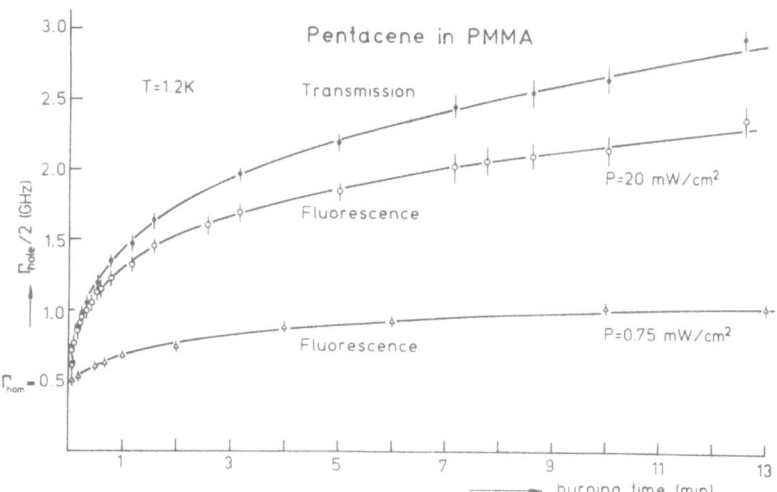

Figure 11. $\frac{1}{2}\Gamma_{hole}$ as a function of burning time for pentacene in PMMA, at T = 1.2 K for two burning powers. At P = 20 mW cm^{-2} holes were detected simultaneously via the fluorescence excitation signal, and the transmission signal through the sample. At P = 0.75 mW cm^{-2} holes were observed in fluorescence only. For t → 0 $\frac{1}{2}\Gamma_{hole}$ ≃ Γ_{hom} = 0.50 GHz [18].

        Furthermore, a critical point in this type of experiments is the hole detection technique. If holes are probed in transmission by scanning with a monochromator (in combination with a lamp) [6-8,23,25,31,36,45], they can be observed only if very high burning powers have been used. These powers are orders of magnitude larger than those needed when one detects via the fluorescence excitation method. In addition, the resolution is much higher with the latter technique, because it is given by the bandwidth of the laser (of a few MHz) instead of being limited by the slit width of the monochromator (about 3-20 GHz).
        The discrepancy between the photon echo and NPHB data of ref. [23] and our NPHB results [18] may principally be attributed to burning power and temperature. The experiments of ref. [23] were performed in a conduction cryostat (sample in contact with helium gas), and the signals were measured in transmission through the sample. Thus a considerable energy was dissipated in the bulk of a sample of poor thermal conductivity, and the temperature may well have been higher than assumed. Further, the laser fluence used for NPHB in ref. [23] was three orders of magnitude higher than that used by us for the same system (see fig. 9). Finally, holes in ref. [23] were probed in transmission by using a lamp and a monochromator (resolution of 4.5 GHz), in the way mentioned above.

3. SURVEY OF MODELS FOR OPTICAL DEPHASING IN GLASSES

In the last two years there have been various theoretical attempts to explain the broading of Γ_{hom} in glasses with a non-integral power dependence on temperature. These new models have in common with earlier ones [2,6,9-11] that they assume the existence of low-frequency modes in the glass (two-level tunneling systems, TLS [1]), but they differ from each other in the way the coupling between the optical transition and the TLS is introduced. One of these models (Jackson and Silbey [19]), explains the observed $\Gamma_{hom} \propto T^{1.3}$, for T between 2 and 20 K, by a combination of two dephasing processes: one resulting from coupling to TLS in the glass (Γ_{TLS}), and the other from coupling to librational modes of the guest (Γ_{lib}), such that $\Gamma_{hom} \propto \Gamma_{TLS} + \Gamma_{lib}$. For the term Γ_{lib} the "exchange model" [46] was used as applied to crystals, but the librational modes in the glass are assumed to have a Gaussian distribution.

In a second model (Lyo and Orbach [20]) the phonons are replaced by "fractons" [47], and an electrostatic dipole-quadrupole interaction is postulated to account for coupling between impurity and TLS. When assuming that polymers are "fractals" with a spectral dimension $\bar{d} = 4/3$ [47], it appears that $\Gamma_{hom} \propto T^{4/3}$.

A third model (Hunklinger and Schmidt [21]) is based on "spectral diffusion" of the electronic transition due to elastic dipolar coupling to "flipping" TLS [9,48]. With the assumption of a constant density of TLS-states, it yields a T^{α}- dependence of Γ_{hom}, where α is related to the dominant relaxation mechanism. If the latter is a Raman process then $\Gamma_{hom} \propto T^{1.3}$, at least between 0.4 and 5 K.

Two other models have recently appeared in the literature, both related to specific experiments. One of them [49] explains photon-echo data for an inorganic glass [22] in terms of "spectral diffusion of TLS" [48]. Assuming an elastic dipole-dipole interaction in a similar way as in ref. [21], but with a density of TLS-states that varies with energy as E^{μ}, where $\mu = 0.3$ [50], $\Gamma_{hom} \propto T^{1+\mu} = T^{1.3}$ is obtained. The other model is based on optical Redfield theory [46b], and explains photon-echo dephasing of an organic glass [23]. An approach similar to that of Lyo [51] was taken here, but instead of postulating a dipole-quadrupole interaction between impurity and TLS with a constant density of TLS-states [51], an electrostatic dipole-dipole interaction in combination with a density of TLS-states varying as $E^{0.3}$ [49,50] was assumed, from which $\Gamma_{hom} \propto T^{1.3}$ results.

Another model, by Pietronero [52], assumes that the TLS have hierarchical constraints. The consequence of this assumption is that the available degrees of freedom for low temperature properties of glasses increase with T with an effective density of TLS equal to $\rho(T) = \rho_0(1 + \rho_0 BT)$, where B may be taken as a proportionality constant, and ρ_0 is the density of TLS in the limit $T \to 0$. With the further assumption that relaxation occurs via dipole-dipole interaction [51], the linewidth then becomes $\Gamma_{hom} \propto \rho T \simeq \rho_0 T + B \rho_0^2 T^2$, which could explain the experimentally observed T^{ν}, with ν ranging from 1.3 for organic glasses to 2 for inorganic glasses. A critical test of these parameters can only be done when data of specific heat measurements are available.

Finally, a model being developed by Maynard [53], considers a dipole-quadrupole interaction between guest molecules and TLS. With the assumption that an elastic quadrupole is associated with a librational mode of the guest [19,37,46c),54], and that the density of thermally excited TLS is $\rho_0 kT$, , he obtains $\Gamma_{hom} \propto (\rho_0 kT)^{4/3}$. For constant ρ_0, $\Gamma_{hom} \propto T^{1.3}$, in formal agreement with our results.

Although these models account for the observed $T^{1.3}$-behaviour, they do not provide parameters from which Γ_{hom} or T could be calculated

in order to make quantitative predictions of the results. It is therefore difficult to decide which model is the correct one, and more work is needed.

3.1 Optical dephasing in semi-crystalline polymers.

We have performed a study aimed at finding out which dephasing model for glasses applies best. For this purpose, we chose semi-crystalline systems with polyethylene (PE) as host, between 0.3 and 4.2 K.

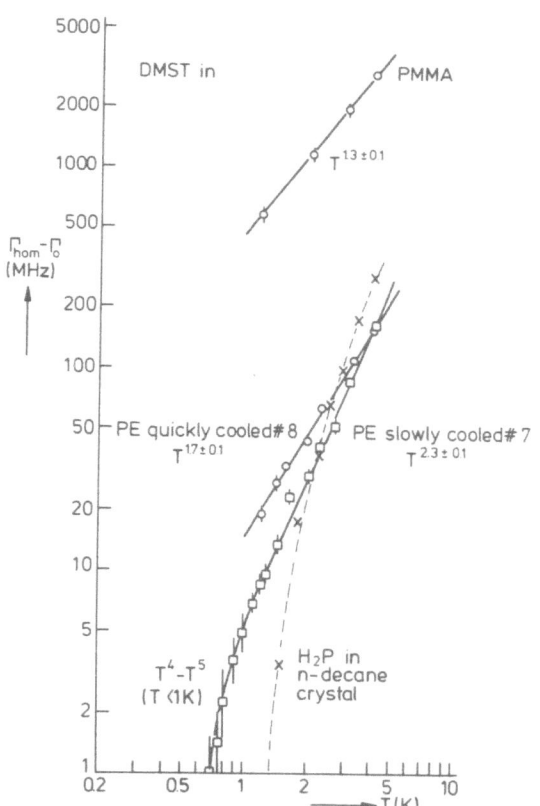

Figure 12. Top: DMST diffused and/ or dissolved into amorphous PMMA, $\Gamma_{hom} - \Gamma_0 \propto T^{1.3\pm0.1}$. Bottom: DMST diffused into semi-crystalline PE. Sample # 8 (quickly cooled) follows $\Gamma_{hom} - \Gamma_0 \propto T^{1.7\pm0.1}$ between 1.2 and 4.2 K, whereas sample #7 (slowly cooled) follows $\Gamma_{hom} - \Gamma_0 \propto T^{2.3\pm0.1}$ for $T > 1$ K, and $\propto T^4 - T^5$ for $T < 1K$ [43,55]. Data for H_2P in an n-decane crystal are plotted for comparison [37].

The results show that the power dependence on temperature of Γ_{hom} is much higher than $T^{1.3}$ (see fig. 12), and increases with the degree of crystallinity of the host [43,55]. In fact these systems behave as intermediate cases between crystals and glasses.

We have tried to fit three of the theoretical models existing in the literature [19,46c,52] to our experimental data for amorphous and semi-crystalline systems. In particular, we wanted to estimate the relative influence of librational modes as compared to TLS modes on the dephasing processes of the various materials studied. As regards the three theories examined, neither of them seems to offer a simple physical explanation for the $T^{1.3}$-dependence of Γ_{hom}. It appears, however, that low frequency librational modes [19] may contribute to the dephasing, although as yet there is no independent spectroscopic evidence for such librations from site-selection or hole-burning experiments. From the results we conclude that two-level systems (TLS) [1], if present at all in semi-crystalline solids, are much more diluted than in amorphous materials [55].

4. CONCLUSIONS

The experimental results presented here prove that spectral hole-burning, whether photochemical (PHB) or non-photochemical (NPHB), is a reliable technique to study optical dephasing processes in glasses, provided holes are burnt and probed with a narrow-band laser ($\Gamma_{laser} \ll \Gamma_{hom}$), at very low fluences. A systematic analysis of the holewidth dependence on laser power and burning time is necessary to get the true value of Γ_{hom}. We have further found that detection of holes by fluorescence excitation spectroscopy is more sensitive than probing the holes in transmission, because lower laser fluences can be used.

In all organic amorphous systems studied, the homogeneous linewidth Γ_{hom} follows a $T^{1.3}$-dependence between ~ 0.3 and 10-20 K, which seems characteristic for relaxation in glasses and perhaps indicates a fundamental property of disordered systems at low temperature. Further, Γ_{hom} was found to extrapolate or actually reach the fluorescence lifetime-limited value of the guest molecule when $T \to 0$.

Recently developed models to explain optical dephasing in glasses have briefly been described. Neither of them, however, seems to offer a simple physical explanation for the $T^{1.3}$-dependence of Γ_{hom} observed. Experiments on amorphous and semi-crystalline materials, aimed at fitting some of the models to our data, suggest that low frequency librations may contribute to the dephasing, while two-level systems (TLS) in semi-crystalline materials only have a small contribution and may be neglected.

REFERENCES

[1] a) For a review see: Amorphous Solids. Low Temperature Properties,
     ed. W.A. Phillips (Springer, Berlin, 1981).
     b) W.A. Phillips, J. Low Temp. Phys. 7, 351 (1972); c) P.W.
     Anderson, B.I. Halperin and C.M. Varma, Phil. Mag. 25, 1 (1972).
[2] P.M. Selzer, D.L. Huber, D.S. Hamilton, W.M. Yen and M.J. Weber,
     Phys. Rev. Letters 36, 813 (1976).
[3] P. Avouris, A. Campion and M.A. El-Sayed, J. Chem. Phys. 67, 3397
     (1977).
[4] J. Hegarty and W.M. Yen, Phys. Rev. Letters 43, 1126 (1979).
[5] J.R. Morgan and M.A. El-Sayed, Chem. Phys. Letters 84, 213 (1981).
[6] J.M. Hayes, R.P. Sout and G.J. Small, J. Chem. Phys. 74, 4266
     (1981).
[7] E. Cuellar and G. Castro, Chem. Phys. 54, 217 (1981).
[8] J. Friedrich, H. Wolfrum and D. Haarer, J. Chem. Phys. 77, 2309
     (1982).
[9] T.L. Reinecke, Solid State Commun. 32, 1103 (1979).
[10] S.K. Lyo and R. Orbach, Phys. Rev. B 22, 4223 (1980).
[11] P. Reineker and H. Morawitz, Chem. Phys. Letters 86, 359 (1982).
[12] H.P.H. Thijssen, A.I.M. Dicker and S. Völker, Chem. Phys. Letters
     92, 7 (1982).
[13] H.P.H. Thijssen, S. Völker, M. Schmidt and H. Port, Chem. Phys.
     Letters 94, 537 (1983).
[14] H.P.H. Thijssen, R.E. van den Berg and S. Völker, Chem. Phys.
     Letters, 97, 295 (1983).
[15] H.P.H. Thijssen, R.E. van den Berg and S. Völker, Chem. Phys.
     Letters 103, 23 (1983).
[16] H.P.H. Thijssen and S. Völker, Chem. Phys. Letters 120, 496 (1985).
[17] H.P.H. Thijssen, R. van den Berg and S. Völker, Chem. Phys. Letters
     120, 503 (1985).
[18] R. van den Berg and S. Völker, Chem. Phys. Letters (in print).
[19] B. Jackson and R. Silbey, Chem. Phys. Letters 99, 331 (1983).
[20] S.K. Lyo and R. Orbach, Phys. Rev. B 29, 2300 (1984).
[21] S. Hunklinger and M. Schmidt, Z. Phys. B 54, 93 (1984).
[22] J. Hegarty, M.M. Broer, B. Golding, J.R. Simpson and J.B.
     MacChesney, Phys. Rev. Letters 51, 2033 (1983).
[23] L.W. Molenkamp and D.A. Wiersma, J. Chem. Phys. 83, 1 (1985).
[24] F.A. Burkhalter, G.W. Suter, U.P. Wild, V.D. Samoilenko, N.V.
     Rasumova and R.I. Personov, Chem. Phys. Letters 94, 483 (1983).
[25] T.P. Carter, B.L. Feary, J.M. Hayes and G.J. Small, Chem. Phys.
     Letters 102, 272 (1983).
[26] J. Fünfschilling and I. Zschokke-Gränacher, Chem. Phys. Letters
     110, 315 (1984).
[27] A.A. Gorokhovskii, R.K. Kaarli and L.A. Rebane, J. Exp. Theor.
     Phys. Letters 20, 216 (1974); Opt. Commun. 16, 282 (1976).
[28] H. de Vries and D.A. Wiersma, Phys. Rev. Letters 36, 91 (1976).
[29] S. Völker and J.H. van der Waals, Mol. Phys. 32, 1703 (1976).
[30] B.M. Kharlamov, R.I. Personov and L.A. Bykovskaya, Opt. Commun. 12,
     191 (1974).
[31] J.M. Hayes and G.J. Small, Chem. Phys. 27, 151 (1978).

[32] A. Szabo, Phys. Rev. B 11, 4512 (1975).
[33] R.M. Shelby and R.M. Macfarlane, Chem. Phys. Letters 64, 545 (1979).
[34] A.I.M. Dicker, L.W. Johnson, S. Völker and J.H. van der Waals, Chem. Phys. Letters 100, 8 (1983).
[35] R. van den Berg and S. Völker, to be published.
[36] F. Drissler, F. Graf and D. Haarer, J. Chem. Phys. 72, 4996 (1980).
[37] A.I.M. Dicker, J. Dobkowski and S. Völker, Chem. Phys. Letters 84, 415 (1981).
[38] S. Völker and R.M. Macfarlane, J. Chem. Phys. 73, 4476 (1980).
[39] D.M. Burland, F. Carmona and J. Pacansky, Chem. Phys. Letters 56, 221 (1978).
[40] H.P.H. Thijssen, R. van den Berg and S. Völker, in: Photoreaktive Festkörper, ed. H. Sixl (Wahl-Verlag, Karlsruhe, 1984), p.p. 763-782.
[41] A.P. Marchetti, M. Scozzafava and R.H. Young, Chem. Phys. Letters 51, 424 (1977).
[42] A.F. Childs and A.H. Francis, J. Phys. Chem. 89, 466 (1985).
[43] H.P.H. Thijssen, R. van den Berg and S. Völker, J. de Physique C7-363 (1985).
[44] R. Jankowiak and H. Bässler, Chem. Phys. Letters, 95, 310 (1983).
[45] W. Breinl, J. Friedrich and D. Haarer, J. Chem. Phys. 81, 3915 (1984).
[46] a) R. Kubo and T. Tomita, J. Phys. Soc. Japan 9, 888 (1954);
     b) A.G. Redfield, IBM J. Res. Dev. 1, 19 (1957);
     c) R.M.Shelby, C.B. Harris and P.A. Cornelius, J. Chem. Phys. 70, 34 (1979).
[47] S. Alexander and R. Orbach, J. Phys. (Paris) 43, L-625 (1982).
[48] J.L. Black and B.I. Halperin, Phys. Rev. B. 16, 2879 (1977).
[49] D.L. Huber, M.M. Broer and B. Golding, Phys. Rev. Letters 52, 2281 (1984).
[50] J.C. Lasjaunias, A. Ravex, M. Vandorpe and S. Hunklinger, Solid State Commun. 17, 1045 (1975).
[51] S.K. Lyo in: Electronic Excitations and Interaction Processes in Organic Molecular Aggregates, eds. P. Reineker, H. Haken and H.C. Wolf (Springer, Berlin, 1983, pp. 215-226.
[52] L. Pietronero, preprint.
[53] R. Maynard, J. de Physique, C7-325 (1985).
[54] S. Völker, R.M. Macfarlane and J.H. van der Waals, Chem. Phys. Letters 53, 8 (1978).
[55] H.P.H. Thijssen and S. Völker, J. Chem. Phys. (in print).

COHERENT VOLTAGE OSCILLATIONS IN ULTRA-SMALL CAPACITANCE STRUCTURES

K. Mullen

E. Ben-Jacob

Department of Physics and Astronomy, Tel-Aviv University
Tel-Aviv, Israel

Department of Physics, The University of Michigan
Ann Arbor, Michigan 48109

Z. Schuss

Department of Applied Mathematics, Tel-Aviv University
Tel-Aviv, Israel

ABSTRACT: New coherent dynamics are predicted in the process of charge transfer in ultra-small capacitance structures. For structures with a capacitance of 10^{-15}F driven by a 1nA current source, voltage oscillations at a frequency of $\sim 10^{10}$Hz with an amplitude of $\sim$ .1mv are predicted. The analysis is presented in an inter-disciplinary spirit, rather than one of mathematical rigor.

1. Introduction

Attempts at understanding the interplay between quantum and classical mechanics have been essential to the development of modern physics. However there lie an enormous number of systems that are neither properly macroscopic nor microscopic but somewhere in between. This intermediate 'mesoscopic' domain includes such diverse systems as biological macromolecules and submicron electronic circuits. The problems with describing the dynamics of such structures is two-fold. First, their size puts them well below the thermodynamic limit, thereby ruling out the use of ensemble averaging. Secondly, much of the interest is in the non-equilibrium properties of these mesoscopic systems, and nonequilibrium effects are notoriously more difficult to describe than the simpler equilibrium ones. Understanding the behavior of the these systems may involve the introduction of new physical principles and reveal new phenomena that are without a classical analog.

Charge transfer, a fundamental process in both biology and submicron electronics, often takes place on the mesoscopic level. It is important to note that since the structures

J. Jortner and B. Pullman (eds.), Tunneling, 245–259.

carrying the charge are quite small, their capacitance is quite small also, and therefore the energy required to transfer an electron on to them, $e^2/2C$, where C is the capacitance, may be relatively large. It is precisely this effect, the dominance of the charging energy, which makes the mesoscopic case so different from the macroscopic case. While admittedly the examples of charge transfer in biology (i.e. the activities of proteins or of neurotransmitters) are much more complicated than electronic circuits, understanding the latter may be a good first step towards understanding the former. Analysis of such ultra-small capacitance circuits indicates the possibility of coherent dynamics that can lead to a new form of quantum oscillation applicable to all such systems.

In this paper we provide a simplified description of sub-micron Josephson junctions written for a reader who is not necessarily a physicist but has a knowledge of the bacic principles of quantum mechanics; more detailed reports are available in the literature.[1-3] A Josephson junction is a sandwich of two superconductors separated by an extremely thin insulating oxide layer, only 10 Å thick (figure (1)). Unlike conventional Josephson juctions, the ones we shall consider are extremely narrow as well, only $.01\mu$ on a side. The concommitant capacitance of such a 'quantum dot' is quite small — on the order of 10^{-15} F or 1 femtofarad (fF). The charging energy of a single electron is $\sim 10^{-4}$ eV, corresponding to an equivalent thermal energy of $\sim 1°$K. We demonstrate that if the junction is operated below this temperature, a new dynamism is possible, producing coherent voltage oscillations. If the junction is driven by an ideal current source, it will produce an alternating voltage of extremely high frequency, typically on the order of 10^{10} Hz. For simplicity our results are derived for Josephson junctions, but they in no way require a superconducting medium, and can be generalized to normal tunnel junctions.

This paper is divided into five sections. In section 2 we derive our results for a mesoscopic Josephson junction and show the difference between it and the conventional Josephson junction. The fluctuations of the volatge about this coherent oscillatory state are described in section 3. We generalize the above results to normal tunnel junctions in section 4. We conclude with a discussion, including possible applications to biological systems in section 5.

2. Mesoscopic Josephson Junctions

We start by considering a Josephson junction, a sandwich of a thin insulating oxide layer between two superconductors. Unlike conventional Josephson junctions, we require our junction to be extremely narrow — on the order of $.01\mu$ on a side, so small that the term 'quantum dot' has been coined to describe it. Since we are dealing with superconductors the fundamental charge carrier is the *Cooper pair*, two electrons bound together by means of their interaction with the crystal lattice through which they move.[4]

There are three energies which serve to characterize the way in which we describe the junction. The first is the *charging energy*, given by

$$E_C = \frac{2e^2}{C} \tag{1}$$

where C is the capcitance, which is the electrostatic energy assosciated with a charge difference of one pair across the junction. This energy is analogous to the the energy stored in a macroscopic parallel plate capacitor when a charge difference is maintained

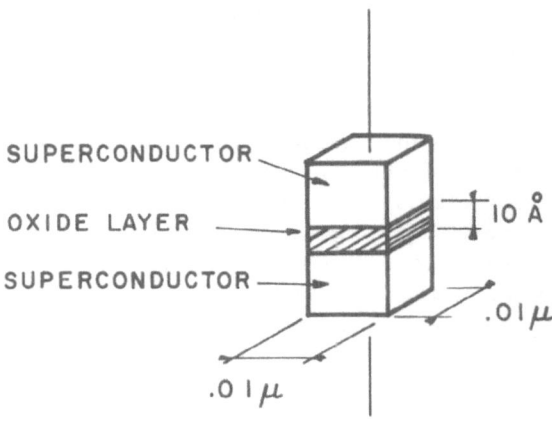

Figure (1): A schematic diagram of a mesoscopic Josephson Junction.

between its two plates. The second energy is the *tunnelling energy* or *coupling energy*, given by

$$E_J = \frac{\hbar I_J}{2e} \tag{2}$$

where I_J is the maximal pair current. The tunnelling energy is a measure of the ease with which pairs tunnel through the junction. The third energy is the *thermal energy*,

$$E_{thermal} = k_B T \tag{3}$$

where k_B is Boltzmann's constant and T is the temperature. In all cases discussed below we assume that $E_{thermal} << E_J, E_C$ so that the behavior of the junction does not depend upon thermal effects. In a conventional Josephson junction, E_C is very small so that the energy of the junction does not change significantly when a Cooper pair tunnels across the the junction and changes the net charge across the junction. However, because the 'quantum dot' is so small, it will have an extremely small capacitance, and by equation (1), a large charging energy, so that transferring one pair across the junction changes the energy of the junction by a significant amount. It is this difference that causes a sub-micron junction to behave differently than a conventional junction.

If the interaction between the junction and the heat bath is neglected, the Hamiltonian of the system consists of just two terms:

$$H = H_C + H_T \tag{4}$$

The first term H_C is the charging Hamiltonian, and is the electrostatic energy stored in the junction when there is a charge difference across the junction as described above. The second term is the tunnelling Hamiltonian; without it our description of the junction would have no proviso for transferring pairs through the junction and so an electrical current could never flow.

The charging Hamiltonian depends upon two important quantities. Imagine we start with a neutral junction and then transfer one pair across the insulating barrier by tunnelling. We now have a charge separartion of $+2$ electrons on one side of the junction and -2 electrons on the other, increasing the electrostatic energy by $(2e)^2/2C = 2e^2/C$. The number of pairs, n, which have tunnelled through the junction is thus an important quantity; we may denote the wave functions of the junction by $|n\rangle$ *i.e.* $|1\rangle, |2\rangle \ldots$. Then if $\hat{n}$ is the operator that counts the number of pairs transferred across the junction,

$$\hat{n}|m\rangle = m|m\rangle \tag{5}$$

We may write the charging Hamiltonian as

$$H_C = \frac{2e^2}{C}\hat{n}^2 = E_C \hat{n}^2 \tag{6}$$

so that, in the absence of tunnelling,

$$E = \langle l|H_C|m\rangle = E_C m^2 \delta_{l,m}. \tag{7}$$

However there is a second way to develop a charge difference across the junction. We may drive a d.c. current through the junction by connecting it to an ideal current source. An ideal current source is a device that transfers a fixed amount of charge per unit time from one of its terminals to the other. There is some question as how to model such a driving force in a quantum mechanical system. Our approach shall be to view it as suppling a source of continuous charge to the junction, so that the charging Hamiltonian including the current source is written as:[5]

$$H_C = E_C(\hat{n} - q)^2 \tag{8}$$

$$\dot{q} = I_{dc}/2e \tag{9}$$

where I_{dc} is the d.c. current and q is the amount of external charge, measured in pairs, brought to the junction by the external source. The current source is thought of as bringing in the external charge from infinity, so that in the absence of H_T the energy of the junction changes continuously in time rather than in small jumps. The energy as a function of q for various fixed values of n is plotted in figure 2.

For the tunnelling Hamiltonian we need a function that will connect states of different n, allowing the junction to change $\langle \hat{n} \rangle$, the expectation value of $\hat{n}$, and thereby pass a current. In the simplest case we want the Hamiltonian to connect states that differ by $\Delta n = 1$. The tunnelling Hamiltonian therefore satisfies:[5]

$$\langle l|H_T|m\rangle = \frac{1}{2}E_J(2\delta_{l,m} - \delta_{l+1,m} - \delta_{l-1,m}) \tag{10}$$

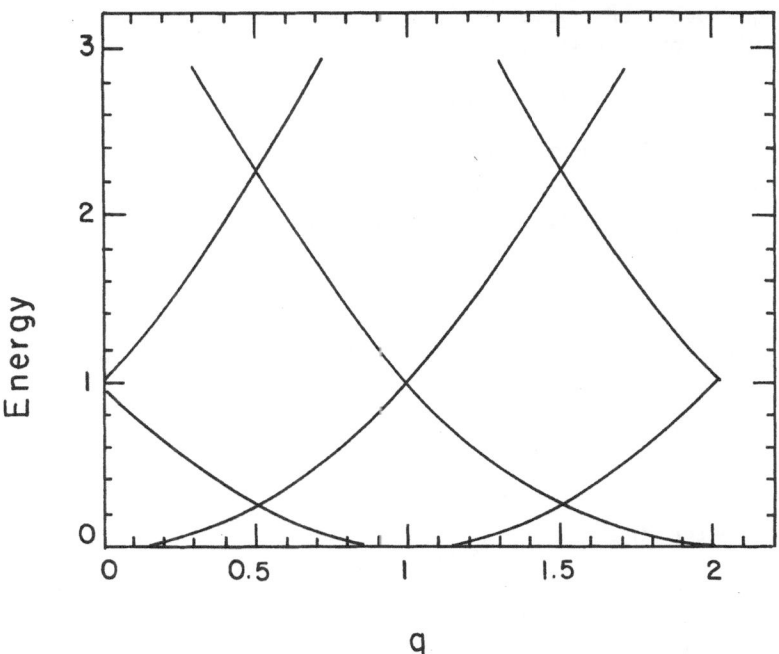

Figure (2): The charging energy as a function of q for different values of n.

The tunnelling Hamiltonian can also be written in terms of $\hat{\theta}$, the quantum mechanical variable conjugate to $\hat{n}$. The value of θ corresponds to the phase difference between the wavefunctions of the pairs on the two sides of the junction. Since $\hat{n}$ and $\hat{\theta}$ are conjugate operators then $\hat{n} = i\partial/\partial\theta$ in the θ representation and $\hat{\theta} = -i\partial/\partial n$ in the n representation, and

$$[\hat{\theta}, \hat{n}] = -i \tag{11}$$

Then H_T can be expressed in terms of the the the operator $\hat{\theta}$ as:

$$H_T = E_J(1 - \cos\hat{\theta}) \tag{12}$$

This form of the Hamiltonian is valid in the limit $E_J >> E_C$, whereas the mesoscopic Josephson junction works in the opposite limit. However, since we intend to treat H_T as a perturbation, we assume its exact form is not important, so long as it serves to connect states that differ by $\Delta n = 1$. One possible modification of H_T would be an enhancement of higher order tunnelling terms corresponding to $\Delta n = 2, 3 \ldots$ due to the coherent nature of the tunnelling dynamics in mesoscopic junctions. These terms would then be added on to H_T as $\cos 2\theta$, $\cos 3\theta \ldots$ but we do not consider such modifications in this paper.

The full Hamiltonian is thus of the form

$$H = E_C(\hat{n} - q)^2 + E_J(1 - \cos\hat{\theta}). \tag{13}$$

If we treat q as a fixed parameter then equation (13) can be used in the time independent Schrödinger equation for different values of q. In order to better understand equation (13), first consider the simpler case where $E_J = 0$.

$$H'\psi = E_C(\hat{n} - q)^2\psi = E\psi \tag{14}$$

The variable q can be eliminated frome equation (14) by performing the transformation $\varphi = e^{iq\theta}\psi$ so that

$$H'\psi = E_C(i\frac{\partial}{\partial\theta} - q)^2 e^{iq\theta}\varphi =$$

$$-\frac{\partial^2}{\partial\theta^2}\varphi = E\varphi \tag{15}$$

The solution to this equation is $\varphi = e^{\pm in\theta}$, subject to the boundary conditions:

$$\psi(0) = \psi(2\pi) \tag{16}$$

or

$$\varphi(0) = e^{i2\pi q}\varphi(2\pi) \tag{17}$$

which leads to the condition

$$e^{i2\pi(n-q)} = 1 \tag{18}$$

We are now in a better position to understand figure (3). At $q = 0$ the states of $\pm n$ are degenerate in energy. A slight increase in q breaks this degeneracy: if $n > 0$ it must decrease as q increases so as to satisfy equation (18); if $n < 0$ then n must become more negative as q increases. At $q = \frac{1}{2}$ we obtain a new set of degeneracies with the state $|n\rangle$ having the same energy as $|-(n-1)\rangle$, and again as q increases the degeneracy is resolved as explained above.

Reintroduction of the tunnelling Hamiltonian at this point helps to clarify the importance of the ratio E_J/E_C. If $E_J/E_C > 1$ then equation (13) corresponds to a particle in a box with a steep well at the center. The lowest energy states will be 'bound' states, strongly localized near the center of the well. In this limit the uncertainty in θ will be very small and perforce the uncertainty in n will be large. Indeed, in the semi-classical description of the conventional Josephson junction the operator $\hat{\theta}$ can be replaced by its expectation value $\langle\hat{\theta}\rangle$ because $\Delta\theta$ is extremely small. The effect of increasing q will be minimal since it is tantamount to adjusting the boundary conditions, and ψ will be almost zero at the boundaries.

In the opposite limit, $E_J/E_C < 1$, the effect of the the tunnelling Hamiltonian is a small perturbation of a free plane wave. Therefore the uncertainty in θ is at a maximum and the uncertainty in n is minimized. The problem is very similar to the nearly-free electron model of solid state physics. The effect of the the perturbation is to resolve the degeneracies at the point of intersection in figure (3) to produce the band structure shown in figure (3). To illustrate this we look at the specific case of $q = \frac{1}{2}$ at the intersection of the $|0\rangle$ and $|1\rangle$ states. Without the tunnelling Hamiltonian the states

$$\psi_+ = \frac{1}{\sqrt{2}}(|0\rangle + |1\rangle))\psi_- = \frac{1}{\sqrt{2}}(|0\rangle - |1\rangle) \tag{19}$$

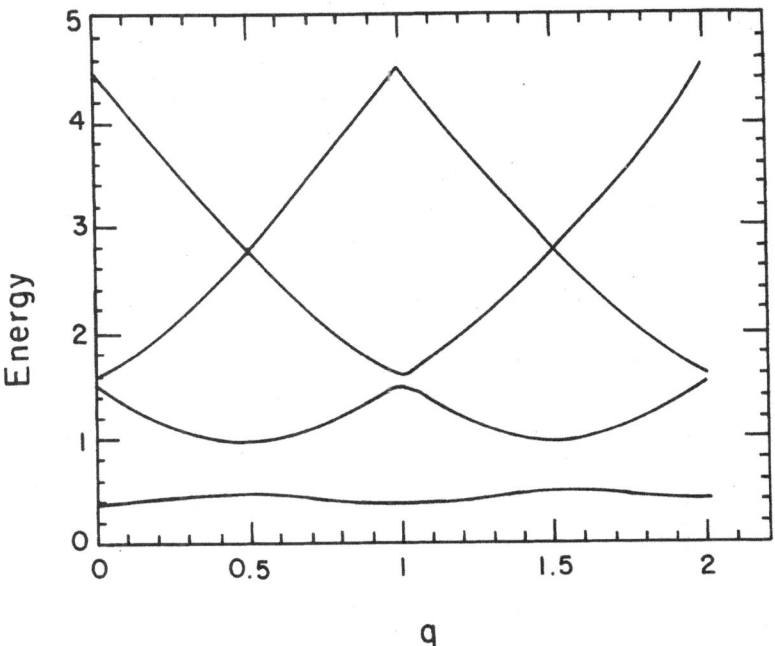

Figure (3): The band structure of a Josephson junction for $E_J/E_C = \frac{1}{2}$.

have the same energy:

$$\langle \psi_+ | H' | \psi_+ \rangle = \langle \psi_- | H' | \psi_- \rangle = \frac{1}{4} E_C$$

However, with the tunnelling Hamiltonian there is a difference in energy:

$$\langle \psi_+ | H | \psi \rangle = \frac{1}{4} E_C + \delta E$$

$$\langle \psi_- | H | \psi \rangle = \frac{1}{4} E_C - \delta E$$

$$\delta E = E_J \langle 0 | \cos \theta | 1 \rangle^2$$

so that the symmetric and antisymmetric combinations of $|0\rangle$ and $|1\rangle$ have different energies. The symmetric combination, with the higher energy, corresponds to the point in the upper band while the antisymmetric one is the lower.

At $q = 0$ the ground state is $|0\rangle$. As the value of q is increased and equation (13) is again solved to find the lowest energy state the ground state changes to an admixture of $|0\rangle$ and an increasing amount of $|1\rangle$, until at $q = \frac{1}{2}$ the amplitudes of the two states are equal. As q increases still further the amplitude of $|0\rangle$ decreases until at $q = 1$ the junction

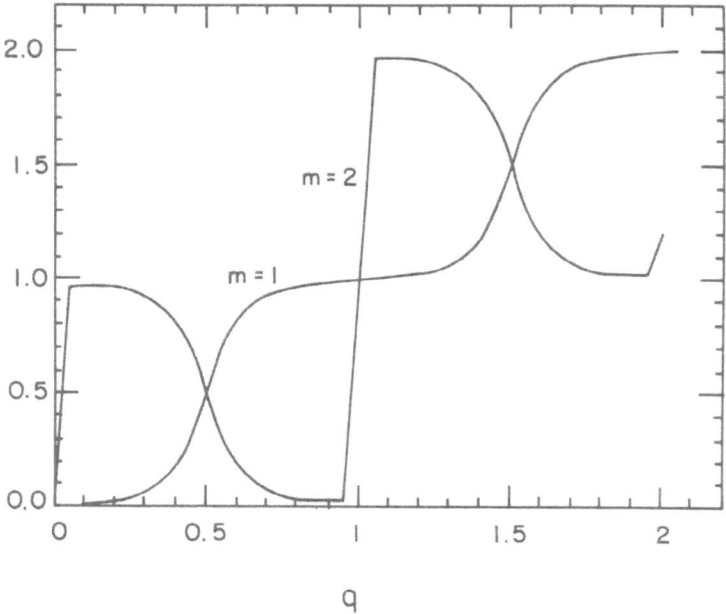

Figure (4): The expectation value of $\hat{n}$ as a function of q for $E_J/E_C = \frac{1}{2}$.

is in a state of pure $|1\rangle$. This can be neatly portrayed by calculating the expectation value of $\hat{n}$ and plotting how the $\langle\hat{n}\rangle$ changes with increasing q (figure (4)).

All of the above calculations were done for fixed q and the solutions were derived by solving the time-independent Schrödinger eqaution. At this point we make the strong assumption that if we change q slowly enough, the junction will always be in an eigenstate of equation (13), the time-independent equation. In that case the energy of the junction will follow one of the bands in figure (2) and will oscillate as q increases, with a frequency $f = \dot{q} = I_{dc}/2e$. The voltage across the junction is given by:

$$V = \frac{1}{2e}\frac{\partial\langle E\rangle}{\partial q}$$

and plotted in figure (5). Thus by driving the junction with an ideal dc current the net result is an alternating voltage with frequency f. The amplitude of the oscillations depends upon which band number, with higher bands corresponding to larger numbers of pairs transferred back and forth across the junction, and therefore a larger swing in the voltage. Although in each band the amount of current passed through the junction is the same, the oscillations about this average value are larger for the higher bands. The current is given by rate of change of $\langle\hat{n}\rangle$ with time; figure (5) indicates that the average increase of $\langle\hat{n}\rangle$ is proportional to q so that the average value of the current is I_{dc}. For a 1 fF junction the voltage oscillations in the lowest band are $\sim .1$ mV.

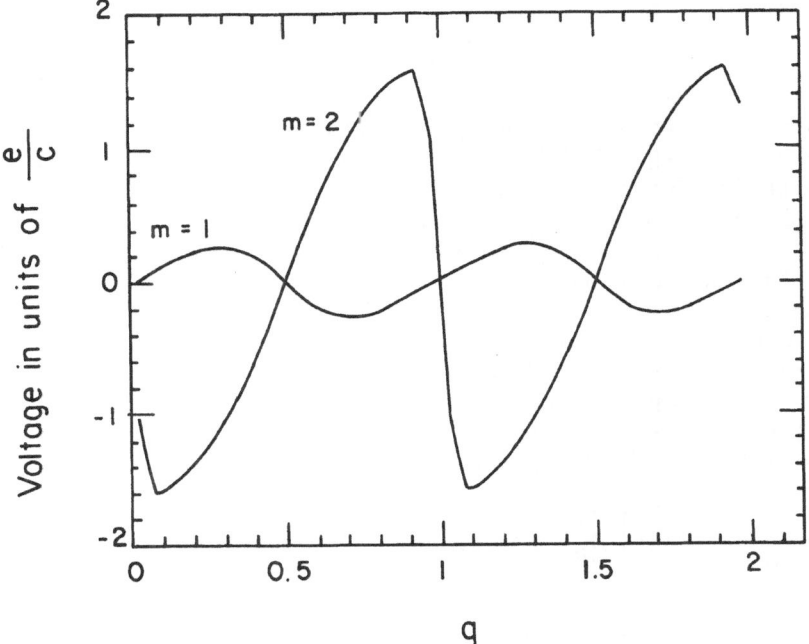

Figure (5): The voltage $V = \partial \langle E \rangle / \partial q$ as a function of q for $E_J/E_C = \frac{1}{2}$.

3. Zener Tunneling

The above treatment is a simplification of the dynamics contained within the time dependent Schrödinger equation because we required that $\dot{q}$ be so small that we could treat the problem adiabatically. A more complete treatment is in progress, but as a first approximation we can include the interband transitions due to the interaction between the junction and the current source. Even in the absence of dissipation and at zero temperature it is possible for the junction to absorb energy from the external driving force so as to transfer to a different band, a phenomenon known as *Zener tunnelling*.[7] If the junction Zener tunnels to a higher band then its energy continues upward along one of the branches of a parabola shown in figure (2), rather than switching direction and transferring to a different parabola so as to stay in the same band as is the case in figure (3). The junction is charging, just as a normal capacitor would when driven by an external source, so Zener tunnelling can be looked at as a failure of a pair to tunnel. The importance of such an effect must be stressed: moving along a branch of the parabola implies that the voltage, $\partial \langle E \rangle / \partial q$, grows larger at a rapid rate. This implies that we can see a measureable, macroscopic effect (a rapid change in the voltage), due to a quantum mechanical effect (Zener tunnelling).

The probability of such an event depends strongly on two quantities, the current and the energy gap between the two bands. The larger the current the greater is the probability; the larger the gap the smaller is the probability. The probability for Zener tunnelling from one band to another is given by[1]

$$P_{m+1,m} = P_{m,m+1} \approx \exp[-\frac{\pi}{4} \frac{E(m)_{gap}^2}{h\dot{q}E_C} \frac{1}{m}]$$ (20)

Since the gap decreases with increasing band number (and hence the Zener tunnelling probability increases), at high band numbers, once the system tunnels upward it has a higer probability to continue upward rather than to stay in a given band. Similarly, at high band numbers, once the system tunnels downward it has a high probability to continue moving down in band number. Hence we expect to observe long runs up and down in band number. These long runs will contribute to the low frequency part of the power spectrum of the noise.

We can then simulate the time evolution of the state of the junction as a random walk in the two parameters: the band number m and in the spin s which is ± 1 indicating whether $\partial \langle E \rangle / \partial q$ is positive or negative (i.e. whether we are going up or down in band number). If at time n the junction is in band $m(n)$ with spin $s(n)$ then the band number and spin at the next time step are determined by:

$$\text{if } s(m) = +1 \begin{cases} \begin{aligned} m(n+1) &= m(n) \\ s(n+1) &= -s(n) \end{aligned} & \quad \text{with probability } 1 - P_{m,m+1}; \\ & \quad or \\ \begin{aligned} m(n+1) &= m(n) + 1 \\ s(n+1) &= s(n) \end{aligned} & \quad \text{with probability } P_{m,m+1}. \end{cases}$$

$$\text{if } s(m) = -1 \begin{cases} \begin{aligned} m(n+1) &= m(n) \\ s(n+1) &= -s(n) \end{aligned} & \quad \text{with probability } 1 - P_{m,m-1}; \\ & \quad or \\ \begin{aligned} m(n+1) &= m(n) - 1 \\ s(n+1) &= s(n) \end{aligned} & \quad \text{with probability } P_{m,m-1}. \end{cases}$$

The probability of of the junction to Zener tunnel at a given instant is calculate from equation (20) and compared to a computer generated random number; if the random number is the lower of the two then the junction Zener tunnels to a new band; if not, it continues to oscillate in the same band. However in using this approach we have made a very strong assumption. A more exact approach would be to keep track of the quantum mechanical amplitudes of occupying different bands, and then forcing the wave function to 'collapse' after some characteristic time to a state selected with a probability distribution given by these amplitudes. Our simpler method assumes that we are always in an eigenstate of $\hat{n}$, never in a superposition of states. In figure(6) we show a plot of band number as a function of q for some of these simple simulations. Note the presence of the large runs up and down in band number predicted above. At non-zero temperature the Zener tunnelling dominates the noise spectrum as long as the gap energy is smaller than the

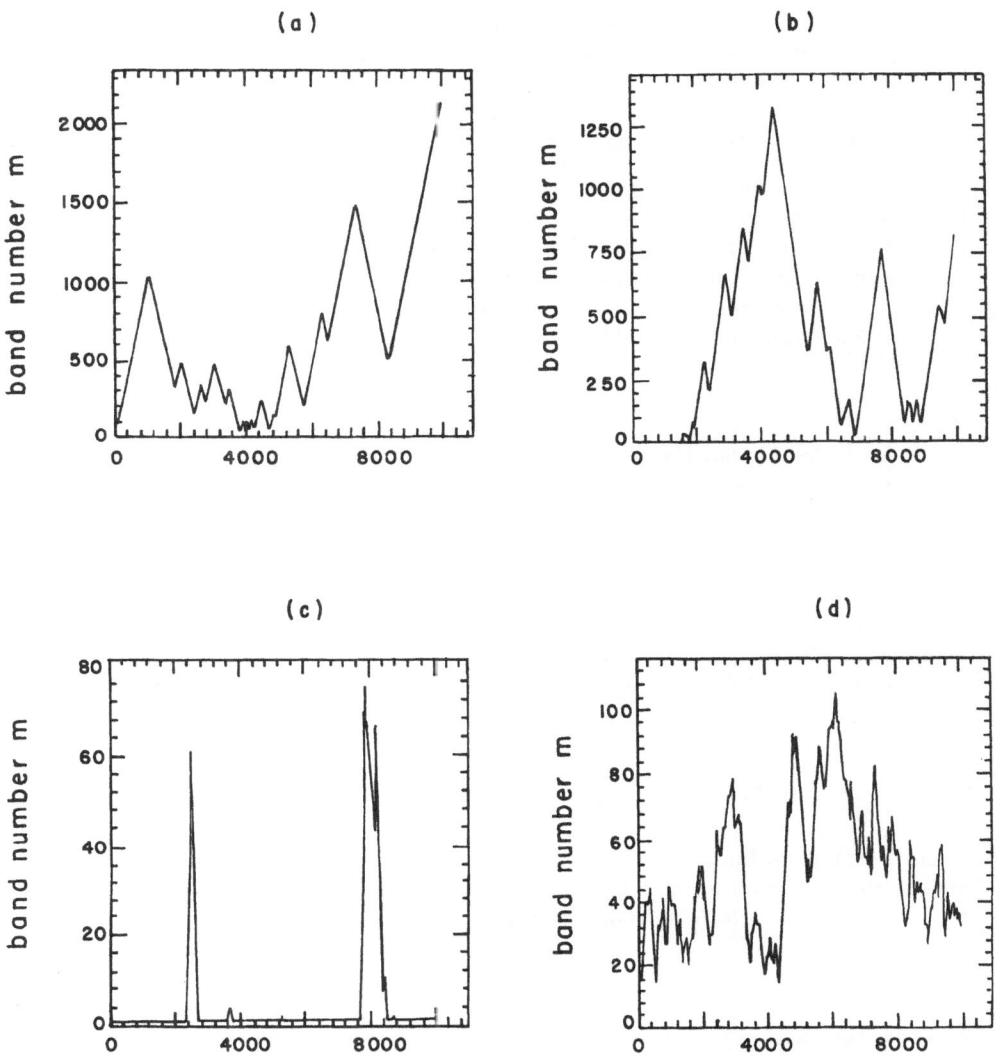

Figure (6): Plots of band number versus as a function of q from the simulations of elastic Zener tunneling. The transition probability is given by $P \sim \exp[-\frac{\pi}{4} \cdot \gamma \cdot \frac{1}{m}]$. Figures (a), (b), and (c) correspond to $\gamma = 1.0$, $\gamma = 2.63$, $\gamma = 4.0$ respectively. Figure (d) shows brownian motion for fixed transition probability $P = \frac{1}{2}$ which corresponds to the high temperature limit.

thermal fluctuations, $E(m)_{gap} >> 2k_BT$. When $E(m)_{gap}$ is of the order of k_BT or smaller, the transitions up and down in band number become almost equally likely, so that the interband random walk is essentially a Brownian motion in the band number. These fluctuations are due to elastic Zener tunnelling at zero temperature and thus this noise does not satisfy the fluctuation-dissipation principle.

4. Coherent Voltage Oscillations in Normal Tunnel Junction

As mentioned before, the coupling of q to the number operator $\hat{n}$ leads to oscillations in normal tunnel junctions. There are two basic approaches to this problem. The first is the Coherent Junction Approach, which is developed much along the lines we used above for Josephson junctions and which we explain below. The second one is the Classical Junction Approach, which we discuss subsequently.

The Hamiltonian for an unbiased normal tunnel juction can be written as

$$H = H_C + H_T$$

where H_C is the charging Hamiltonian and H_T is the tunnelling Hamiltonian. The charging Hamiltonian for the unbaiased junction is

$$H_C = \frac{e^2}{2C}\hat{n}^2 = E_C\hat{n}^2$$

It is important to note that while in a normal tunnel junction the phase difference θ is not a well defined quantity, n, the number of electrons that have tunneled through the junction, is a well defined quantity. The quasi-particle tunnelling across the junction is described by[7]

$$H_T = \int d^3x \int d^3x' \, T(x,x')\psi_{left}^{\dagger}(x,\sigma)\psi_{right}(x',\sigma) + h.c. \qquad (21)$$

where $T(x,x')$ is a function expressing the overlap of the two wave functions. Again, we assume that the exact form of H_T is unimportant, since we treat it as a perturbation. In the presence of the external driving field, q is defined such that

$$H_C = E_C(\hat{n} - q)^2$$

As above the effect of the tunnelling Hamiltonian is to remove the degeneracies present in the energy spectrum at the points $q = m/2$, producing a band structure picture for the energy spectrum as in figure (4). This characteristic tunnelling energy is given by

$$E_J = \frac{\hbar}{RC}$$

where R is the normal resistance of the junction, and C its capacitance. In the limit $E_C >> E_J$ oscillations can appear, provided that $\dot{q}$ is sufficiently small and that the inelastic interactions within the junction are sufficiently weak. Since the driving force is coupled to the number operator the effect is insensitive to interactions which destroy the phases of the electron wave functions.

A key similarity between the Josephson and normal tunnel junctions is that the voltage oscillates; there is no d.c. component to the voltage. While this is not surprising in the case of the Josephson juntion, it is remarkable in the case of the normal tunnel junction because this implies that there is no dissipation in the process. Therefore we can transfer charge across the system with an efficiency unobtainable in normal macroscopic systems.

In addition, all of the analysis we used in section 3 for Zener tunnelling applies in this case as well. Therefore, if the coherent junction approximation is correct, we should see the same type of fluctuations, evidenced by an enhancement of the low frequency power spectrum.

A second approach, the Classical Junction Approach, is based upon different physical assumptions and produces very similar results. We may assume that the electrons on either side of the juction cannot be characterized by a single quantum mechanical parameter such as the phase or number operator, but that the electrons act independently and have independent probabilities to tunnel across the junction. Then the statistical distribution of electrons will be almost a Fermi distribution and we can then calculate the tunnelling rates from one side to another, bearing in mind that the fermi energy is shifted by an amount E_C every time an electron tunnels across the junction. The net result is that this shift in the fermi energy causes the voltage across the junction to alternate at the same frequency as in the coherent junction approach. However, ther Zener tunnelling analysis of section 3 would not apply; the fluctuations will be due to shot noise. Thus the fluctuations give an important method to distinguish between the two approaches, when these predictions are compared to future experiments.

5. Discussion

We have predicted that a Josephson juntion driven by a direct current source I_{dc} will show voltage oscillations, provided

$$\frac{2e^2}{C} > \frac{\hbar I_J}{2e} >> 2k_B T \tag{23}$$

In order to avoid Zener tunneling we need the exponent in equation(20) to be small or

$$I_{dc} << I_J$$

Using current junction fabrication techniques, Josephson junctions with a capacitance of 10^{-15} fF can be contructed. Thus according to equation(23) the charge oscillations can observed for

$$I_J << 10^{-7} A$$
$$T << 1^{\circ} K$$

If we choose $I_J = 10^{-8}$A then in order to reduce the probability of Zener tunnelling we need $I_{dc} \simeq 10^{-9}$A which yields oscillations at a frequency of $\sim 10^{10}$Hz.

Returning to normal tunnel juctions, we have predicted that charge oscillations can be observed in the limit

$$\frac{e^2}{2C} > \frac{\hbar}{RC} >> 2k_BT$$

which implies $R > \hbar/e^2$. In order to neglect Zener tunnelling we need

$$I_{dc} << \frac{e}{RC}$$

which for the above parameters means $I_{dc} << 10^{-7}$A. When the quasi-static limits above are not satisfied we expect to see the Zener noise discussed in section 3. This noise is characterized by the enhancement of the low frequency power spectrum.

It may be possible to apply the above analysis of coherent voltage oscillations to charge transfer between or along macromoleculs. If the macromolecules are anchored to some larger structure then we can view each one as a junction; their small size ensuring the requisite charging energy. The molecules could bucket charge from one to another thereby passing a current. Alternatively, we can imagine that each side arm of a branched macro-molecule plays the role of the junction, with charge tunnelling from branch to branch. As explained above, the transfer of charge would be without dissipation, reaching efficiencies impossible in macroscopic systems.

In addition, the analysis used throughout this paper to study the dynamics of charging can be extended to other dynamical systems. For example a Hamiltonian for magnetic interactions can be written in much the same way as equation(13), where the driving force is no longer an external current source but an external magnetic field. In cyclic molecules it should be possible to produce a coherent oscillation in the current flowing around the molecule by steadily increasing the magnetic field.

A second possible extension is to generalize the analysis to mechanical systems. It may be possible to write the interactions between sections of a large molecule so that we view the transfer of 'kinks' along the molecule as a mechanical 'current'. The molecule, when stressed, would transfer mechanical energy along its length in a coherent fashion, with no loss of energy. The dynamics of the molecule would then be described in terms of the motion of these coherent excitations.

In conclusion we would like to stress that it is very important that much of the activity in biological systems occurs in this intermediate domain of physics. This allows a large flexibility in the dynamics of these systems, for it may be possible that they decouple from the external world for brief periods during which they behave in a quantum mechanical fashion, allowing it to perform processes that are energetically unfavorable and otherwise impossible. Yet, they are also large enough that they are constantly interacting with the external heat bath and so their dynamics are almost classical. There is some promise that the key to many such processes lies in mesoscopic physics.

Acknowledgements: One of us (Z.S.) acknowledges the Israel Academy of Science Grant and the U.S.–Israel Binational Science Foundation Grant for partial support.

References

1 E. Ben-Jacob, Y. Gefen, K. Mullen, Z. Schuss, *Proceedings of the Third international Conference on Superconducting Quantum Interference Devices, Berlin, West*, H. Ha-halbohm, H. Lübbig eds., (Walter de Gruyter, Berlin, New York) 1985; T. P. Spiller, J. E. Mutton, H. Prance, R. J. Prance and T. D. Clark *ibid.,*; E. Ben-Jacob, D. Berg-man, B. Matkowsky, Z. Schuss, "A Master Equation Approach to Shot Noise in Josephson Junctions", (to be published); E. Ben-Jacob, Y. Gefen, Phys. Lett. **108A** (1984) 289 Y. Ivaanchenko and L. Zilberman, Zh. Eksp. Teor. Fiz **55** (1968) [JETP **28** (1969) 1272].

2 K. K. Likharev, A. R. Averin, preprint # 7 (1984)

3 F. Guinea, G. Schön, "Coherent Oscillations in Tunnel Junctions", (preprint)

4 D. Tinkham, *Superconductivity*, McGraw Hill, N.Y. (1975)

5 E. Ben-Jacob, E. Mottola, G. Schön, Phys. Rev. Lett. **51** 2064, (1983)

6 D. J. Scalapino, *Tunneling in Solids*, E. Burstein, S. Lundquist, eds., Plenum, N.Y. (1968)

7 A. Messiah, *Quantum Mechanics*, North Holland, Amsterdam (1969)

SURFACE STATES OF ELECTRONS ON ALKALI-HALIDE CLUSTERS

Dafna Scharf,[†] Uzi Landman[#] and Joshua Jortner[†]
[†]School of Chemistry, Tel Aviv University
69978 Tel Aviv, Israel
[#]School of Physics, Georgia Institute of Technology
Atlanta, Georgia 30332, USA

ABSTRACT. The energetic and structural properties of a surface state of an excess electron localized on a $[Na_{14}Cl_{13}]^{+}$ cluster were established on the basis of quantum path integral molecular dynamics calculations.

1. INTRODUCTION

The physical and chemical phenomena associated with the attachment of an excess electron to a molecular cluster [1] are of considerable interest because of two reasons. Firstly, quantum phenomena are expected to be pronounced in such finite systems, where the electron wavelength is comparable to the cluster size. Secondly, the excess electron can serve as a probe for the nuclear motion and dynamics of the cluster. Non-reactive electron attachment to the clusters, which is not accompanied by the breaking of either intermolecular or intramolecular bonds, can result in either bulk or surface states of the excess electron. From the point of view of general methodology, the formation of electron surface states on clusters is facilitated by the large surface to volume ratio of clusters. From the technical point of view, the formation of such surface states is dominated by the structure of the cluster and by the nature of the electron-cluster interaction. We have explored the compositional, structural and size dependence of electron localization modes in an alkali-halide cluster (AHC), which provide novel information on surface states of an excess electron in these systems [2].

2. ELECTRON LOCALIZATION IN ALKALI-HALIDE CLUSTERS

The alkali-halide clusters (AHCs) were chosen because of three reasons. Firstly, the nature of interionic interactions in these clusters is well understood. Secondly, there exists an abundance of model calculations on both neutral and charged ACHs. Thirdly, quite extensive information is available on electron-alkali cation (M^{+}) and electron-

261

J. Jortner and B. Pullman (eds.), Tunneling, 261–267.

halide anion (X^-) interaction. The structure, energetics and dynamics of an e-AHC has been explored using the quantum path-integral molecular dynamics (QUPID) method. This approach, which rests on a discrete version of Feynman's path integral method [3-7], provides a powerful method for the study of these systems. In this scheme the quantum problem is isomorphic to an appropriate classical problem, with the excess electron being mapped onto a closed flexible polymer of P points. The isomorphism becomes exact as $P \to \infty$. The practical applicability of the computational method rests on the development of numerical algorithms, which achieve convergence with manageable values of P. From the point of view of general methodology, the QUPID method rests on the application of computers for the simulations of the properties of complex systems.

The QUPID method was applied to a system of an electron interacting with an AHC, which is specified by the Hamiltonian,

$$H = - (\hbar^2/2m) \nabla^2 + V_e(r) \quad , \tag{2.1}$$

with $V_e(r)$ being the e-AHC potential. The partition function for such a system is

$$Z = \text{Tr}[\exp(- \beta H)] \quad , \tag{2.2}$$

where $\beta = 1/kT$ is the inverse temperature, while the energy of the system is

$$E = - \frac{\partial}{\partial \beta} \ln Z \quad . \tag{2.3}$$

The quantum-mechanical-classical-isomorphism is established by (a) the factorization of the partition function into P contributions, and (b) by invoking the high temperature, i.e., a small β/P, expansion for the matrix elements. This procedure allows for the expression of the partition function

$$Z \simeq (Pm/2\pi\hbar^2\beta)^{3P/2} \int \vec{dr}_1 \ldots \vec{dr}_p \, \exp[-\beta V_{eff}(\vec{r}_1 \ldots \vec{r}_p)] \tag{2.4}$$

in terms of the effective potential

$$V_{eff} = \sum_{i=1}^{P} \left[\frac{Pm}{2\hbar^2\beta^2} - (\vec{r}_i - \vec{r}_{i+1})^2 + \frac{1}{P} V_e(\vec{r}_i) \right] \quad , \tag{2.5}$$

which consists of a superposition of a harmonic potential and the cluster potential.

Equations (2.4) and (2.5) establish an approximate isomorphism between the quantum problem characterized by the Hamiltonian H, Eq. (2.1),

and the classical problem defined by the effective potential, Eq.
(2.5). In this isomorphism the quantum particle is mapped onto a
closed polymer, which is also referred to as a necklace of P pseudo-
particles (beads). Each point of the necklace exerts two types of
interactions (Fig. 1), that is (i) nearest neighbour interactions between the beads in the chain, which are characterized by a harmonic
potential with a force constant $Pm/\hbar^2\beta^2$, and (ii) interactions with the
AHC through the scaled potential $V_e(\vec{r})/P$.

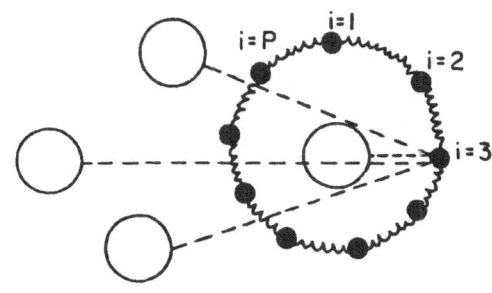

Fig. 1: A schematic
representation of
the electron neck-
lace.

∿∿∿ HARMONIC PSEUDOINTERACTIONS

- - - CONTRIBUTIONS TO e-AHC POTENTIAL

◯ CLASSICAL IONS

● POINTS OF NECKLACE

At this stage it will be convenient to provide an explicit description of the e-AHC system. The AHC, which is treated as a classical system, consists of N ions (N_1 cations and N_2 anions) with masses
M_1 and M_2, respectively. The interionic potential within the AHC is
given

$$V_{AHC} = \sum_{I \neq J} \sum \Phi_{IJ}(R_{IJ}) \qquad (2.6)$$

with the interionic potential $\Phi_{IJ}(R_{IJ})$ being given by the Born-Meyer
potential [8]. The e-AHC potential is

$$V_e(\vec{r}) = \sum_I \Phi_{eI}(\vec{r} - \vec{R}_I) \quad , \qquad (2.7)$$

consisting of a sum of electron-ion potentials, which are described by
a purely repulsive pseudo-potential

$$\Phi_{eI}(r) = e^2/r \qquad (2.8)$$

for electron-anion interaction and by the local pseudo-potential [9]

$$\Phi_{eI}(r) = - e^2/R_c \quad ; \quad r \leq R_c$$

$$\Phi_{eI}(r) = - e^2/r \quad ; \quad r \geq R_c \tag{2.9}$$

for the electron-cation interactions. The average energy of the system is now given from Eqs. (2.3) and (2.7) in the explicit form

$$E = 3N/2\beta + <V_{AHC}> + KE + P^{-1} < \sum_{i=1}^{P} V_e(\vec{r}_i) > . \tag{2.10}$$

The first two terms in Eq. (10) correspond to the kinetic and potential energies of the classical AHC system, respectively. The electron kinetic energy is [6]

$$KE = 3/2\beta + \sum_{i=1}^{P} <\partial V_e(\vec{r}_i)/\partial \vec{r}_i \cdot (\vec{r}_i - \vec{r}_p)>/2P \quad , \tag{2.11}$$

consisting of a free-particle term, $3/2\beta$, and the contribution K_{int} from the interaction with the ions. The indicated statistical averages < > are over the Boltzmann distributions as defined in Eq. (2.4). This formalism is converted into a numerical algorithm by noting [7] the equivalence of the sampling described above to that over phase space trajectories generated via molecular dynamics by the classical Hamiltonian

$$H = \sum_{i=1}^{P} m^* \dot{\vec{r}}_i^2/2 + \sum_{I=1}^{N} M_I \dot{\vec{R}}_I^2/2 +$$

$$+ \sum_{i=1}^{P} [Pm(\vec{r}_i - \vec{r}_{i+1})^2/2\hbar^2\beta^2 + V_e(\vec{r}_i)/P] + V_{AHC} \tag{2.12}$$

the mass m^* being arbitrary [7] and taken as $m^* = 1$ a.m.u.

   Numerical simulations were performed for an electron interacting with sodium chloride clusters. Based upon the examination of the stability of the variance of the kinetic energy contribution K_{int}, the number of "electron beads" was taken as $P = 399$. Employing an integration step of $\Delta t = 1.03 \times 10^{-15}$ sec, long equilibration runs were performed ($1-2 \times 10^4 \Delta t$). The reported results were obtained via averaging over $8 \times 10^3 \Delta t$ following equilibration. The temperature of the system was $T = 300$ K. The electron-cation pseudo-potential parameters were varied by changing the cut-off radius R_c in Eq. (2.9). The value chosen for Na^+ was $R_c = 3.22$ a_0 [9], which reproduces well the atomic ionization potential [10]. Weaker electron-cation interactions, which

are appropriate for heavier cations, were mimicked by increasing the
value of R_c.

We have studied the interaction of an electron with positive
clusters $[Na_{14}Cl_{13}]^+$ and $[Na_{14}Cl_{12}]^{++}$, which exhibit the crystallographic structure of the NaCl crystal, without and with an anion vacancy, respectively (Fig. 2). The energetics of the $[Na_{14}Cl_{12}]^{++}$ —e
system [2] demonstrate the stabilization of an internally localized
excess electron state induced by the presence of the anion vacancy and
corresponds to a bulk F centre [11] within a finite system. A drastically different situation is exhibited for the localization of an
excess electron in a $[Na_{14}Cl_{13}]^+$ cluster, which constitutes a "filled"
3x3x3 ionic cube. The $[Na_{14}Cl_{13}]^+$—e system exhibits a novel surface
state. In Fig. 3 we present the results for the equilibrium charge
distribution obtained from two-dimensional projections of the necklace
edge points, and for the nuclear configuration of the clusters. The

$$98.88 \qquad\qquad 91.35$$
$$Na_{14}Cl_{13}^+ \qquad\qquad Na_{14}Cl_{12}^{++}$$

Fig. 2: Configuration of $[Na_{14}Cl_{12}]^{++}$ and $[Na_{14}Cl_{13}]^+$
clusters. The numbers refer to ground-state energies
in eV units.

Fig. 3: Ionic configuration and "electron necklace"
distribution for an excess electron interacting with
the $Na_{14}Cl_{13}^+$ cluster ($R_c = 3.22$ a.u.). Note electron
localization in a compact surface state.

energetics of the $[Na_{14}Cl_{13}]^+$–e surface state is characterized by:

electron binding energy	$E_B{}^e = -0.1594$ Hartree
cluster nuclear reorganization energy	$E_C = 0.1055$ Hartree
cluster electron affinity	$E_A = -0.0539$ Hartree

The characteristics of the external localization of an electron on an NaCl cluster are:
   (a) It is exhibited in a vacancy-free moderately large AHC.
   (b) In such an AHC, internal localization is prohibited by a large value of E_C.
   (c) The excess electron in $[Na_{14}Cl_{13}]^+$ localizes around the Na^+ surface ion leaving a neutral $Na_{13}Cl_{13}$ cluster interacting with a partially neutralized Na atom.
   (d) In this system a surface localized electron state is exhibited.
   (e) For clusters containing heavier M^+ ions, the surface state will become more extended.
   (f) The excess electron surface state is expected to be exhibited also in moderately large neutral AHCs.
The theoretical prediction for the existence of energetically stable surface states of an excess electron on an AHC is of considerable interest. Such surface states were considered for macroscopic alkali-halide crystals by Tamm [11] about fifty years ago, but were never experimentally documented.
   The observation of the localization of an excess electron around a single surface Na^+ ion in the centre of the face of the $[Na_{14}Cl_{13}]^+$ cluster is intriguing. This cluster is characterized by six face-centered Na^+ ions, which provide equivalent binding sites for the electron. The O_h symmetry of the system implies the existence of six delocalized electronic states of symmetries A_{1g}, E_g and T_{1u} for this system. At 0 K a delocalized state of the excess electron is expected to be exhibited. At finite temperatures severe symmetry breaking effects are expected to prevail in view of the negligibly small transfer integral between different Na^+ face-centered ions. Such symmetry breaking will be induced by cluster vibrational excitations, which will result in the localization of the excess electron around a single Na^+ ion. The results of our QUPID calculations at 300 K concur with the notion that the cluster acts as its own heat bath inducing localization, which is due to symmetry breaking.

REFERENCES

[1] J. Jortner, Ber. Bunsengess. Phys. Chem. 88, 188 (1984).
[2] U. Landman, D. Scharf and J. Jortner, Phys. Rev. Lett. 54, 1860 (1985).
[3] R.P. Feynman and A.R. Hibbs, Quantum Mechanics and Path Integrals, McGraw Hill, New York, 1965.
[4] D. Chandler and P.G. Wolynes, J. Chem. Phys. 74, 4078 (1981).

[5] M. Parrinello and A. Rahman, J. Chem. Phys. 80, 860 (1984).
[6] M.F. Herman, E.J. Bruskin and B.J. Berne, J. Chem. Phys. 76, 5150
     (1982).
[7] D. Calloway and A. Rahman, Phys. Rev. Lett. 49, 613 (1982).
[8] F.G. Fumi and M.P. Tosi, J. Phys. Chem. Solids 25, 31, 45 (1984)
[9] R.W. Shaw, Phys. Rev. 174, 769 (1968).
[10] D. Scharf, U. Landman and J. Jortner, to be published.
[11] N.F. Mott and R.W. Gurney, Electronic Processes in Ionic Crystals,
     Oxford University Press, Oxford, 1946.

THE CHEMISORBED STATE OF HYDROGEN ON A METAL SURFACE STUDIED VIA
QUANTUM PATH INTEGRAL MOLECULAR DYNAMICS SIMULATIONS

Uzi Landman, R. N. Barnett, and C. L. Cleveland
School of Physics
Georgia Institute of Technology
Atlanta, Georgia 30332

P. Nordlander
IBM Thomas J. Watson Research Center
P. O. Box 218
Yorktown Heights, New York 10598

ABSTRACT. The states of hydrogen and tritium chemisorbed on a Ni(100)
surface are investigated, at various temperatures, using quantum path-
integral molecular dynamics simulations. Following a brief description
of the method we present results which demonstrate clearly the quantum
nature of adsorbed hydrogen, and it's isotopes, at all temperatures.

1. INTRODUCTION

The nature of chemisorbed hydrogen, and it's isotopes, on metallic
surfaces and the mechanism of transport in these systems are subjects
of coupled technological and fundamental interest. These systems are
particularly intriguing since we expect them to exhibit marked quantal
effects and thus they can be used as prototype systems for testing of
fundamental physical concepts both theoretically and experimentally.
Indeed, recent experimental studies of the diffusion of atomic hydrogen
and it's isotopes on tungsten surfaces, using the field-emission current fluctuation technique [1] and measurements of the vibrational
dynamics of hydrogen adsorbed on metals [2], aroused a great deal of
interest and theoretical work [3-9]. Among the issues of current
interest we mention:
>   (i)   the nature of chemisorbed hydrogen, and it's isotopes, i.e.,
>   the degree of quantal character and spatial localization,
>   and the effect of substrate dynamics on these properties;
>   (ii) the transport mechanisms, distinguishing band motion, ther-
>   mally activated hopping and direct or phonon activated tunnel-
>   ing regimes, and the dependence of these on the adsorbed
>   species, the coverage, the substrate metal surface and the
>   ambient temperature;
>   (iii) the energetics of the adsorbed state and calculations of
>   potential energy surfaces and interaction potentials, parti-

J. Jortner and B. Pullman (eds.), Tunneling, 269–279.
© *1986 by D. Reidel Publishing Company.*

cularly forms which could be employed in molecular dynamics
simulations;
(iv) the dynamics of the adsorption system.

In this study we focus on the first issue listed above. Customar-
ily, in the chemisorbed state the adsorbed atoms are regarded as point
masses localized and vibrating about well defined adsorption sites.
However, quantum mechanical calculations of the energy levels and wave
functions for the motion of chemisorbed hydrogen or Nickel surfaces
revealed significant delocalization [9], suggesting that a proper description can be given only in terms of hydrogen energy bands. Moreover, calculations of diffusion coefficients of hydrogen on metal
surfaces which have included quantal effects in transition state theory
[3,4] have predicted large quantal contributions, especially at low
temperatures. In addition, it has been found that diffusion on nonrigid surfaces is enhanced compared to that on rigid ones, [4,5] via
participation of the surface atoms' dynamics in the reaction coordinate.
The coupling to the substrate dynamics introduces a shift in the effective potential for the tunneling process, as well as dissipative effects. The calculated enhancement of the tunneling contribution to the
diffusion is not inconsistent with phenomenological studies which have
concluded that frictional effects decrease tunneling rates [10], since
in these investigations only the dissipative effects were considered.

This paper is organized as follows: the quantum path-integral
molecular dynamics (QUPID) simulation method is described briefly in
Section 2. Results of simulations of adsorbed hydrogen (H) and tritium
(T) whose chemisorption to a Ni (100) surface is described by a full
adiabatic potential-energy surface calculated via the effective medium
theory, are shown in Section 3. These results show pronounced quantum
character for both isotopes even at room temperature.

2. QUANTUM PATH-INTEGRAL MOLECULAR DYNAMICS (QUPID)

The Feynman path integral formulation of quantum statistical mechanics
[11] allows a derivation of an approximate expression for the partition
function, Z, for a system consisting of a quantum particle (mass m and
coordinate $\vec{r}$), interacting with a set of N classical particles (whose
phase space trajectories are generated by classical equations of motion)
via a potential $V(\vec{r}, \vec{R}_1, \ldots, \vec{R}_N)$,

$$Z = \left(\frac{mP}{2\pi\hbar^2\beta} \right)^{3P/2} \int d\vec{r}_1 \ldots d\vec{r}_P \int d\vec{R}_1 \ldots d\vec{R}_n \ e^{-\beta V_{eff}}, \qquad (1a)$$

where

$$V_{eff} \equiv \sum_{i=1}^{P} \frac{mP}{2\hbar^2\beta^2} (\vec{r}_i - \vec{r}_{i+1})^2 + \frac{1}{P} \sum_{j=1}^{N} \sum_{i=1}^{P} V(\vec{r}_i, \vec{R}_j)$$

$$+ V_c(\vec{R}_1, \ldots, \vec{R}_N). \qquad (1b)$$

V_c is the interaction potential between the classical particles, and $\beta = (k_B T)^{-1}$.

Eqs. (1) establish an isomorphism [12] between the quantum problem and a classical one in which the quantum particle is represented by a flexible periodic chain (necklace) of P pseudoparticles (beads) with nearest-neighbor harmonic interactions with a temperature dependent spring constant, $Pm/\hbar^2\beta^2$. The above expression for the partition function is exact as $P \to \infty$. In practice, the finite value of P employed in a calculation is chosen to yield convergent results and depends upon the temperature and characteristics of the interaction potential, V.

The average energy of the system at equilibrium is given by

$$E = \frac{3N}{2\beta} + <V_c> + K + \frac{1}{P} < \sum_{i=1}^{P} V(\vec{r}_i)>, \tag{2a}$$

where

$$K = \frac{3}{2\beta} + \frac{1}{2P} \sum_{i=1}^{P} < \frac{\partial V(\vec{r}_i)}{\partial r_i} \cdot (\vec{r}_i - \vec{r}_p)> , \tag{2b}$$

and the angular brackets indicate statistical averages over the probability distribution as defined in Eq. (1). The first two terms in Eq. (2a) are the mean kinetic and potential energies of the classical component of the system. The quantum particle kinetic energy estimator [13], K, consists of the free particle term ($K_f = 3/2\beta$) and a contribution due to the interaction (K_{int}). Finally, the last term in Eq. (2a) is the mean potential energy of interaction between the quantum particle and the classical field.

A measure of the quantum character of the system may be inferred from the relative contribution of K_{int} to the kinetic energy K. The magnitude of K_{int} depends on two factors, the gradients of the potential at the location of the beads and the bead spatial distribution. For a classical particle whose thermal wavelength, $\lambda_T = (\beta\hbar^2/m)^{1/2}$, approaches zero, the Gaussian factor in Eq. (1a) (see first term in Eq. (1b)) reduces to a delta function ($\delta(\vec{r}_i - \vec{r}_{i+1})$, for all i) and the necklace collapses to a single point, representing a classical particle. Under this circumstance $K_{int} = 0$. Generally, the degree of quantal character depends upon the interplay between the spatial extent of the particle (which in the above formulation is related to the spatial distribution of the pseudo-particles) and a characteristic length associated with the rate of change of the potential (for a similar discussion in the context of the classical limit of the partition function see [14,15]).

Another convenient measure of the quantum character of the system and the degree of localization of the quantum particle is provided by the complex time correlation function [15],

$$R^2(t-t') = <|\vec{r}(t) - \vec{r}(t')|^2> , \tag{3}$$

where $|\vec{r}(t) - \vec{r}(t')|$ is the distance between two points (beads) on the electron path separated by a "time" $(t-t')$ ε $(0,\beta\hbar)$. The value at $t-t' = \beta\hbar/2$, $R \equiv R(\beta\hbar/2)$, characterizes the breadth of the bead distribution, and yields the correlation length. Note that for a free particle, where the bead distribution obeys Gaussian statistics, $R_f = \sqrt{3}\ \lambda_T/2$. For a localized state, when fluctuations in $R(t-t')$ are inhibited, $R(t-t')$ is dominated by the ground state [15], and as a function of $(t-t')$, $R^2(t-t')$ starts from zero at $t-t' = 0$ and rapidly achieves a constant value, with $R < R_f$ (i.e., it is independent of "time" except for $t-t'$ close to 0 and βh). The rise time, τ, is a (rough) measure of the excitation energy $\Delta E \equiv E_1 - E_0$, (where E_1 is a Boltzmann weighted average of the manifold of excited states), given by $\Delta E = \hbar/\tau$. A delocalized state, on the other hand, is characterized by a dependence of $R^2(t-t')$ on $(t-t')$ over the whole range $(0,\beta\hbar)$.

The formalism described above is converted into a numerical algorithm by noting the equivalence [16,17] between the equilibrium statistical averages over the probability distribution given in Eqs. (1), and sampling over phase space trajectories, generated by a classical Hamiltonian

$$H = \sum_{i=1}^{P} \frac{m^* \dot{\vec{r}}_i^2}{2} + \sum_{I=1}^{N} \frac{M_I \dot{\vec{R}}_I^2}{2} + \sum_{i=1}^{P} [\frac{Pm}{2\hbar^2\beta^2} (\vec{r}_i - \vec{r}_{i+1})^2 + \frac{V(\vec{r}_i)}{P}]$$
$$+ V_c(\vec{R}_1,\ldots,\vec{R}_N) \ , \qquad (4)$$

where m^* is an arbitrary mass chosen such that the internal frequency of the necklace, $\omega = [mP/(m^*\beta^2\hbar^2)]^{1/2}$, will match the other frequencies in the system, and M_I is the mass of a classical particle. For a description of the application of the method to a wide variety of quantum many-body systems of chemical and physical interest the reader is referred to a recent review [18].

3. HYDROGEN AND TRITIUM ADSORBED ON $N_i(100)$

The starting point for our calculation is a specification of the interaction potentials, or a potential energy surface. Such information is available only in approximate forms for a small number of chemisorption systems. For the purpose of this investigation we have chosen to use a potential energy surface for chemisorbed hydrogen on a Nickel surface, which was obtained via the effective medium theory [19]. This potential was shown to yield good chemisorption energies and bond lengths in comparison to other theoretical estimates and experiment. In addition the quantum mechanical energy levels and wave functions for this system were calculated via a numerical solution of the Schrodinger equation, revealing a significant delocalization of the adsorbed hydrogen [9].

Potential energy contours (energy in eV, distance in units of the nickel nearest-neighbor distance, $4.7a_0$), are shown in Figs. (1a), 2(a) and 2(e) for a cut vertical to the surface plane along the <110> direction through the fourfold center position (Fig. (1a)) and in planes parallel to the surface plane, at two heights above it,

Figure 1. Potential energy contours (a) and bead distributions
for H/Ni(100) at 300K and 80K
(b,d) and for T at 300K (b), on
a vertical cut along the <110>
direction through the 4-fold
site. Distance in units of
4.7 a_0 and energy in eV.

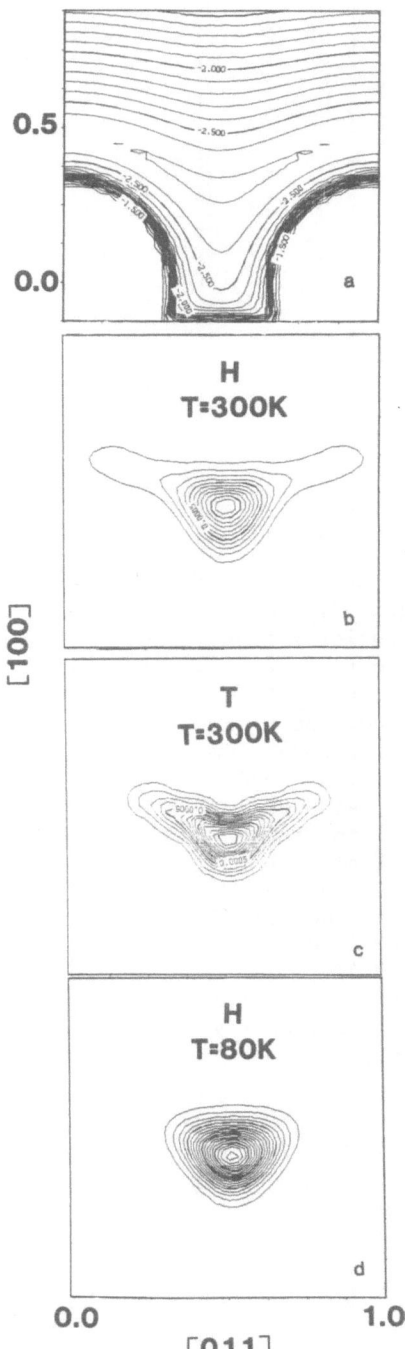

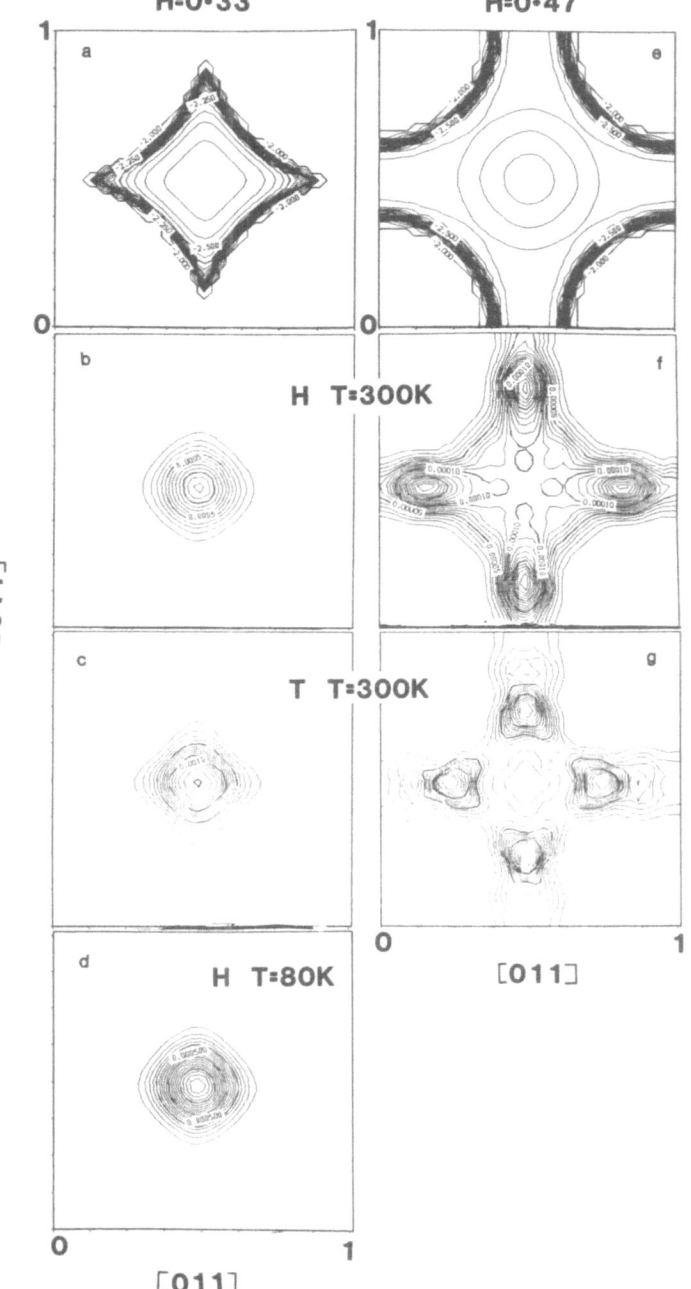

Figure 2.
Potential energy
contours (a,e)
and bead distri-
butions for H/Ni
(100) at 300K and
80K (b,c,d) and
for T at 300K
(c,g) on planes
parallel to the
surface plane
and at two heights
above it H = 0.33
and 0.47.
Distance in units
of 4.7 a_0, energy
in eV.

H = 0.33 and 0.47, Figs. (2a) and (2e) respectively. In the minimum
energy configuration the hydrogen is adsorbed at the fourfold site at
a height of $\sim$ 0.33 x 4.7 a_0 above the metal surface plane.

The simulations discussed here were performed for a canonical
(constant temperature) ensemble [20], using a stochastic collision
frequency parameter ν = 0.001, and the integration time steps at
300°K, 80°K and 20°K were 0.125 tu, 0.5 tu and 1 tu, respectively,
with tu = 1.46 x 10^{-15} sec. The number of beads, P (see Eq. (1)),
which yielded convergent results were found to be 512, 256 and 256 in
order of descending temperature, and m^* was given the value of 1 amu.
To achieve good equilibrium statistics (i.e., exploration of all of the
accessible phase space, including saddle points), 4 x 10^5, 2 x 10^5 and
10^5 integration steps were required for simulations at 300, 80 and 20 K
respectively.

Equilibrium averaged pseudoparticle density contours on the
vertical plane, through the fourfold site in the <110> direction, for
H and T at 300K and for H at 80K (the results at 20K are very similar
to those at 80), are shown in Figs. (1b)-(1d), respectively. In all
the density contours shown, the outermost contour corresponds to the
lowest value, beyond which the density vanishes. We observe that at
T = 300K the bead distribution extends over a significant portion of
the surface unit cell, for both H and T. A more spatially localized
state is observed at T = 80K. Comparison of the probability density
shown in Figs. (1b) and (1c) with the calculated wave functions for the
ground and excited state (see figure 1 in reference [9]) indicates
that while at T = 80K the ground state dominates, at room temperature
the state is a superposition (with Boltzmann weights) of the ground
and first, and perhaps a small contribution from an even higher,
excited state. Similar conclusions are drawn from analysis of the
probability density contours in planes parallel to the surface drawn
at different heights, H, above the surface plane. It is seen that at
T = 300K the state extends to a further distance from the surface (see
contours for H = 0.47 x 4.7 a_0, Figs. (2f) and (2g)), whereas at T =
80K the state is confined to the vicinity of the minimum of the potential energy surface.

Further information about the nature of the adsorbed particle is
provided by the correlation function R(t), see Eq. (3). In Fig. 3 we
show R(t) versus t ϵ (0,$\beta\hbar$) for H and T at 300K (Figs. 3(a) and 3(b))
and for H for 80 and 20K (Figs. 3(c) and 3(d)). First we observe that
at 300K the R(t) depends upon t over the entire interval indicating an
extended state in the well, and excited state(s) participation,
(slightly more so for tritium). The curves for 80K and 20K are of a
different character corresponding to a more confined state dominated
by the ground state (from the curve at 20K, we estimate an excitation
energy E $\sim$ 30 meV).

The values for K_f and K_{int} given in Table 1 for H and T chemisorbed at various temperatures on Ni(100) and for H on Cu(100) at 300K
corraborate our previous results. As discussed in Section 2, in the
classical limit $K_{int} \to 0$. As seen from our results that limit is not
achieved for chemisorbed H even at room temperature, and not even for
the heavy isotope, T at 300K. In addition we provide values for

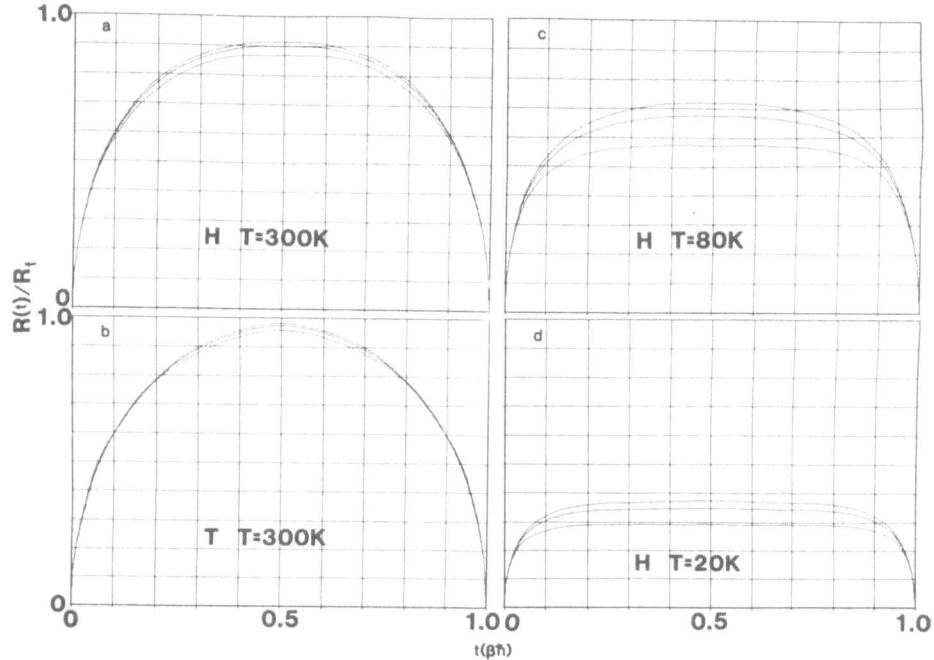

Figure 3. $R(t)/R_f$ versus $t \in (0,\beta\hbar)$ for $H/N_i(100)$ at 300, 80 and 20K (a,c,d) and for $T/N_i(100)$ at 300K.   Ground state dominance and increased confinement at the lower temperatures is evident.

$R(\beta\hbar/2)/R_f(\beta\hbar/2)$ and $R_T = [(P/P-1) \sum_{i=1}^{P} <(\vec{r}_i - \vec{r}_{i+1})^2>]^{1/2}$, which for a free particle yields $R_T = \sqrt{3} \lambda_T$, (for free H at 300K, $R_T = 1.304\ a_o$, and for T, $R_T = 0.753\ a_o$).

The marked quantum nature of chemisorbed hydrogen, even at room temperature revealed by our calculations, and the delocalized character of the adsorbed atom (more pronounced at high temperature due to an increased contribution from the excited states) are in accord with previous calculations [9] (performed at zero temperature) and experimental observations, such as the absence of ordered structures of chemisorbed H up to close to a full monolayer coverage [6,21] where the interaction with neighboring hydrogen atoms restricts the mobility. Additionally, our calculations provide further evidence that theoretical treatments of dynamical and transport phenomena in hydrogen chemisorption systems should incorporate quantal effects.

In the simulation presented above the substrate was kept static. An interesting question is the effect of substrate dynamics on the nature of the adsorbed particle, particularly since studies of hydrogen diffusion indicate a marked increase in diffusion rates when compared to those calculated for rigid surfaces [4,5]. We have attempted to

TABLE I

The free interaction kinetic energies, K_f and K_{int}, potential energy of the chemisorbed particle $<V>$, R_T, $R \equiv R(\beta\hbar/2)$ normalized by the free particle value R_f, for chemisorbed H on Ni(100) at three different temperatures and for T at 300°K. Values for H chemisorbed on Cu(100) are also given. Energy in Hartrees, length in Bohr radii. Variances are given in parentheses.

	H 300K	T 300K	H 80K	H 20K	H/Cu(100) 300K
K_f	1.425×10^{-3}	1.425×10^{-3}	3.800×10^{-4}	9.500×10^{-5}	1.425×10^{-3}
K_{int}	1.239×10^{-3}	5.017×10^{-4}	1.752×10^{-3}	2.019×10^{-3}	1.193×10^{-3}
	(1.352×10^{-3})	(6.845×10^{-4})	(1.028×10^{-3})	(6.225×10^{-3})	(1.123×10^{-3})
$<V>$	-9.760×10^{-2}	-9.835×10^{-2}	-9.811×10^{-2}	-9.812×10^{-2}	-9.252×10^{-2}
	(1.149×10^{-3})	(1.022×10^{-3})	(4.522×10^{-4})	(2.286×10^{-4})	(1.384×10^{-3})
R_T	1.31	7.63×10^{-1}	2.53	4.99	1.32
R/R_f	8.99×10^{-1}	9.73×10^{-1}	6.75×10^{-1}	3.5×10^{-1}	9.11×10^{-1}

---

address this question using the Morse potential models used in previous studies [3,4]. However, our results showed that unlike the results obtained by using the effective medium (EMT) potentials, for these semi-empirical potentials the adsorbed hydrogen exhibits an almost classical character at room temperature (K_{int}= 6.43 x 10^{-5} a.u. for a dynamical Cu substrate, and K_{int} = 7.06 x 10^{-5} a.u. for a static substrate, compared to the value of 1.19 x 10^{-3} a.u. obtained with static substrate EMT potentials, see Table 1). From these results and a comparison of the potential energy surfaces we conclude that prior to further studies of substrate dynamical effects in these systems improved potential energy models, suitable for molecular dynamics simulations, are required.

Finally we comment on simulations of transport processes in quantum systems. It is important to recognize that the dynamics in the classically isomorphic system (see Eqs. (1) and (4)), solely provides an algorithm for the determination of equilibrium properties. The simulation of real time dynamics of a quantum system is a most difficult problem, for which no satisfactory solution is currently available.

An estimation of quantal effects in investigations of diffusion phenomena, via an effective potential approximation, was suggested by Doll and his coworkers [22,3]. This method is based on Feynman and Hibbs (FH) who have shown that leading quantal corrections to the classical partition function can be obtained by replacing the classical potential by a temperature dependent effective potential which is obtained as a convolution of the classical potential with a Gaussian whose width is given by $\lambda_T^2/12$. For a generic potential energy function, $V(x)$, with a harmonic frequency ω_b near the bottom of the well, and $|\omega_t|$ in the vicinity of the top of the barrier separating the well from a neighboring one, the effective potential obtained by such a convolution, is characterized by a barrier height reduction $U^2 = \hbar^2(\omega_b^2+|\omega_t|^2)/24kT$. Interestingly, this factor when incorporated in the customary transition state theory rate expression yields, to lowest order, the Wigner tunneling correction to the rate of chemical reactions [23]. While suggestive, one should be cautious that the FH effective potential was developed in the context of a calculation of the partition function, and that the justification and utility of that effective potential to generate dynamical information is not obvious.

Alternatively, one may take a totally different approach and attempt to solve the time-dependent Schrodinger equation [24] for the quantum component of the system, in conjunction with classical dynamical evolution of the rest of the system, generated by a classical Hamiltonian. The two time evolutions are then coupled self-consistently.

We gratefully acknowledge useful conversations with J. W. Gadzuk and R. F. Fox. This work was supported by the U. S. DOE, Grant DE-FG05-86 ER45234.

REFERENCES

1. R. Difoggio and R. Gomer, Phys. Rev. B25, 3490 (1982); S. C. Wang and R. Gomer, J. Chem. Phys. 83, 4193 (1985); see also article by R. Gomer et al. in this volume.
2. S. Andersson, Chem. Phys. Lett. 55, 185 (1978); W. Ho, N. J. DiNardo and E. W. Plummer, J. Vac. Sci. Technol. 17, 134 (1980); C. Nyberg and C. G. Tenstal, Solid State Commun. 44, 251 (1982); U. J. Chabal in Vibrations At Surfaces 1985, eds. D. A. King, N. V. Richardson and S. Holloway (Elsevier, Amsterdam, 1986), p. 159.
3. S. M. Valone, A. F. Voter and J. D. Doll, Surface Sci. 155, 687 (1985).
4. J. G. Lauderdale and D. G. Truhlar, J. Am. Chem. Soc. 107, 4590 (1985); J. Chem. Phys. 84, 1843 (1986).
5. R. Jaquet and W. H. Miller, J. Phys. Chem. 89, 2139 (1985).
6. K. A. Muttalib and J. Sethna, Phys. Rev. B32, 3462 (1985).
7. K. B. Whaley, A. Nitzan and R. B. Gerber, J. Chem. Phys. 84, 5181 (1986).
8. M. Tringides and R. Gomer, Surface Sci. 166, 419, 448 (1986).
9. M. J. Puska, R. M. Nieminen, M. Manninen, B. Chakraborty, S. Holloway and J. K. Norskov, Phys. Rev. Lett. 57, 1081 (1983).
10. A. O. Calderia and A. J. Legett, Phys. Rev. Lett. 46, 211 (1981).
11. R. Feynman, Statistical Mechanics (Benjamin, Reading, Mass., 1972); R. P. Feynman and A. R. Hibbs, Quantum Mechanics and Path Integrals (McGraw-Hill, New York, 1965).
12. D. Chandler and P. G. Wolynes, J. Chem. Phys. 79, 4078 (1981).

13. M. F. Herman, E. J. Bruskin and B. J. Berne, J. Chem. Phys. 76, 5150 (1982).
14. K. Huang, Statistical Mechanics (Wiley, New York, 1963), pp. 213-220; L. D. Landau and E. M. Lifshitz, Statistical Physics (Pergamon, New York, 1980), Part 1, pp. 98-104.
15. A. L. Nichols III, D. Chandler, Y. Singh and D. M. Richardson, J. Chem. Phys. 81, 5109 (1984).
16. M. Parrinello and A. Rahman, J. Chem. Phys. 80, 860 (1984); D. Callaway and A. Rahman, Phys. Rev. Lett. 49, 613 (1982).
17. U. Landman, D. Scharf and J. Jortner, Phys. Rev. Lett. 54, 1860 (1985).
18. B. J. Berne and D. Thirumalai, "On the Simulation of Quantum Systems: Path Integral Methods", to appear in Annual Rev. Phys. Chem. (1986).
19. P. Nordlander, S. Holloway and J. K. Norskov, Surface Sci. 136, 59 (1984).
20. J. R. Fox and H. C. Andersen, J. Phys. Chem. 88, 4019 (1984).
21. K. Christmann, R. J. Behm, G. Ertl, M. A. van Hove and W. H. Weinberg, J. Chem. Phys. 70, 4168 (1979).
22. J. D. Doll, J. Chem. Phys. 81, 3536 (1984).
23. E. Wigner, Z. Physik Chem. B19, 203 (1932).
24. See article by Z. Kotler, A. Nitzan and R. Kosloff, this volume.

TUNNELING IN SURFACE DIFFUSION

Assa Auerbach, Karl F. Freed and Robert Gomer
The James Franck Institute and Department of Chemistry
The University of Chicago
Chicago, Illinois 60637, U.S.A.

ABSTRACT. A description is provided of experimental data for tunneling
diffusion of atoms on single crystal planes as obtained using field
emission-fluctuation spectroscopy. A focus is made upon the
significantly weaker than anticipated isotope effect for H/W(110)
tunneling diffusion and on the coverage dependence which suggests an
influence of the nuclear statistics on the diffusion. A unified
analysis of both the tunneling and thermally activated regions provides
insight into the nature of the interactions between the adsorbed
hydrogen atom and the tungsten lattice, and these interactions, in
turn, provide an explanation of the observed anomolous isotope effect
for diffusion of H/W(110).

INTRODUCTION

This paper presents some experimental results of tunneling
diffusion of light adsorbates on metal surfaces as measured by the
fluctuation method. We then discuss the implications of these results,
particularly the smallness of the observed isotope effect and present
theoretical approaches to understanding the experiments.

FLUCTUATION METHOD

The experimental method [1] requires a brief description. It
takes advantage of the very high linear magnification (10^5 - 10^6) of
the field emission microscope in which an electron emission map of a
hemispherical single crystal surface is projected onto a fluorescent
screen [2]. The projected surface consists of regions of varying
orientation and some well developed flats, corresponding to low index
planes. Emission is exponentially dependent on work function and the
latter in turn on adsorbate coverage.

By cutting a small hole in the screen it is possible to examine
emission from a region of 50-100 Å in linear dimension, involving a
single crystal plane. If the emitter is uniformly covered with
adsorbate and then kept at a temperature where diffusion occurs, small
fluctuations in emission current from the probed region can be

281

J. Jortner and B. Pullman (eds.), Tunneling, 281–295

observed, corresponding to fluctuations in the number of adsorbed atoms
diffusing into and out of the probed region. Measurements may be made
of the time autocorrelation function f_i of the current fluctuations
which are simply related to the fluctuation in the number $\delta n(t)$ of
adsorbed atoms in the probed region. For a circular probed region of
radius r_0 and area A a natural time scale is provided by

$$\tau_0 = r_0^2/4D. \tag{1}$$

The number autocorrelation function f_n is given by

$$f_n(t) = \langle \delta n(0)\delta n(t) \rangle. = \frac{\langle (\delta n)^2 \rangle}{A} \int_A d^2\vec{r} \int_A d^2\vec{r}\,' \, \frac{\exp[-|\vec{r}-\vec{r}\,'|^2/4Dt]}{4\pi Dt} \tag{2}$$

and can easily be expressed in units of t/τ_0. This is equally true of
the actually measured quantity $f_i(t) = \langle \delta i(0)\delta i(t) \rangle / \langle i \rangle^2$ with $\delta i(t)$ the
tunneling current, so that a comparison of the experimental function
with the theoretical one yields τ_0 and thus, to the accuracy of the
experiment, in the determination of r_0 and D. The experiments can then
be repeated at different temperatures and coverages.
      If the probed region has the form of a long narrow rectangular
slit of dimensions 2a x 2b with b >> a and if the slit is aligned along
the principal axes of the diffusion tensor (say, with the 2a dimension
parallel to the x direction), it can be shown [3] that the correlation
function decomposes into a product of two one-dimensional ones
$f(t/\tau_x)f(t/\tau_y)$, each with its own relaxation time given by

$$\tau_x = a^2/D_{xx} \tag{3a}$$

$$\tau_y = b^2/D_{yy}, \tag{3b}$$

where D_{xx} is the component of the diffusion tensor along the x
direction. Unless $D_{yy} >> D_{xx}$, the decay of the function $f(t/\tau_x)$ is so
much faster than that of $f(t/\tau_y)$, and consequently only D_{xx} is
measured. Rotation of the slit by 90° then allows measurement of D_{yy}.
      The resolution of the field emission microscope is $\lambda = 30\text{-}40\text{Å}$ [2]
which can be comparable to the narrow dimension of a rectangular probe.
However, this complication can be handled straightforwardly. In
essence, the result is an increase in the effective probe dimensions by
an additive amount $\sim 0.75\lambda$ [4].
      The fluctuation method requires for its implementation fields of
the order of 0.3-0.5 volt/Å. Adsorbates with high polarizability α
and/or high dipole moment μ have their activation energies (or barrier
heights) at these fields differ from those at zero field by

$$\Delta E = (1/2)\alpha(F_s^2 - F_o^2) + \vec{\mu}\cdot(\vec{F}_s - \vec{F}_o), \tag{4}$$

where F_s and F_o are the fields experienced by the ad-particle at saddle points and stable binding sites respectively. The effects for adsorbates like H, O and CO are essentially negligible, particularly on close packed planes where $F_s/F_o \sim 1$. It can also be demonstrated experimentally that the field emission current seems to have no effect on diffusion over at least two orders of magnitude, the experimentally accessible range [5].

RELATION OF D TO MICROSCOPIC QUANTITIES

The diffusion coefficient D, which is measured by the fluctuation method, is the Fick's law, or "chemical", rather than the single atom displacement or tracer diffusion coefficient D^*. D and D^* differ from each other except at zero coverage, and the relation between them is not analytically accessible except for the simplest kind of interaction, i.e., delta function repulsion which produces site exclusion. Direct and substrate mediated adsorbated-adsorbate interactions enter in a complicated way into the determination of the tracer coefficient D^* and in even more complicated ways into D.

Both D and D^* are often forced into the form

$$D = a^2 \nu P, \qquad\qquad (5)$$

where $\underline{a}$ is a mean jump length and νP an effective jump frequency; ν can be thought of as an attempt frequency and P an effective jump probability. In thermally activated diffusion P is a Boltzmann average which is complicated because the actual activation energies for a given jump vary with local adsorbate configuration because of ad-ad interactions. (Incidentally, the same seems to be true of ν as well as for hydrogen on W(110).) In tunneling diffusion P is a similarly weighted tunneling probability. Thus the connection between an overall D and individual atomic events is not as direct as might be hoped. (D^* cannot be measured presently, except on the computer.) An important aid in interpretation is provided by computer simulations, even at the Monte Carlo level, where P and ν can be assigned values based on local ad-ad configurations and interaction energies both in activated and tunneling diffusion. By working backward and experimenting with various interactions in the simulation or by calculating microscopic jump rates theoretically, a fit to experiment can then be attempted.

A simple but quite useful example of this approach is provided by diffusion anisotropy. Ratios of D_{xx}^*/D_{yy}^* can be calculated for any surface structure if the actual jump directions are known and if all ad-ad interactions leave the surface symmetry intact. It then turns out, both from Monte Carlo simulations and from quasi-analytical theory [6] that

$$D_{xx}^{*}/D_{yy}^{*} = D_{xx}/D_{yy}, \tag{6}$$

so that a measurement of anisotropy can often be interpreted straightforwardly

TUNNELING DIFFUSION

We now present experimental results for the diffusion of hydrogen and its isotopes (^{m}H, m = 1,2,3) on the W(110) plane of tungsten [5]. Figure 1 shows plots of log D vs. 1/T for ^{1}H. Similar sets of curves are found for ^{2}H and ^{3}H. The data exhibit two markedly different temperature regimes, a thermally activated region above 140K and a rather temperature independent one between 27K and 140K with a rather sharp crossover between them. Analysis of the data in the thermally activated regime produces activation energies of 5-6 Kcal/mole, depending on coverage and very slightly on mass. This section concentrates on the temperature independent regime, while the following one considers some aspects of the thermally activated range in order to provide an explanation of the low coverage data ($\theta \leq 0.1$) for both regions using the same model of the hydrogen-tungsten lattice interactions and their effects on the diffusive processes.

The temperature independent regimes cover a wide range of T in which D is also sensitively to coverage. Figure 2 shows log D vs. coverage θ for all three isotopes at 27K. θ is deduced from work function measurements and assumes the change in work function Δφ to vary linearly with θ. It is also believed that θ = 1 corresponds to a ratio H/W = 1; this is based on relative saturation coverages on the (110) and (100) planes and absolute measurements on (100) [7].

The most striking feature of the temperature independent diffusion data is the relatively small isotope effect seen in tunneling. If we assume that D is approximately given by

$$D = a^{2}\nu P, \tag{7}$$

with the typical values $\underline{a} \sim 2 \times 10^{-8}$ cm, $\nu = 10^{12} - 10^{13}$ sec^{-1} and the conventional tunneling probability

$$P = \exp[-2(2m_{H}/\hbar^{2})^{1/2}\int_{x_{1}}^{x_{2}}(V-E)^{1/2} dx] \tag{8}$$

$$\approx \exp[-2(2m_{H}/\hbar^{2})^{1/2}V_{0}^{1/2}\ell],$$

where V_{0} is barrier height, m_{H} is the hydrogen mass and ℓ an effective barrier length which can be adjusted according to the barrier shape, we are immediately faced with a dilemma, even allowing for the fact that P is an average value (since V_{0} depends on ad-ad interactions and hence on the local environment of the tunneling atom): If the smallness of the isotope effect is to be explained on the basis of the ratio $m_{1} : m_{2} : m_{3} = 1:2:3$, $V_{0}^{1/2}\ell$ must be extremely small, so small in fact

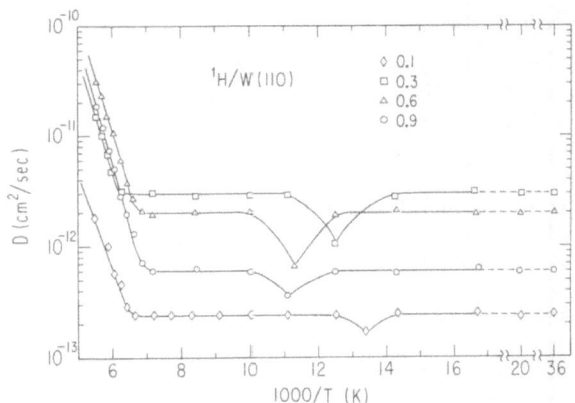

1) Log D vs.1/T for ¹H diffusion on W(110). Relative coverages marked
on figure. The dips in the otherwise temperature independent parts
result from a first order phase transition (from Ref. 5).

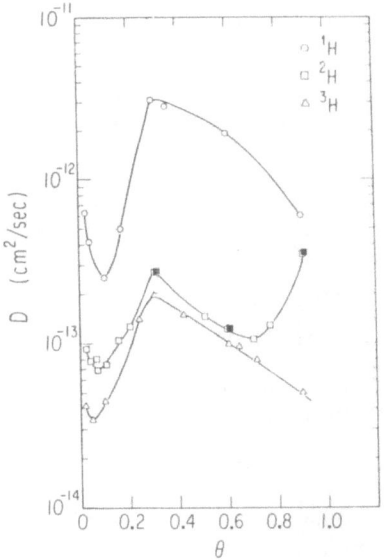

2) Log D vs. θ for ¹H, ²H and ³H at 27K (from Ref. 5).

that D with this predicted isotope effect would be some 10^6 times larger than observed unless $a^2\nu$ were smaller than $\sim 10^{-3}cm^2sec^{-1}$ by a factor of $\sim 10^6$. If V_o is approximately given by the activation energy of thermally activated diffusion, i.e., 5-6 Kcal/mole, there is another problem: ℓ would then have to be a small fraction of an Angstrom, which seems physically unreasonable on the basis of the surface structure discussed below. We are, therefore, forced to assume that there is a contribution which enters in the form of mass renormalization, involving additive mass corrections which arise from lattice distortion by the adsorbate. This distortion accompanies the tunneling H atom and thus leads to increased effective mass. This idea was first suggested in this context by Muttalib and Sethna [8].

Analysis of Figure 2 suggests that even with this approach some difficulties remain, namely the corrections themselves must be slightly mass dependent. Figure 3 plots the product of effective mass M^* in amu times ℓ^2 in $(\text{Å})^2$ vs. θ, assuming no change in V_o. The latter is probably an incorrect assumption, but the approximation serves to give a feeling for the magnitude of the corrections involved. The experimental data provide effective masses of all three isotopes as roughly 10-12 amu (depending on the value of $\ell \leq 0.8$ Å chosen for a triangular barrier). The next section of this paper provides attempts to explain this behavior.

The second point of note involves the coverage dependence, namely the minimum near $\theta = 0.1$ and the maximum near $\theta = 0.3$, as well as the fact that D increases again for ^{2}H at $\theta \geq 0.6$. The initial minimum is not well understood. It may have to do with the asymmetry of forces at low θ which leads to extra mass renormalization. Before describing attempts to explain the maximum near $\theta = 0.3$ and the differences at $\theta \geq 0.6$ between ^{1}H and ^{3}H, on the one hand, and ^{2}H, on the other, some anisotropy results [7] must be presented and analyzed.

Figure 4 shows a picture of the W(110) plane. If the actual jump directions are along the (111) directions labelled x_1 and x_2 in Figure 4, it can readily be seen that D_{110}^*/D_{100}^* (i.e., D_{yy}^*/D_{xx}^*) = 2. Monte Carlo calculations [6] confirm that this is so for D_{yy}/D_{xx} for any set of interactions. This result is in fact observed for oxygen on W(110) [7]. For ^{1}H and ^{2}H, however, (^{3}H was not available in this apparatus) it is found that $D_{110}/D_{100} \sim 1$, even in the tunneling regime at essentially all coverages, although D_{yy}/D_{xx} increases slightly at $\theta \leq 0.1$ [7]. (See Fig. 5.)

Figure 6 again shows the (110) surface with various positions labelled. It seems reasonable that 1 or 5 represent the binding sites of H with probably shallow intervening maxima at locations like 4 and the main diffusion barriers at locations like 2, 6, etc. In order to explain the observed D_{110}/D_{100} ratio, it is therefore necessary to postulate that jumps like 1→7 occur in tunneling diffusion with appreciable probability relative to "ordinary" jumps like 1→5. This suggests that most of the barrier opacity comes from the large

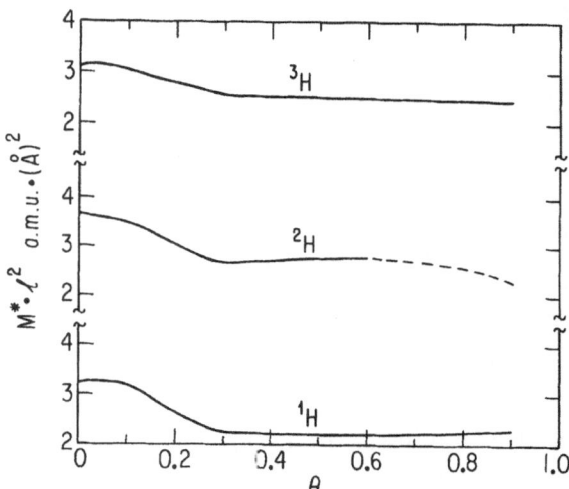

3) Effective mass M^* (in amu) times (barrier length in Å)^2, assuming constant barrier height, as function of θ for ^{1}H, ^{2}H and ^{3}H in the tunneling regime (from Ref. 5).

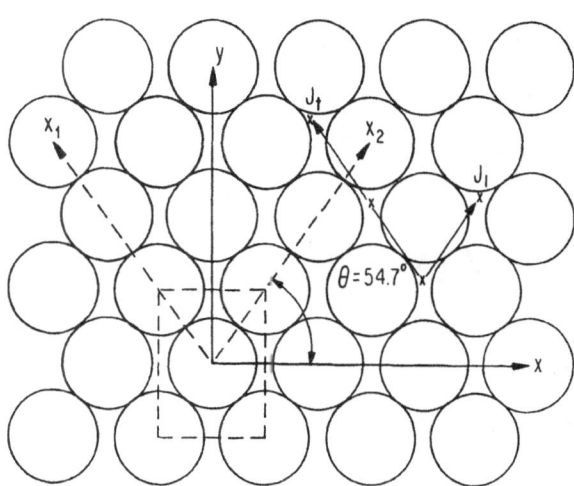

4) Schematic diagram of a (110) plane of a bcc metal. y corresponds to the (110) direction in the plane, x to the (100) direction. x_1 and x_2 correspond to (111) directions and are assumed to represent the actual jump directions in the surface diffusion. Also indicated are the nearest neighbor interaction J_1 and the linear triplet interaction J_t.

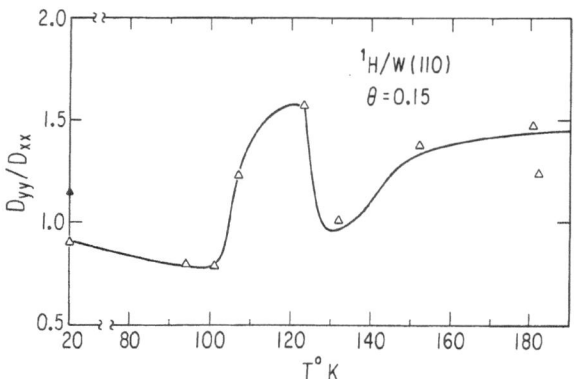

5) D_{yy}/D_{xx} for ^2H at θ = 0.15 as function of temperature (from Ref. 7).

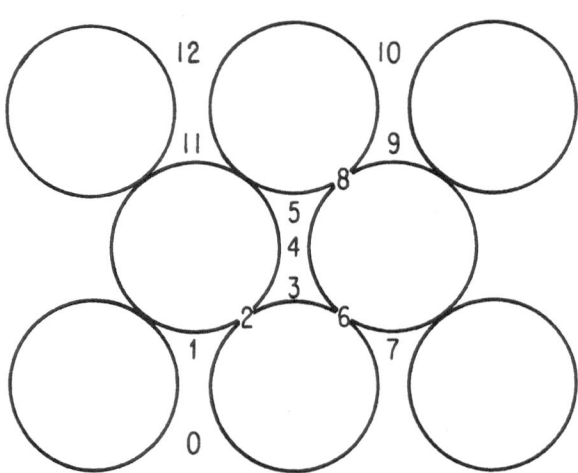

6) Schematic diagram of the (110) plane indicating possible ad-sites and diffusion barriers.

effective mass and that the contributions from integrals over phonon coordinates from events like 1→7 and 1→5 may not be too different.

A possible reason why such events occur for H diffusion is provided by the observation of Estrup [9] that hydrogen adsorption on W(110) shifts the entire top layer of W atoms along the (110) (y in Fig. 4) direction, thus removing the degeneracy of the two ends of the hour-glasses in Figs. 4 and 6 and making sites like say 1, 5 and 7 binding but 3,9, etc., non-binding. This would give a tunneling or thermally activated atom a choice on "reaching" position 3 whether to go to 5 or 7.

We consider next the behavior of ^{2}H at $\theta \geq 0.6$. We hypothesize that the difference between ^{2}H and the other isotopes at high θ has to do with nuclear spin in the following sense. If the motion of bonding electrons is much faster than that of hydrogen nuclei, the exchange of two H atoms involves only the nuclear statistics. Then adsorbed ^{1}H and ^{3}H must be regarded as spin 1/2 fermions, ^{2}H as a spin 1 boson. Since the spin functions are randomly oriented at and above the lowest temperature 27K used, the symmetric spin triplet has a weight of 3/4, and the antisymmetric singlet a weight of 1/4 for ^{1}H and ^{3}H. Thus the space part of the wave function is antisymmetric in 3 out of 4 encounters on average in order to obey Fermi statistics. For ^{2}H, on the other hand, similar reasoning shows that the space part of the wave function is symmetric in 2 out of 3 encounters. Thus two ^{1}H or two ^{3}H atoms can approach each other closely only in 1 out of 4 encounters, while two ^{2}H atoms can do so in 2 out of 3 cases. We now assume that tunneling from 1 to 7 is possible if site 5 is occupied, but that there is sufficient amplitude at 3 from the atom at 5 to affect the tunneling probability of an atom originally at 1. This modulation is assumed to take the form of allowing a 1→7 event only 1 out of 4 times for ^{1}H or ^{3}H, but 2 out of 3 times for ^{2}H.

In order to prevent D_{100}/D_{110} from exceeding unity when 1→7 events are allowed, it is additionally necessary to postulate that concerted double tunneling events occur with finite probability. Such events are assumed to consist of an initial configuration with sites 1 and 5 filled going to a final one with sites 1 empty and 5 and 10 (or 12) filled. The probability of these events is assumed to be finite relative to single tunneling events because the overall distortions involved in the double jump may be much less than that for two non-concerted events. The effect of the statistics postulated for 1→7 jumps is assumed to hold also for concerted double tunneling, i.e., it is allowed 1 out of 4 times for ^{1}H and ^{3}H and 2 out of 3 times for ^{2}H.

We return now to the maximum in D at $\theta = 0.3$. As indicated by Fig. 3, this could simply be the result of small changes in effective mass, but it seems unlikely that these would lead to maxima for all three isotopes at the same value of θ. It is much more likely that the variation in D with θ is due to variations in barrier height, resulting from the opposing effect of attractive and repulsive H-H interactions.

Monte Carlo simulations have been performed in which the local tunneling probability P is a function of such interactions and in which the postulated quantum statistical effects as well as jumps like 1→7 and 1→12 are included. These simulations are, in fact, able to reproduce at least qualitatively the observed behavior, namely a single maximum for ^{1}H and ^{3}H and an upturn in D for ^{2}H at θ > 0.6 [6]. The interactions are chosen also to reproduce the observed D_{yy}/D_{xx} in the activated and tunneling regimes and to consist of repulsive nearest neighbor J_1 and attractive triplet interactions J_t. If J_1 = 1.6 Kcal/mole and J_t = -2.4 Kcal/mole, the results shown in Fig. 7 are found. These have qualitative rather than quantitative significance; they do not quite reproduce the trend in D_{yy}/D_{xx}, and they lead to a p(1x1) ground state; the latter, however, may not be in accord with the phase diagram of H/W(110) which is very incompletely known [6]. If slightly different values of J_1 and J_t are used, a different phase diagram results, the anisotropy can be matched better, the peak for ^{1}H and ^{3}H at θ = 0.3 can be retained, but the primary maximum for ^{2}H is shifted to θ = 0.6. Thus a quantitative fit to experiment is not possible with the simple models used, but the qualitative agreement suggests that both the 1→7 and double tunneling events as well as the invoked quantum statistical effects seem to be pointing in the right direction.

LATTICE-HYDROGEN INTERACTIONS AND MODELS OF TUNNELING DIFFUSION

The same interactions between the adsorbed hydrogen atom and the substrate tungsten lattice must govern the diffusion dynamics in both the tunneling and thermally activated regions. A consistent theory should, therefore, explain the data in both the tunneling and thermally activated regions. We confine attention to the limit of vanishing coverage where complications due to adatom-adatom interactions and statistics are not present. Our discussion begins with a semiquantitative explanation of some of the striking aspects of the observed data in the thermally activated regime as this provides important constraints on the new interpretation of the tunneling data which follows thereafter.

The thermally activated diffusion constant is in agreement with the Arrhenius form

$$D(T) = A(m)\exp(-\overline{V}/T), \quad \overline{V} = 185 \pm 15\text{meV}, \tag{9}$$

where m = 1, 2 or 3 for ^{m}H and where $\overline{V}$ is only slightly isotope dependent in a fashion consistent with small zero point energy differences. However, the prefactor A exhibits unusual exponential mass dependence which is shown in Fig. 8 to vary rather roughly as

$$A(m) \sim \exp(-30m^{-1/2}) \tag{10}$$

and which leads to the enormous inverse isotope effect wherein A(3) is approximately five orders of magnitude larger than A(1). Clearly, such a huge isotope effect must have quantum mechanical origins. [10]

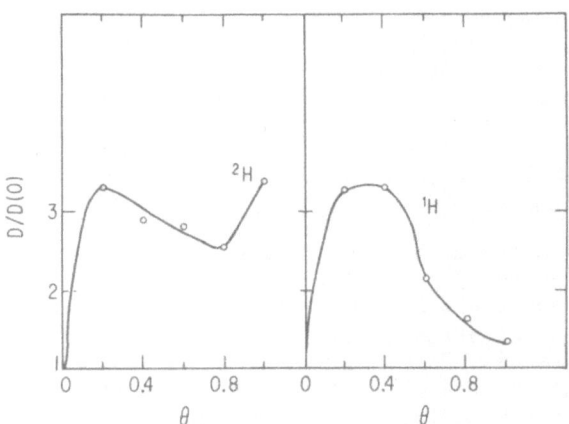

7) Monte Carlo results for D vs. θ in the tunneling regime for ^{1}H (and ^{3}H) and for ^{2}H, using the interactions and postulated events discussed in the text (from Ref. 6).

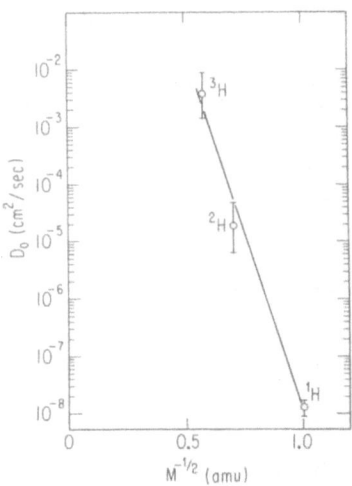

8) Data (extrapolated to θ =0) log D versus $m^{-1/2}$ with straight line designed to guide approximate fit to (11) (from Ref. 10).

An analysis has been made of the dynamics of this thermally activated diffusion by treating the activation process involved in the diffusion as occuring by a series of incoherent jumps between consecutive vibron levels for hydrogen motion in a single well [10]. This model describes a single jump as arising when energy is transfered from a highly excited local phonon mode of frequency ω_ℓ to the hydrogen adatom which is thereby promoted by a vibronic quantum of energy $\hbar\,\omega_H \gg \hbar\,\omega_\ell$ with ^{m}H having the frequency $\omega_{m_H} = \omega_H m^{-1/2}$. The great disparity between ω_H and ω_ℓ leads naturally to the use of an adiabatic approximation where the hydrogen motion occurs adiabatically as a function of fixed substrate lattice position q. The vibron energy $E_n(q)$ provides a contribution to the potential energy determining the slower lattice dynamics. The characteristic timescale for an activation step between the nth and n+1st vibron levels depends on the matrix element of the nonadiabatic hydrogen-tungsten coupling and on a lattice overlap factor for the local phonon mode.

This type of model calculation yields [10]

$$A^f \propto \omega_H \ell^2 (\frac{\omega_H}{\omega_\ell})^{3/2} \exp\{-X_{01} - m^{-1/2}(\frac{\omega_H}{\omega_\ell})\,(\ln[\frac{m^{-1/2}}{X_{01}}(\frac{\omega_H}{\omega_\ell})]-1)\}, \quad (11)$$

where ℓ is the intersite distance $\sim 2\text{Å}$, m is the isotope mass in amu, and the parameter X_{01} is discussed below. (We neglect the weak m dependence in the prefactor.) The ratio ω_H/ω_ℓ equals the number of phonons needed to promote the hydrogen to a higher vibronic state, and the ratio is also measure of the degree of adiabaticty or separation of time-scales which underlies this analysis.

It is natural to assume that the vibron and local phonon motions have effective masses which are, respectively, proportional to that of hydrogen (m_H) and tungsten (m_W). Since values of ω_H/ω_ℓ deduced by comparing (11) with the experimental data yield values which are within an order of magnitude of the reduced mass ratio $(m_W/m_H)^{1/2} = 13.5$, this strongly suggests that pairwise W-W and H-W force constants of the local and vibron modes, respectively, are of similar orders of magnitudes.

The parameter X_{01} in (11) provides a measure of the quantum nature of the hydrogen-lattice forces, and it is given by [10]

$$X_{01} = (m_W \omega_\ell/2\hbar)\,(\bar{q}^{(1)} - \bar{q}^{(0)})^2 \qquad (12)$$

where $\bar{q}^{(n)}$ is the position of the minimum of the adiabatic potential for the local mode with the vibron in the nth state. If the vibron contribution to this potential is $[n + (1/2)]\hbar\,\omega_H$, then X_{01} is proportional to m^{-1} [10]. Comparison of (11) with experimental data provides the rough estimate that X_{01} O(1). The fact that X_{01} is non-zero allows us to anticipate that the adiabatic ^{m}H-W local mode potentials are isotope dependent! This same kind of isotope dependence appears also to be required in order to explain the low temperature data.

The low temperature diffusion coefficient is dominated by
intersite quantum tunneling. The lowest temperature of 27K is still
very high compared to the measured inverse hopping time (D/ℓ^2). The
energy associated with this hopping time is $\hbar \, D\ell^{-2} \approx 10^{-9}$eV, making it
appear impossible to justify a coherent band description, such as that
for electrons in metals, for the motion of the hydrogens. Thus, we are
led to assume that low energy inelastic processes of energy $\Delta \ll 27$K
destroy the phase coherence between consecutive hops. Such processes
are easily explained in terms of a coupling of the hydrogen motion to
electron-hole excitation of the metals and to low lying acoustical
phonons.

This model of the hydrogen tunneling implies that D is therefore
described by the small polaron hopping rate. A detailed calculation by
Flynn and Stoneham [11] using Holstein's treatment [12] gives in this
case

$$D = \ell^2 J^2 / \hbar \, \tilde{\Delta} \, , \tag{13}$$

where J is the transition matrix element which in a semiclassical
approximation is [13]

$$J = m^{1/4} \omega_H \left(\frac{2m_H \omega_H}{\pi \hbar \, \ell_T^2} \right)^{1/2} \exp[-m^{1/2} S_H - S_W]. \tag{14}$$

Here S_H is the classical "action" for the hydrogen tunneling a distance
ℓ_T under the potential barrier $V_H(x)$,

$$S_H = \frac{1}{\hbar} \int_{-\ell_T/2}^{\ell_T/2} [2m_H V_H(x)]^{1/2} \, dx, \tag{15}$$

and S_W is the overlap factor, discussed below in (19), which is
associated with the motion that the tungsten atoms undergo due to the
lattice distortion (polaron) that accompanies the hydrogen tunneling.
Using a smooth potential of the form

$$V_H(x) = \overline{V}[(2x/\ell_T)^2 - 1]^2 \tag{16}$$

with values of $\overline{V}$ as in (9) and ℓ_T of order 0.5 - 0.8 Å, the magnitude
of $D(^1H)$ in the limit of zero coverage (Fig. 3) is fitted by the set of
parameters in the range

$$S_H \approx 3.7 \pm 0.6 \quad , \quad S_W \approx 14 \pm 2, \tag{17}$$

$$\hbar \, \omega_H = 100\text{meV} \quad , \quad \text{and } \tilde{\Delta} \approx 0.1\text{-}2.0 \text{ meV},$$

where the large range for $\tilde{\Delta}$ leads to the quoted uncertainty in the
value of S_W which is fit to experiment. However the smaller than
expected isotope effect cannot be explained by the polaron model (14)

unless S_W is allowed to vary with m, so as to partially compensate for the large isotopic variations of $m^{1/2}S_H$. The required mass variation is roughly given by

$$\frac{\partial S_W(m)}{\partial m} \approx -1.5 \pm 0.3 .$$
(18)

Two important conclusions emerge from (17) and (18). The first is that the lattice participation in the tunneling is very large because $S_W \gg S_H$. This has earlier been described by Muttalib and Sethna [8] as a large "mass renormalization" effect, but our analysis here suggests a separation of timescales which leads to a different interpretation. Secondly, the local mode H-W force constants, which determine S_W, are influenced by the quantum motion of the hydrogen. This influence, manifested by the mass dependence of these force constants, is a natural consequence of the adiabatic approximation for the more rapid hydrogen motion.

A qualitative estimate of the magnitude of X_{01} in (12) can be made by using a model with a single linearly coupled phonon-hydrogen potential. Such a model yields

$$S_W = (m_W \omega_\ell / \hbar) \bar{q}^{(0)2} ,$$
(19)

where $\bar{q}^{(0)}$ is the polaron distortion of the lattice mode which minimizes the sum of W-W and H-H potential energies. Equation (12) contains the lattice distortion $\bar{q}^{(n)}$ defined for each adiabatic vibron potential. We assume here that $\bar{q}^{(n)}$ depends solely on the harmonic vibron state, and this provides a relation between the isotope and n-dependence which is given by

$$\bar{q}_m^{(n)} = \bar{q}[(n + \tfrac{1}{2}) \hbar\, \omega_H m^{-1/2}].$$
(20)

(A specific example of such an interaction has been given by Emin [14].) Thus, combining (12), (19) and (20) leads to the estimate of

$$(\frac{\partial S_W}{\partial m})^2 = S_W X_{01} / 4m,$$
(21)

which in turn gives $X_{01} \sim 0.5-2$. The latter is in accord with the estimates provided on the basis of the thermally activated diffusion data. Although (21) depends on a simplified two-dimensional model for the H-W potential, the agreement of the estimates for X_{01} from the activated and thermal data verifies the consistency of our approach and assures that the relation (20) is approximately valid.

In summary, our analysis of the data suggests a sizeable isotope dependence of the adiabatic H-W local mode potentials as well as a large participation of the lattice in the diffusion process in both the low and high temperature regimes.

We are grateful to D. Emin for a useful discussion. This research is supported by MRL(NSF) facilities at the University of Chicago.

References

[1] R. Gomer, Surf. Sci. 38, (1973) 373; G. Mazenko, J.R. Banavar, and R. Gomer, Surf. Sci. 107, (1981) 459.
[2] R. Gomer, Field Emission and Field Ionization, Harvard University Press (1960).
[3] D.R. Bowman, R. Gomer, K. Muttalib and M. Tringides, Surf. Sci. 138, (1984) 581.
[4] R. Gomer and A. Auerbach, Surf. Sci., in press.
[5] S.C. Wang and R. Gomer, J. Chem. Phys. 83, (1985) 4193.
[6] M. Tringides and R. Gomer, Surf. Sci. 166, (1986) 419; M. Tringides and R. Gomer, Surf. Sci. 166, (1986) 440.
[7] M. Tringides and R. Gomer, Surf. Sci. 155, (1985) 254.
[8] K. Muttalib and J. Sethna, Phys. Rev. B32, (1985) 3462.
[9] P. Estrup, Phys. Rev. Lett., in press.
[10] K.F. Freed, J. Chem. Phys. 82, (1985) 5264.
[11] C.P. Flynn, A.M. Stoneham, Phys. Rev. B10, (1970) 3966.
[12] T. Holstein, Ann. Phys. 8, (1959) 343.
[13] A. Auerbach and S. Kivelson, Nucl. Phys. B257, [F514] (1985) 799.
[14] D. Emin, private communication.

LOW TEMPERATURE ATOM DIFFUSION ON SURFACES: TUNNELING AND ENERGY BAND STRUCTURE

K. B. Whaley,*†‡, A. Nitzan† and R. B. Gerber‡
†Department of Chemistry   ‡Department of Physical Chemistry
Tel Aviv University   Hebrew University of Jerusalem
Ramat Aviv 69978   Jerusalem 91904
Israel   Israel

ABSTRACT.   Low temperature diffusion of atomic hydrogen on metal surfaces is discussed in terms of a band model, applicable to diffusion of many interacting particles obeying quantum statistics. The coverage dependence of diffusion of H, D and T on W(110) is found to be consistent with collision limited band propagation at low coverages, within very narrow bands. At higher concentrations the hydrogen-hydrogen interactions affect both energetics and dynamics of the adsorbed atoms and the coverage dependent diffusion only be quali-tatively analyzed in terms of an effective band picture. The usefulness and limitations of a band treatment for heavy particle diffusion are discussed in the light of these results.

1.   INTRODUCTION

Recent experiments by Gomer and coworkers[1,2,3] on the low temperature diffusion of atomic hydrogen and its isotopes on W(110) surfaces have raised a number of intriguing questions about the mobility of tunnel-ing systems. In their experiments, the two dimensional diffusion is observed to be nonactivated below 100-150 K (depending on isotope and coverage), and shows a complex non-monotonic coverage dependence (see Ref. 2 and contribution by R. Gomer in this volume). In addition, 1) the mass dependence is much weaker than the inverse exponential behavior predicted by a simple one dimensional model of tunneling between adjacent sites,[1] and 2) deuterium shows an anomalous increase at high coverages which was suggested to be a consequence of quantum statistics.[1]

Such behavior suggests that diffusion proceeds by a tunneling mechanism at low temperatures for all coverages and raises the possibility of a band motion of the atomic particles. Energy bands are of course always possible when a particle is situated in a periodic potential, to an extent dependent upon the other interactions present. However, because of their narrow width for high mass particles, bands are rarely seen for systems other than electrons. Some evidence does exist for energy bands of helium,[4] the positive muon,[5] and hydrogen,[6] usually at much lower temperatures. Both the

J. Jortner and B. Pullman (eds.), Tunneling, 297–304

mobility of the particles in the band states and the width of the energy bands depend on interparticle and particle-lattice interactions. Here we shall investigate the possibility and consequences of band motion for atomic diffusion, taking the hydrogen/W(110) system as an example. In particular, we analyze the effects of interparticle interactions and quantum statistics on the coverage dependence of diffusion.

The question of a proper theoretical description of low temperature tunneling diffusion at finite concentrations is related to the more general question of tunneling of mutually interacting particles. With the exception of electron motion in metals, where the interaction is strongly shielded, this has not been extensively investigated. Our study of band motion for atomic systems with strong interparticle interactions also provides a starting point for this problem.

2. BAND PROPAGATION MODEL FOR SURFACE DIFFUSION

Our starting point is the two dimensional energy band structure of a single particle in the surface periodic lattic, assumed to be square for simplicity. In the tight binding approximation this is given by

$$\varepsilon(\underline{k}) = \varepsilon^0 - 2 \beta \cos (k_x \ell) - 2 \beta \cos (k_y \ell) \tag{1}$$

where ε° is the local site energy, ℓ is the lattice spacing, and β the resonance integral between adjacent sites, which we evaluate by relating it to the one dimensional semiclassical tunneling rate

$$\beta = 4\left(\frac{\hbar\omega}{2\pi}\right) \exp \left[- \frac{\sqrt{2m}}{\hbar} \int_{x_1}^{x_2} \sqrt{V(x) - \hbar\omega/2} \, dx\right] \tag{2}$$

Here m is the particle mass and ω a vibrational frequency parallel to the surface. At low coverage, or if no interparticle interactions are present, at equilibrium the adsorbate is distributed over the band states according to the quantum distribution function

$$f^0(\varepsilon(\underline{k})) = \frac{1}{e^{(\varepsilon(\underline{k}) - \mu)/kT}\pm 1} \tag{3}$$

which enables the chemical potential μ to be found from the density $n(\theta)$ at given coverage θ. The adsorbate mobility can then be derived using the semiclassical treatment of transport in energy bands in which the time evolution of the non-equilibrium distribution function is given by the Boltzman equation.[7] A semiclassical approximation is naturally appropriate for atomic particles since their wavelength is short relative to the length scale of interactions. When collisons are treated within a relaxation time approximation, the particle mobility γ is readily derived as

$$\underline{\gamma} = \frac{g_s}{n(\theta)} \int \frac{d\underline{k}}{(2\pi)^2} \, \underline{v}(\underline{k})\underline{v}(\underline{k})\tau(\varepsilon(\underline{k})) \left(\frac{-\partial f^0}{\partial\varepsilon}(\varepsilon(\underline{k}))\right) \tag{4}$$

where $\tau(\varepsilon(k))$ is the relaxation time, $\underline{v}(\underline{k})$ the velocity and g_s the spin degeneracy. The mobility $\underline{y}$ is related to the chemical diffusion coefficient $\underline{D}$ by the Stokes-Einstein relation, generalized to quantum distributions as

$$\underline{D} = \underline{y} \left(\frac{1}{n(\theta)} \frac{\partial n(\theta)}{\partial \mu} \right)^{-1} \qquad (5)$$

This is the diffusion corresponding to collision limited band propagation; the effects of collisions are entirely described by the single parameter τ, the relaxation time. The relaxation time approximation implicitly assumes that collisions are random, uncorrelated events which are of short duration relative to the natural lifetimes of the band states. This is valid as long as

$$\Delta\tau \geqslant \hbar \qquad (6)$$

where $\Delta = 8\beta$ is the bandwidth. Several sources of scattering may contribute to τ: i) impurities or defects, ii) particle-phonon interactions, and iii) particle-particle interactions. We neglect impurity scattering since it will not in general affect the coverage dependence of τ. Particle-phonon interactions are much more important but unfortunately depend crucially on both the particle-lattice coupling and the surface phonon spectrum, neither of which are known. Coupling to lattice nuclear motions can produce both static effects such as mass renormalization and dynamic effects resulting in scattering of the band states. The latter usually results in strong temperature dependence[8] and since this is not seen for the hydrogen/W(110) system below 100 K[1-3] we assume that the nuclear lattice distortions cause only a temperature and concentration independent mass renormalization. The relaxation time τ then depends only on the particle-particle interactions, and is derived from the collision integral in the Boltzman equation as

$$\frac{1}{\tau(\varepsilon)} = n(\theta)v(\varepsilon) \int d\phi \, (1 - \cos\phi) \, |F(k, \phi)|^2 \qquad (7)$$

where ϕ is the scattering angle, $F(k, \phi)$ the scattering amplitude at wave vector k and $v(\varepsilon)$ the velocity. Equation (7) assumes isotropic energy surfaces and neglects relative motion of the adsorbate. A more general expression which takes the relative motion into account can be derived from the general Boltzman equation.[7]

From eqs. (4), (5) and (7) the diffusion coefficient is thus seen to depend inversely on concentration $n(\theta)$ for narrow bands and also implicitly on the quantum statistics by virtue of $f^0(\varepsilon(\underline{k}))$ and $F(k, \phi)$. We expect that this band picture based on independent particle states with binary interparticle collisions providing a finite relaxation time, and hence diffusive motion, will be valid at low coverages but become less so as θ increases. Not only is the criterion (6) invalidated as collisions further reduce τ, but the interparticle interactions induce fluctuations in the local site energies and remove the lattice periodicity. Thus we expect breakdown of the band picture above a certain coverage and a transition from extended to localized

states. Both the location of this transition and the nature of the
nonactivated transport at higher coverages are difficult properties to
study in these systems where all particles are tunneling simultaneously.

To investigate the concentration dependence of diffusion at
higher concentrations we use two approaches. In the first, the band
structure is retained and the effects of interactions on the energy
levels is treated approximately by replacing the periodic potential
$V(x)$ in eq. (2) by an effective potential $\bar{V}(x)$ which contains a static
average of the particle-particle interactions in addition to the
underlying periodic term. In this mean field approximation the
effective band structure is still constructed from independent
particle states, and the effect of higher concentration is merely to
modify the effective potential $\bar{V}(x)$. The second approach is to
consider diffusion as occurring by a series of disconnected tunneling
events between adjacent sites which fluctuate in energy, i.e.,
coherence is destroyed between successive tunneling events. We define
an averaged transition rate between adjacent sites by

$$W = \int P(\delta)W(\delta)\,d\delta \qquad (8)$$

where δ is the energy level asymmetry arising from fluctuations, $P(\delta)$
a distribution of asymmetries and $W(\delta)$ the tunneling rate for a
particle in an asymmetric minimum. Then the (isotropic) diffusion
coefficient in two dimensions is given by

$$D = \tfrac{1}{4}\,\ell^2 W \qquad (9)$$

Both of these approaches are qualitative and represent the
particle-particle interactions in highly averaged ways, but both can
qualitatively represent the concentration dependence at higher
coverages when long range interparticle interactions are taken into
account. We now apply the band model to the low temperature diffusion
of hydrogen isotopes on W(110).

3.   APPLICATION TO LOW TEMPERATURE DIFFUSION OF H, D AND T ON W(110)

When strongly chemisorbed, adsorbed hydrogen may be regarded as
effectively ionized and calculations based on jellium models[9] then
give screened oscillatory long range interactions between adsorbed
hydrogen atoms. Such a potential appropriate to hydrogen adsorbed on
W(110) is shown in Figure 1. This is used to construct an additional
effective potential $\bar{V}(x)$ for tunneling of a hydrogen atom between two
sites when surrounded by other hydrogens, which modifies the band
width.[7] For simplicity, however, the scattering amplitudes $F(k,\phi)$
(eq.(7)) which are not very sensitive to the form of the potential,
are calculated with an effective hard core repulsive potential.
Additional modification of the band model at high coverages was made,
namely introduction of a factor $(1-\theta)$ in the density of states (eq.
(4)) to account for the decrease in density of delocalized states

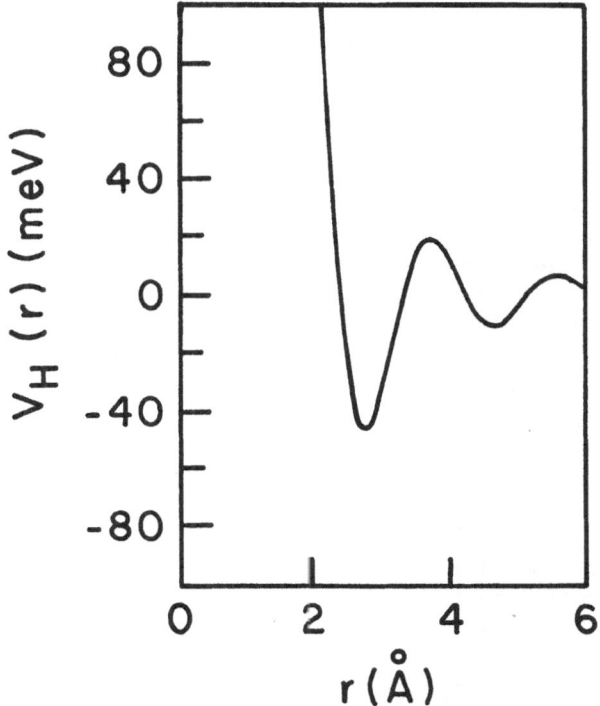

Figure 1.   Interaction potential $V_H(r)$ between two chemisorbed hydrogen atoms on tungsten.[7]

resulting from lack of multiple occupation of states as a result of short range hydrogen-hydrogen repulsions.

The diffusion coefficient ($D=D_{xx}=D_{yy}$) is then calculated for the hydrogen isotopes as a function of coverage θ, with the parameters shown in Table I. The mass renormalization is chosen to fit the diffusion coefficient for all three isotopes at low coverage to the experimental values [2]. An additive renormalization of $m^*=5.5$ amu is found, in good agreement with the path integral estimate of Muttalib and Sethna [10]. Very narrow band widths result, e.g. $\Delta \sim 10^{-9}$ meV for ^{1}H. One consequence of this is that since $\theta \ll kT$, all quantum statistical factors cancel and no difference is seen for bose-einstein or fermi-dirac distributions, eq. (3). This is appparent in Figure 2 which shows calculated $D(\theta)$ for all three isotopes with ^{2}H treated as a boson of spin 1 and ^{1}H, ^{3}H as fermions of spin $1/2$. This result compares very well with the experimental coverage dependence[2,3], the only feature not reproduced being the increase for ^{2}H at high coverage. At low concentrations we see the inverse dependence on concentration expected from the relaxation time τ of eq. (7). The same result is found when relative motion of the colliding hydrogens is included in τ.[7] As the concentration increases, however, a minimum is reached, after which $D(\theta)$ rises to a maximum at $\theta=0.3$. This is a consequence of the long range oscillatory hydrogen-hydrogen interaction

TABLE I. Parameters for the resonance integral β at zero coverage, eq. (2), calculated for a rectangular barrier. A more realistic barrier function is used in ref. 7.

frequency (^{1}H)	ω=95 meV
barrier height	V_B=300 meV
barrier width	x_B=0.95 Å
lattice spacing	ℓ=2.65 Å
hard core radius	a=1.5 Å

band width Δ(meV) with and without mass renormalization m^* (amu)

	$m^*=0$	$m^*=5.5$
Δ(^{1}H)	1.35 (-2)	1.19 (-9)
Δ(^{2}H)	8.34 (-4)	5.35 (-11)
Δ(^{3}H)	1.79 (-6)	4.51(-12)

(Figure 1): in our mean field treatment[7] the attractive component of $V_H(r)$ causes a reduction in the effective barrier potential $\bar{V}(x)$ and hence an increase in bandwidth. The interactions were taken to be additive at low coverage and saturated at θ=0.3, in a simple approximate representation of the non-linearity of these interactions.[7] Beyond saturation of $\bar{V}(x)$ at θ=0.3, D(θ) again decreases, due to both concentration dependence of τ and the empirical modification of the density of states by (1-θ).

Figure 2 shows that all the coverage dependence of D(θ) except the increase for ^{2}H at θ>0.7 may be qualitatively understood within a tunneling model based on band propagation. For θ<0.1 where the long range interaction is unimportant and $\bar{V}(x) \sim V(x)$ we expect this to be also quantitatively correct. At higer coverages although the concept of an effective periodic potential is not very appropriate, we expect the behavior of D(θ) to be the same, if it is determined essentially by the hydrogen-hydrogen interactions. A simple estimate based on the tunneling rate between localized sites, eq. (8), assuming that both the width of the asymmetry distribution P(δ) and the tunneling rate W(δ) increase with coverage, also shows a rise of D(θ) to a maximum, followed by a monotonic decrease at higher coverages.[7] (W(δ) increases because of the decrease in $\bar{V}(x)$ with coverage.) However, the initial decrease in D(θ) is not reproduced, as was also found in Monte Carlo simulations of tunneling diffusion which employed tunneling between localized sites.[11] This feature appears to be only explicable in terms of the band model.

4.   ENERGY BANDS IN ATOMIC DIFFUSION

We have seen that the coverage dependence of low temperature diffusion of atomic hydrogen on W(110) is consistent with a collision limited band propagation at low concentrations, while for higher coverage the

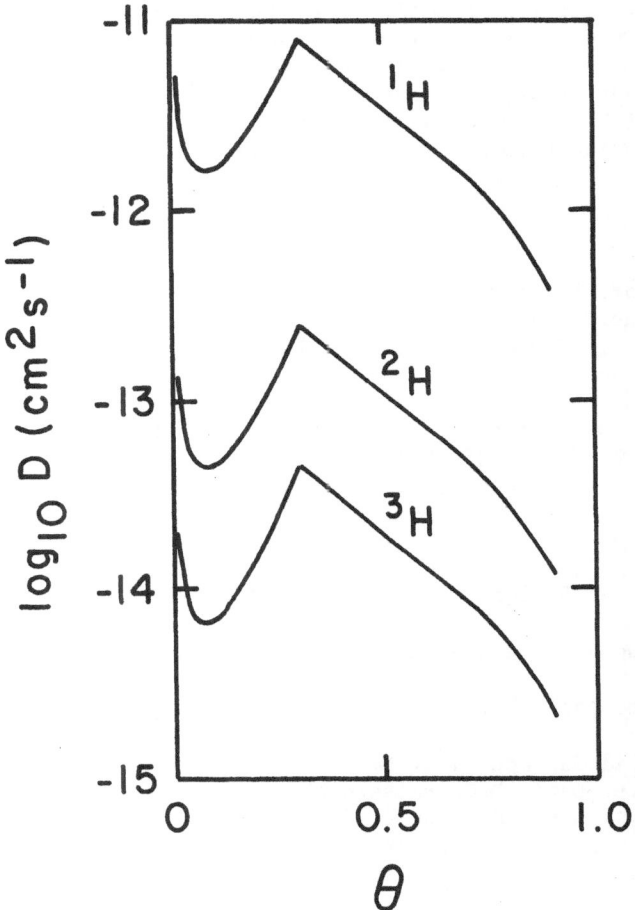

Figure 2.   Diffusion coefficient vs. coverage θ for ^{1}H, ^{2}H and ^{3}H on W(110) calculated in the energy band model.

behavior is explicable in terms of hydrogen interactions mediated by the metal electron distribution, within either an effective band or a hopping tunneling model of the atomic motion. The resulting energy bands for H/W(110) appear to be very narrow due to mass renormalization. This raises a serious question of consistency, namely whether inelastic scattering by phonons should not destroy the bands at temperatures near 100K. Lack of temperature dependence suggests that inelastic phonon scattering is weak or absent, but in addition the lack of intraband and interband transitions induced by phonons may also be a consequence of the low phonon density of states for the former, and a high interband gap for the latter. Finally, at low temperatures the interactions would be limited to long wavelength acoustic phonons whose wave length may be too long to destroy coherence within the small domains probed in the H/W(110) experiments (~100 Å).[1]

These results clearly show the usefulness and applicability of
the band model for mass transport of atomic particles on periodic
surfaces. To go beyond the limitations of an effective band model
requires a rigorous theory of the tunneling dynamics of many strongly
interacting particles in periodic lattices. This remains a major
problem for diffusion behavior at low temperatures.

ACKNOWLEDGMENTS

    We thank Prof. R. Gomer for stimulating discussions and providing
preprints of unpublished work. This work received support from the
Minerva Gesellschaft Füdie Forschung, FRG and from the Israel National
Academy of Sciences.

✱Current address: Department of Chemistry, University of California
Berkeley, Berkeley, CA 94720, USA.

REFERENCES

1.   R. DiFoggio and R. Gomer, Phys. Rev. B25, 3490 (1982).

2.   S.C. Wang and R. Gomer, J. Chem. Phys. 83, 4193 (1985).

3.   R. Gomer, see contribution in this volume.

4.   Films on Solid Surfaces J.G. Dash, Academic Press 1975, p. 26.
      M. Bretz, J.G. Dash, D.C. Hickernell, E.O. McLean and O.E.
      Vilches, Phys. Rev. A8, 1589 (1973).
      H. Chow and E.D. Thompson, Surf. Sci. 59, 225 (1976).
      M.G. Richards, J. Pope and A. Widom, Phys. Rev. Lett. 29, 708
      (1972).

5.   O. Hartmann, E. Karlsson, L.O. Norlin, D. Richter and T.O.
      Niinikoski, Phys. Rev. Lett. 41 1055 (1978).

6.   K. Christman, R. Behm, G. Ertl, M.A. van Hove and W.H. Weinberg,
      J. Chem. Phys. 70 4168 (1979).
      R.C. Casella, Phys. Rev. B27 5943 (1983)

7.   K.B. Whaley, A. Nitzan and R.B. Gerber, J. Chem. Phys. 84, 5151
      (1986).

8.   C.P. Flynn and A.M. Stoneham, Phys. Rev. B1, 3966 (1970).

9.   K.H. Lau and W. Kohn, Surf. Sci. 75, 69 (1978).
      A.G. Eguiluz, D.A. Campbell, A.A. Maradudin and R.F. Wallis,
      Phys. Rev. B30, 5449 (1984).

10.   K.A. Muttalib and J.P. Sethna, Phys. Rev. B32, 3462 (1985).

11.   M. Tringides and R. Gomer, Surf. Sci. 166, 448 (1986).

AB INITIO MODELS FOR ELECTRON TUNNELLING BETWEEN TRANSITION METAL
COMPLEXES

Marshall D. Newton
Chemistry Department
Brookhaven National Laboratory
Upton, NY 11973 USA

ABSTRACT. Electron transfer matrix elements are calculated and analyzed
for several pairs of transition metal complexes of the type ML_6^{2+}-ML_6^{3+},
chosen so as to allow a comparison of $t_{2g}(\pi)$ vs $e_g(\sigma)$ transfer (M =
Fe, Co, and Ru), and NH_3 vs H_2O ligands, based on an apex-to-apex
orientation of the reactant complexes. The magnitude of the calculated
matrix elements, which span the range from non-adiabatic to strongly
adiabatic coupling, are found to exhibit a pronounced dependence on
transfer type ($e_g > t_{2g}$) and ligand type ($NH_3 > H_2O$), an effect
which is correlated with the degree of mixing of the metal and ligand
orbitals, using a simple LCAO-type analysis of the ab initio molecular
orbitals.

1. INTRODUCTION

The mechanism of electron transfer processes associated with transition
metal complexes has been the object of intensive study in recent years
[1,2]. The crucial elements of these mechanistic analyses have included
quantum mechanical aspects of the various electronic and nuclear degrees
of freedom, which, for the sake of brevity, we denote, respectively, as
electronic and nuclear tunnelling. In the present study, we focus on
electronic tunnelling for the following class of electron exchange
processes,

$$ML_6^{2+} + ML_6^{3+} \rightleftharpoons ML_6^{3+} + ML_6^{2+} \qquad (1)$$

where the ML_6 species are octahedral transition metal ion complexes.
Specifically, we wish to understand the manner in which the electronic
transmission factor (i.e., the probability of "reaction" occurring when
the activated reactant system passes through the "crossing region" [2])
depends on the electronic structure of the transition metal redox system
[1-5]. For this purpose we consider two different ligands, H_2O and NH_3,
which by virtue of their highest filled molecular orbitals are facile
$\pi(t_{2g})$ and $\sigma(e_g)$ electron donors, respectively. In addition, both

J. Jortner and B. Pullman (eds.), Tunneling, 305–314.

ligands can interact with the t_{2g} components of the transition metal
d-manifolds via hyperconjugation [5,6]. In the case of the water ligand
we emphasize the difference between the "out-of-plane" π character of
the highest filled orbital, which does not involve the hydrogen atoms,
and the "in-plane" π-character of the orbital which may participate in
hyperconjugative ligand-metal interactions. In terms of the local c_{2v}
symmetry of the water ligand, we label these two types of π-orbitals as
b_1 (out-of-plane) and b_2 (in-plane). In the NH_3 ligand both the sigma
type (or a_1 in local c_{3v} symmetry) and the π-type (or e in local c_{3v}
symmetry) orbitals involve participation of the hydrogen atoms. (The a_1
orbital is primarily a nitrogen lone pair, but also involves appreciable
admixture with the hydrogen s-orbitals.) Since the hydrogen atoms form
the periphery of the hexa-aquo and hexammine ML_6 complexes,
delocalization of the "transferring" electron onto the hydrogen atoms
can have a significant bearing on the strength of the electronic
interaction between redox partners. Thus the b_2 orbital of the water
ligand, and the a_1 and the e orbital of the ammine ligand are expected
to be more important than the b_1 water orbital as contributors to the
overlap between redox partners (and hence to the extent of "electronic
tunnelling"). The systems chosen to exemplify eq (1) include the
exchange of both t_{2g} (high-spin $Fe^{2+/3+}$ and low spin $Ru^{2+/3+}$) and
e_g (high-spin ligand field excited $Fe^{2+/3+}$ and low spin $Co^{2+/3+}$)
electrons [5].

It is convenient to express the electronic transmission factor
$\kappa_{e\ell}$, according to the Landau Zener model, as [2]

$$\kappa_{e\ell} = 2P/(1 + P) \tag{2}$$

$$P = 1 - \exp(-2\pi\gamma) \tag{3}$$

and $\qquad 2\pi\gamma = (H'_{if})^2 (\pi^3/E_\lambda RT)^{1/2}/h\nu_n \tag{4}$

The crucial features of electronic overlap, which dominate the behavior
of $\kappa_{e\ell}$, are contained in H_{if}, the Hamiltonian matrix element
coupling the initial (ψ_i) and final states (ψ_f) in the electron
exchange process:

$$H'_{if} = (H_{if} - S_{if}H_{ii})/(1 - S_{if}^2) \tag{5}$$

where $\qquad H_{if} = \int \psi_i H \psi_f , \quad S_{ij} = \int \psi_i \psi_f$

and $\qquad H_{ii} = \int \psi_i H \psi_i ,$

with H being the total electronic Hamiltonian operator for the system of
interest [2]. The other parameters necessary for specifying $\kappa_{e\ell}$ are
E_λ, the reorganization energy and ν_n, the effective nuclear
frequency associated with the reaction coordinate [2].

Our primary goal is to calculate H'_{if} elements for the various redox couples described above, using appropriate ab initio electronic structure models, and to analyze their magnitudes in terms of basic electronic structural features. In this manner we attempt to achieve a chemical interpretation of the propensity for electronic tunnelling in the representative transition metal systems chosen for study.

2. METHODS AND MODELS

2.1 Model Complexes

The electron transfer matrix elements, H'_{if}, for four different $ML_6^{2+/3+}$ redox couples, were calculated in terms of model $(M-L\cdots L-M)^{5+}$ clusters corresponding to an apex-to-apex transition state geometry [3,5]: i.e., an orientation of reactants (ML_6^{2+} and ML_6^{3+}) is adopted in which one pair of ligands is in close contact in the region between the metal ions (corresponding to $r_{MM} = 7.1-7.3$ Å [3,5]), with the ML distances for each redox couple assigned the common value appropriate for the activated inner-shells [1-5]. Clearly, in the apex-to-apex orientation, the electronic overlap associated with the close-contact ligand pair can be expected to be of great importance, especially that involving the hydrogen atoms, where the closest-contact is established, as noted above.

2.2 Computational Model

The wavefunctions characterizing the reactants (ψ_i) and products (ψ_f) were obtained as spin-restricted ab initio self-consistent field (SCF) valence-level wavefunctions [3,5] (effects of inner shell electrons were simulated through the use of ab initio effective potentials [3,5]). The matrix elements, H'_{if} were then evaluated according to eq (5), using the "corresponding orbital method" [7] to account for effects of non-orthogonality between molecular orbitals of the initial (ψ_i) and final (ψ_f) states. It should be noted that the spatially unconstrained SCF procedure for the $(ML\cdots LM)^{5+}$ complexes of interest, yields charge-localized (i.e., "broken symmetry" [3]) diabatic states corresponding to the desired valence bond structure for the initial ($ML^{2+}\cdots ML^{3+}$) and final $(ML^{3+}\cdots ML^{2+})$ states, even though the two reactants are symmetry-equivalent in the transition state geometry. This result follows from the relatively weak interaction (compared to the scale of typical bond energies) between reactants. An alternative, albeit indirect, route to H'_{if} is to calculate the SCF energies, $E_\pm$ corresponding to the (constrained) symmetrical states [8]

$$\psi_\pm = (\psi_i \pm \psi_f)/(1 \pm S_{if}) \tag{6}$$

The splitting of these two energies, then provides an estimate for H'_{if},

$$H'_{if} = (E_- - E_+)/2 \qquad (7)$$

The calculations outlined above are based on the super-molecule comprising the two inner-shell reactant complexes. The influence of the polarized medium ("outer-sphere") on the electronic structure of the transition state is not included in the model. While the radial extent (and hence the associated H'_{if}) of weakly-bound electrons involved in long-range electron transfer may be appreciably influenced by the medium [9,10], the matrix element (H'_{if}) for short-range transfer of relatively tightly bound electrons of the type considered in the present study is expected to be much less sensitive to the medium polarization. We have previously supported this contention by determining that the effect of the medium in the region of space most critical for the magnitude of H'_{if} (i.e., the overlap region between the two complexes) is quite small, as indicated by the fact that in this region, the electron field is strongly dominated by the contribution from the inner-sphere complexes [11].

By evaluating H'_{if} for r_{ML} values characteristic of the transition state, we include the influence of inner shell reorganization on the electronic wavefunction. Previous studies of Fe hydrates [8] have indicated that the Condon approximation for H'_{if} is quite accurate near the transition state, as far as motion along the inner-shell component of the reaction coordinate is concerned.

2.3 Orbital Models

To provide a basis for analyzing the calculated H'_{if} values, we proceed to develop some 1-electron orbital models.

At the simplest level we consider a single orbital, M (or M') for each reactant,

$$\psi_i = (M_\ell)^n (M'_r)^{n-1} \qquad (8)$$

$$\psi_f = (M'_\ell)^{n-1} (M_r)^n \qquad (9)$$

where the two reactants are distinguished by the subscripts ℓ, r (for left, right), and the redox-process involves changes in occupation numbers, n and n-1 (n = 1 or 2). In this and the following examples we assume an effective 1-electron Hamiltonian, h, and also assume orthogonality between orbitals on different reactants. If orbital relaxation is suppressed ($\bar{M} = M = M'$), we obtain

$$H'_{if} = h_{\bar{M}_\ell \bar{M}_r} \qquad (10)$$

For the case n = 2, we may expect the form of the orbital to depend on oxidation state (M would typically be somewhat expanded relative to M', due to screening of the nuclear change when the orbital is doubly occupied). This would lead to

$$H'_{if} = (S_{MM'})^2 h_{M_\ell M_r} \tag{11}$$

where $\quad S_{MM'} = \int M_i M'_i \leq 1, \quad i \equiv \ell \text{ or } r$

Relative to the constrained case, the dominant effect of relaxation is expected to be an increase in the magnitude of H'_{if}, since any expansion of M relative to $\overline{M}$ will tend to increase the inter-reactant matrix element, $h_{M_\ell M_r}$, an effect partially offset by the overlap integrals, $S_{MM'}$, between the initial and final state orbitals of a given reactant (see related discussion [12]).

Additional relaxation may be anticipated in terms of dependence of ligand-metal mixing on oxidation state. To explore this possibility, we extend the above model to include a (primarily) metal (M) and a (primarily) ligand (L) orbital for each reactant:

$$\psi_i = (M_\ell)^n (L_\ell)^2 (M'_r)^{n-1} (L'_r)^2 \tag{12}$$

$$\psi_f = (M'_\ell)^{n-1} (L'_\ell)^2 (M_r)^n (L_r)^2 \tag{13}$$

If we assume that the relaxation in this case can be accounted for in terms of mixing between M and L (i.e., M' and L' can be expanded in terms of the orthogonal basis, M and L, or _vice versa_), we then have,

$$H'_{if} = (S_{LL'})^2 h_{M'_\ell M'_r} \tag{14}$$

If the higher oxidation state corresponds to relatively greater delocalization of the metal orbital (M') onto the ligand [3,5], then the dominant effect of this relaxation (relative to a fixed set of orbitals, $\overline{M}$, $\overline{L}$) can be expected to be an enhancement of the magnitude of H'_{if}, since $h_{M_\ell M_r}$ should increase in magnitude as the M' orbitals take on more ligand character. Once again, as for eq (11), this enhancement is somewhat offset by the intra-reactant overlap integrals. As additional orbitals, aside from those directly involved in the inter-reactant overlap, begin to participate in the relaxation process, the intra-reactant overlap effects (connecting initial and final state orbitals) will tend to become the dominant factor, leading to a reduction in the H'_{if} magnitude.

The qualitative considerations presented above are in fact borne out in the results of detailed calculations for model systems, and help to explain why H'_{if} values obtained from the delocalized (symmetrical) representations (eqs (6 and 7)) are often smaller than those based on the localized model, where a higher degree of relaxation in the M, L orbitals (or their counterparts) is achieved. Most of the examples cited in ref. [8] fall into this category, although the $[Fe(H_2O)_3]_2$ cluster, based on a face-to-face orientation of reactants, displays the

opposite trend. However, the role of the ligands in the inter-reactant
overlap may not be as important in the face-to-face orientation [5,8] as
it is in the apex-to-apex case, which forms the basis for the above
analyses.

2.4 LCAO Analysis

In order to characterize quantitatively the degree of ligand-metal
mixing in orbitals M, L, etc., we adopt an LCAO-type model, bearing in
mind, of course, that the ligand orbitals are molecular orbitals (MO's),
not atomic orbitals (AO's):

$$M = N_M(\chi_M - \lambda\chi_L) \tag{15}$$

$$L = N_L(\chi_L + \gamma\chi_M), \tag{16}$$

where N_M and N_L are normalization constants [5]. We assume a common
basis of metal (χ_M) and ligand (χ_L) orbitals, and include
relaxation (upon change in oxidation state) through changes in the
linear coefficeints (i.e., M' and L' are obtained by replacing λ and γ
with λ' and γ', respectively). If H'_{if} is dominated by interaction
between the close-contact ligand pair, we expect (based on eq (14)) the
major contribution to arise from the term $(S_{LL'})^2(\lambda')^2(\int \chi_{L_\ell} h \chi_{L_r})$.

While all of the factors in the latter expression may be expected
to vary to some extent with the nature of the redox partners, it is
significant that the results of detailed calculations, as discussed
below, reveal a monotonic dependence between H'_{if} and $(\lambda')^2$ [5]. The
$(S_{LL'})^2$ factors are found to have a modest quantitative influence,
since individual $S_{LL'}$ elements are generally $\geqslant 0.9$ in magnitude [5].
In the case of ML_6 clusters, the orbitals χ_L refer to group ligand
orbitals appropriate to octahedral symmetry [5].

The necessary λ' values have been obtained by casting the
appropriate ab initio MO's for each ML_6^{3+} complex (this is the
high-oxidation state species corresponding to $(M')^{n-1}(L')^2$ in
eqs (12) and (13)) in the form given by eq (15) or (16), with χ_M and
χ_L being defined "in situ" by the particular linear combination of
metal or ligand basis functions, respectively, characteristic of the MO
under consideration [5]. Of course, such a construction, based upon
ab initio MO's expanded in a flexible basis, is not constrained to yield
a common χ_M, χ_L basis for the set of MO's of interest for a
given redox pair (ϕ_M, $\phi_{M'}$, ϕ_L, and $\phi_{L'}$). However, in practice,
the basis functions, χ, implied by the analysis of eqs (15 and 16) do
not vary greatly from one MO to another, and an apparently meaningful
set of covalency parameters (λ, λ', γ, γ') can be obtained in general,
subject to the usual caution, that mixing between orbitals of identical
occupancy is arbitrary in the Hartree-Fock framework, in the sense that
the many-electron wavefunction and total energy are invariant to such
mixing.

We note that for the case n = 1 (see eqs (12) and (13)), where
$\phi_{M'}$, is unoccupied, we obtain λ' from the orthogonal component of
$\phi_{L'}$ (eq 16), according to

$$\lambda' = (\gamma' + S_{ML})/(1 + \gamma'S_{ML}) \tag{17}$$

where $\quad S_{ML} = \int \chi_{L_i}\chi_{M_i}$, $\quad i = \ell$ or r

3. RESULTS AND DISCUSSION

The results of ab initio calculations are presented in Table I, which
includes values of H_{if} and λ', and also various indices of charge
transfer obtained from a population analysis [13]. The four cases
considered here span the range from non-adiabatic to strongly adiabatic
interactions (the switchover in behavior typically [2] occurs for H_{if}'
magnitudes in the vicinity of KT at room temperature (i.e.,
~ 200 cm^{-1})).

The H_{if}' values for the $ML_6^{2+/3+}$ exchange processes are obtained
by scaling the ab initio values for the $(ML \cdots LM)^{5+}$ clusters, using the
form of the group ligand orbitals for the octahedral complexes, as
described in [5]. The calculated λ' values, based on the $\phi_{M'}$
counterpart of eq 15, are in general agreement with the values obtained
previously from redox orbitals defined on the basis of a "corresponding
orbital" analysis [5,7]. The most important feature of the data in
Table I is the strong dependence of H_{if}' magnitude on the nature of the
transferring electron (t_{2g} vs e_g) and the electron donor properties
of the ligand (H_2O vs NH_3). The role of metal-ligand mixing is evident
in terms of the monotonic correlation between H_{if}' and λ' exhibited by
both the model (M-L) complexes and the full octahedral complexes. The
correlation is most meaningful in the former case, where no scaling of
H_{if}' is involved, and a rough proportionality is observed between
(H_{if}') and $(\lambda')^2$, with coefficient ~ 6 ± 1 x 10^3 cm^{-1}, in accord with
the discussion following eqs (15) and (16). It should be emphasized
that the inferred λ' values do not appear to depend strongly on the size
of the complex from which they were obtained, a satisfying result which
supports the utility of the small model clusters in ab initio studies.

Not surprisingly, the calculated λ' (and H_{if}') values are
correlated with the shift (between the 2+ and 3+ charge states) in the
ligand populations associated with the "redox manifold" (i.e., the
occupied t_{2g} or e_g MO's of the ML_6 complex, depending on the
symmetry of the "transferring" electron). However, the total 2+/3+
shift in ligand populations bears no obvious relation to the H_{if}'
values. This situation results from the fact that the dominant ligand
population shift occurs in the e_g manifold for all cases [3],
including those where electron transfer occurs between t_{2g} orbitals.
In the latter cases, the e_g population shifts correspond to additional

TABLE I. Calculated Properties for $ML_x^{2+/3+}$ Redox Pairs
(Apex-to-Apex Orientation)

Complex M/L	Electronic Configuration[a] (2+/3+)	$\phi_{M'}$[b] symmetry	λ' x=6(x=1)	$\|H'_{if}\|$[c] (cm^{-1}) x=6(x=1)	Population Shift[d] in redox manifold ($q_L^{2+} - q_L^{3+}$)
Fe/H$_2$O	(t^4e^2/t^3e^2)	$t_{2g}(b_2)$[e]	0.04(0.06)	12(25)	0.01(0.33)
Ru/NH$_3$	(t^6/t^5)	t_{2g}	0.21(0.11)	16(66)	0.06(0.66)
Fe/H$_2$O	(t^3e^3/t^3e^2)[f]	e_g	0.34(0.25)	1048[g](312)	0.16(0.33)
Co/NH$_3$	(t^4e^2/t^3e^2)[h]	e_g	0.61(0.71)	940[i](2820)	0.61(0.80)

[a] Metal-ion configuration corresponding to nominal 2+ or 3+ oxidation states ($t \equiv t_{2g}$; $e \equiv e_g$).

[b] The MO associated with the "transferring" electron (eq (15), with λ replaced by λ'), similar to the "redox orbital" defined in [5]. The ab initio λ' values are based on ML_x^{3+} complexes, with x = 6 (or 1).

[c] Electron transfer matrix elements, as defined by eq (5). The values in parentheses (x = 1) are the ab initio values obtained from the (ML···LM)$^{5+}$ clusters, while the x = 6 values are obtained from the x = 1 values by scaling factors (1/4, 1/3, or 1/2) based on the form of the group ligand orbitals for the octahedral complexes [5].

[d] q_L^{n+} is the Mulliken population [12] of the six ligands in the ML_6^{n+} complex (based on the ab initio SCF wavefunction), with the primary quantities referring only to population in the "redox manifold" (i.e., t_{2g} or e_g, according to the symmetry of the "transferring" electron, or the b_2 subset of the t_{2g} population in the case of Fe(H$_2$O)$_6^{2+/3+}$), while the parenthetical quantities refer to the total population shifts.

[e] The λ' value refers to the admixture of the b_2 MO of the water ligand with the metal t_{2g} orbital (see the Introduction).

[f] The Fe^{2+} complex is in a ligand-field excited state.

[g] A value of 170 cm^{-1} was estimated semi-empirically for the analogous hexa-aquo Cr$^{2+/3+}$ electron exchange [18].

[h] Similar H'_{if} values were obtained for other Co$^{2+/3+}$ configurations whose self-exchange involves the change in occupation of a single e_g orbital (e.g., t^5e^2/t^5e). The ground state configurations (t^5e^2/t^6) correspond to ψ_i and ψ_f which differ in the occupation of three orbitals, and yield a very small H'_{if} element ($\lesssim$ 1 cm^{-1}).

[i] Extended-Hückel calculations [17] also yield a relatively large H'_{if} value ($\sim$ 1950 cm^{-1}) for this redox pair.

attenuation factors contributing to H'_{if}, analogous to the $S_{LL'}$ factors discussed in connection with eq (14).

For both t_{2g} and e_g electron transfer, we see that NH_3 is the more effective ligand (based on H'_{if} magnitude). Although H_2O is a good π-donor, this involves primarily the high-lying b_1 (out-of-plane) ligand π orbital, in contrast to the lower-lying b_2 (in-plane) π-orbital whose hyperconjugative ability to mix with metal t_{2g} orbitals appears less effective than that exhibited by the e orbitals of the NH_3 ligand.

As a result of more favorable overlap, e_g transfer is found to be considerably more effective than t_{2g} transfer (as measured, once again, by H'_{if} magnitude), especially when coupled to the strong σ-donor capability of NH_3 ligand. The large H'_{if} value calculated for the $Co(NH_3)_6^{2+/3+}$ system has potentially important consequences for the mechanism of the thermal electron exchange reaction [5,14-17]. If the primary pathway is dominated by the ground state electronic configurations, then one must invoke spin-orbit coupling, in which case the large spatial H'_{if} matrix element would help to offset the attenuation due to the spin-orbit factor. On the other hand, if as suggested in a recent study [17], the process involves thermally excited configurations differing in the occupation of a single orbital, then the large H'_{if} value would imply a significant reduction (by $\sim |H'_{if}|$) of the usual inner-shell barrier, which is based on the crossing point of zeroth-order diabatic potential energy surfaces [2,5].

We note that in accord with the discussion in Section 2, the delocalized representation (eqs (6 and 7)) for the $(Co-NH_3\cdots NH_3-Co)^{5+}$ system yields a reduced magnitude for H'_{if} (~ 1400 vs 2820 cm^{-1}), corresponding to the smaller value of λ' (0.34) inferred from an "LCAO" analysis (eq (15)) of the delocalized MO's, which of necessity are intermediate in character between the limits of the fully-relaxed 2+ and 3+ charge states.

4. CONCLUSIONS

The preceeding analysis has demonstrated that the detailed aspects of electronic overlap important for understanding electron tunnelling in the electron transfer reactions of prototype transition metal systems can be quantitatively evaluated and then systematically analyzed in terms of simple orbital models for ligand-metal interactions. A broad range of interactions, from non-adiabatic to strongly adiabatic regimes, has been spanned by various combinations of ligand electron donating propensity and transferring electron orbital type (t_{2g} or e_g). The apex-to-apex orientation is especially suitable for studying such effects in terms of detailed but manageably-sized super-molecule clusters.

5. ACKNOWLEDGEMENT

This research was carried out at Brookhaven National Laboratory under contract DE-AC02-76CH00016 with the U. S. Department of Energy and supported by its Division of Chemical Sciences, Office of Basic Energy Sciences.

6. REFERENCES

1. Faraday Discuss. Chem. Soc. 1982, **74** (entire issue).
2. Newton, M. D.; Sutin, N. Ann. Rev. Phys. Chem. 1984, **35**, 437.
3. Logan, J.; Newton, M. D.; Noell, J. O. Int. J. Quant. Chem., Symp. 1984, **18**, 213.
4. German, E. D. J. Chem. Soc., Faraday Trans. 1 1985, **81**, 1153.
5. Newton, M. D. J. Phys. Chem. 1986, **90** (in press).
6. Wayland, B. B.; Rice, W. L. Inorg. Chem. 1966, **5**, 54; Wayland, B. B.; Rice, W. L. Inorg. Chem. 1967, **6**, 2270.
7. King, H. F.; Stanton, R. E.; Kim, H.; Wyatt, R. E.; Parr, R. G. J. Chem. Phys. 1967, **47**, 1936.
8. Logan, J.; Newton, M. D. J. Chem. Phys. 1983, **78**, 4086.
9. Kuznetsov, A. M., Ref. 1, pp 31 and 49.
10. Kuznetsov, A. M. Chem. Phys. Lett. 1982, **91**, 34.
11. Friedman, H. L. and Newton, M. D., Ref. 1, p 73.
12. Mulliken, R. S. J. Chem. Phys. 1955, **23**, 1833.
13. (a) Schwarz, W. H. E. and Chang, T. C. Int. J. Quant. Chem. Symp. 1976, **10**, 91; (b) Sawatsky, G. A. and Lenselink, A. J. Chem. Phys. 1983, **78**, 3097.
14. Buhks, E.; Bixon, M.; Jortner, J.; Navon, G. Inorg. Chem. 1979, **18**, 2014.
15. Sutin, N. Prog. Inorg. Chem. 1983, **30**, 441.
16. Hammershoi, A.; Geselowitz, D.; Taube, H. Inorg. Chem. 1984, **23**, 979.
17. Larsson, S.; Stahl, K.; Zerner, M. D. Inorg. Chem. 1986, 25 (in press).
18. Hush, N. S. Electrochim. Acta 1968, **13**, 1005.

Electron Transfer Between Self Trapped (Solitary) States

Sighart F. Fischer and Norbert Heider
Technische Universität München
D-8046 Garching
Germany

Abstract. A model for electron transfer is presented. It is based on a
selfconsistent treatment of the initial (final)-state charge distribution
and the equilibrium arrangements for the molecular coordinates. The
theory comprises the nonadiabatic electron transfer and internal conversion processes as special limits within the same frame. It is shown
that the initial state contains a small degree of charge delocalization,
which is a resonance like function of the free energy change. It peaks
when the free energy change matches the nuclear reorganisation energy.
The physical origin for this resonance like enhancement is a non-Condon-
effect. The analysis of exact eigenstates for an asymmetric dimer is used
to interrelate this soliton approach with others based on the small
polaron transformation.

1. Introduction

Intramolecular processes are usually visualized by the use of adiabatic
potential curves. The dynamic description focuses on the nuclear coordinates and some electronic (nonadiabatic) coupling. For nonadiabatic electron transfer processes it is common to identify the initial (final)
state by complete charge localization at the donor (acceptor) [1]. In
this case an electronic coupling term which may be mediated by a bridging
molecule (superexchange coupling) is treated as the perturbation which
causes the transfer. As a simple model one might consider an asymmetric
dimer with the Hamiltonian:

$$(1) \quad H = E_r a_r^+ a_r + E_l a_l^+ a_l + V(a_r^+ a_l + a_l^+ a_r)$$

$$+ \omega b_r^+ b_r + \omega b_l^+ b_l + \omega g(b_r + b_r)a_r^+ a_r + \omega g(b_l + b_l)a_l^+ a_l$$

We set $\hbar = 1$. E_r and E_l denote diagonal energies of the acceptor and the
donor sites. V is the electronic coupling between them. The coupling
constant of the two local modes with the same frequencies is g. The
vibrations are described by the dimensionless creation and destruction
operators b_r^+ and b_r or b_l^+ and b_l, respectively, and the electrons by
the operators a_r^+ and a_l^+. The nonadiabatic electron transfer rate at
zero temperature may be presented as [2]

315

J. Jortner and B. Pullman (eds.), Tunneling, 315–326.
© 1986 by D. Reidel Publishing Company.

$$(2) \quad k^p = 2 \pi V^2 \, 2^n \, \frac{g^{2n}}{\Gamma(n+1)\omega} \, \exp\{-2g^2\}$$

where we used the gamma function $\Gamma(n+1)$ with $n = (E_l - E_r)/\omega$.
The density of states is $1/\omega$. It is well known how to improve this expression in order to get a more consistent treatment of the Franck-Condon
weighted density of states, to incorporate coupling to low frequency
medium modes and to include temperature effects /3/. Here we use the
simplified expression in order to concentrate on the basic theoretical
concepts that lead to such an expression. The interpretation of (2) is
simple. The transferrable electron induces the transition via V and the
nuclei pick up the energy $E_r - E_l$ as vibrational excitation with the
corresponding Franck-Condon factor.

We like to note already here that this simple description is in conflict
with the Born-Oppenheimer (B.O.)-approximation, where the electronic
couplings would lead to partial delocalization of the electron for any
fixed nuclear arrangement. It is also clear that an B. O.-eigenstate cannot serve in general as initial state for the electron transfer process .
For instance for $E_r = E_l$ the system would be symmetric and the electron
distribution of the B.O.-eigenstates would reflect this symmetry. In
this case one needs symmetry breaking in order to get electronic localization .

In this paper we will see that one can get partial localization, which is
consistent with the electronic charge distribution for certain equilibrium configurations. Our approach is closely related to the concept of
self trapping or solitary states /4/. We want to show that this approach
has a series of interesting implications. In order to illustrate the
theoretical concepts we compare the localization properties of exact
eigenstates with the self trapped solitary states and with polaronic
states, which are commonly used for the deriviation of the rate expression (2).

The need for partial delocalization of the initial and the final state is
examplified by the following examples. Let us consider a charge separation within a molecular complex. The resulting radical pair may exist in
its singlet state or may convert into the triplet. The splitting between
the singlet and the triplet state is determined by an exchange integral
J. It can be measured as splitting in an EPR-experiment /5/ or it can be
extracted from the magnetic field effect of the radical pair recombination /6/. The theory has to interrelate J and V . Our investigation can
shed some new light on this interrelation for fast electron transfer processes. This might be of interest in connection with the primary electron
transfer step in photosynthetic reaction centers , where a seemingly discrepancy between theory and experiment has been found /7/.

2. Polaronic states

We start with the description of the transformation of the "small polaron". We construct the unitary operator

$$(3) \qquad U = \exp [g(b_r - b_r^\dagger) \, a_r^\dagger \, a_r + g(b_l - b_l^\dagger) a_l^\dagger \, a_l]$$

and get for the transformed Hamiltonian H :

$$(4) \qquad H = U^\dagger H U = (E_r - g^2\omega) a_r^\dagger a_r + (E_l - g^2\omega) a_l^\dagger a_l$$
$$+ \omega b_r^\dagger b_r + \omega b_l^\dagger b_l + V \exp [g(b_r - b_r^\dagger) - g(b_l - b_l^\dagger)] a_r^\dagger a_l$$
$$+ V \exp [-g(b_r - b_r^\dagger) + g(b_l - b_l^\dagger)] a_l^\dagger a_r$$

If we choose for the electron transfer as initial and final states the states projected on the $a_r^\dagger | \Omega\rangle$ and the $a_l^\dagger | \Omega\rangle$ state, respectively, we obtain in the transformed basis the polaronic states:

$$(5) \qquad | \psi_f^p\rangle = \exp (g(b_r - b_r^\dagger)) \, a_r^\dagger | \Omega\rangle$$

and

$$(6) \qquad | \psi_i^p\rangle = \exp (g(b_l - b_l^\dagger)) \, a_l^\dagger | \Omega\rangle$$

Here $| \Omega\rangle$ denotes the vacuum state. Using the golden rule for the transfer with the density of states $1/\omega$ one gets the rate expression (2).

An effective electronic Hamiltonian can be obtained by taking the vibrational ground state expectation value of (4)

$$(7) \qquad H_{el} = (E_r - g^2\omega) a_r^\dagger a_r + (E_l - g^2\omega) a_l^\dagger a_l$$
$$+ V \exp (-g^2) \, . \, (a_r^\dagger a_l + a_l^\dagger a_r)$$

The effective interaction V_{ef} is the Franck Condon dressed interaction $V \exp (-g^2)$.

We can diagonalize (7) and get the eigenvalues

$$(8) \qquad E_\pm = 1/2(E_r + E_l - 2g^2\omega) \pm 1/2[(E_l - E_r)^2 + 4V^2\exp(-2g^2)]^{1/2}$$

Of special interest to us will be the charge distribution. For instance, the probability of finding the electron on the right side is given for $E_r > E_l$ by :

$$(9) \qquad | \psi_r^p |^2 = [1 - [1 + 4V^2\exp(-2g^2)/(E_r - E_l)^2]^{-1/2}]/2$$

We refer to $|\psi_r{}^p|^2 = \varrho_r{}^p$ as a dynamic charge distribution for the delocalized polaronic ground state. This equation allows the calculation of the half width Δ_H in the charge distribution as function of $E_l - E_r$ for the polaron model . The result can be generalized to resonances at $E_l - E_r = n\omega$ with $n=1,2...$, if one forms the corresponding nondiagonal matrixelements $H_{0,n}$ to give

(10) $\Delta_H(n) = [4/(3n!)]^{1/2} \, V \, \exp(-g^2)(\sqrt{2}g)^n$

In deriving (10) we introduced a transformation to a symmetric and an asymmetric mode (see also chapter 3). The symmetric mode decouples, so that in Eq. (10) only the quantum number n of the asymmtric mode appears.

3. Solitary states

The solitary states are partially delocalized. The amount of localization is determined in a selfconsistent way together with the equilibrium changes of the localized vibrational modes. We choose as trial functions the following form for the initial state mostly localized at the site 1.

(11) $| \psi_i^s \rangle = \exp[-g_r^i(b_r-b_r^+)-g_l^i(b_l-b_l^+)](\varphi_r^i a_r^+| \Omega\rangle+\varphi_l^i a_l^+| \Omega\rangle)$

For the final state we replace the index i by f. $g_r{}^i$ and $g_l{}^i$ are the displacement parameters and $\varphi_r{}^i$ and $\varphi_l{}^i$ are the electronic amplitudes at the right and the left site, respectively. The polaronic state is obtained as the limit $g_r{}^i = \varphi_r{}^i = 0$. We determine the parameters by a variation of the expectation value $\langle\psi_i{}^s| \, H \, | \psi_i{}^s\rangle$.

(12) $\langle \psi_i^s| \, H \, | \psi_i^s\rangle = E_r\varphi_r^{i2}+E_l\varphi_l^{i2} +2V\varphi_l^i\varphi_r^i +2g\omega(g_r^i\varphi_r^{i2}+g_l^i\varphi_l^{i2})+\omega(g_r^{i2}+g_l^{i2})$

The variation with respect to $g_r{}^i$ and $g_l{}^i$ gives

(13) $g_r^i = -g\varphi_r^{i2}$

and

(14) $g_l^i = -g\varphi_l^{i2}$

From the normalization condition

(15) $\varphi_r^{i2}+\varphi_l^{i2} = 1$

it follows

(16) $g_r^i + g_l^i = -g$

If we introduce a symmetric and an asymmetric mode $b_s = 1/\sqrt{2}(b_r + b_l)$ and $b_a = 1/\sqrt{2}(b_r - b_l)$ and use (15) we can rewrite the trial function (11) as

(17) $| \psi_i^s \rangle = \exp[\frac{g}{\sqrt{2}}(b_s - b_s^+) + \frac{g_i^i - g_r^i}{\sqrt{2}}(b_a - b_a^+)](\varphi_r^i a_r^+ + \varphi_i^i a_i^+)| \Omega \rangle$

That means the symmetric mode couples equally to the initial and the final state and it can be factored out. To complete the variational treatment we substitute (13) and (14) as well as (15) in (12) and obtain our final functional H(x) for the variable $\varphi_i{}^i = x$

(18) $H(x) = E_r + (E_l - E_r)x^2 + 2Vx(1-x^2)^{1/2} - g\omega [x^4 + (1-x^2)^2]$

We set $dH(x)/dx = 0$ and obtain as equation for x

(19) $(E_r - E_r)x + V[(1-x^2)^{1/2} - x^2(1-x^2)^{-1/2}] - 2g^2\omega(x^3 - x(1-x^2)) = 0$

The approximate solution for the charge distribution $\varrho_r{}^s = 1 - x^2$ reads for $\varrho_r{}^s \ll 1$

(20) $\varrho_r^s = V^2/[(E_l - E_r - 2g^2\omega)^2 + 4V^2]$

This result differs essentially from the expression (9). For the localized polaronic initial state (Eq.(6)) the static charge distribution to the right site $\varrho_r{}^p$ is zero. Only after diagonalization of H_{el} (Eq. 7) we got a delocalized charge distribution. We referred to this as the dynamic delocalization. It is screened exponentially by $\exp(-g^2)$. The solitary trial functions on the other hand provide a static delocalization $\varrho_r{}^s$ with the unscreened interaction V. It peaks as function of $E_l - E_r$ at $E_l - E_r = 2g^2\omega$. One could obtain also a dynamic contribution to the delocalization within the soliton approach by diagonalizing H within the basis set of $\psi_i{}^s$ and $\psi_f{}^s$. The corresponding resonances in the charge distribution are approximately given by (10).

4. Comparison with exact eigenstates

We introduced polaronic states, which are completely localized and solitary states which have a small degree of delocalization. We now want to analyze exact eigenstates with regard to their delocalization properties in order to understand the delocalization of the solitary states in more detail.

The eigenstates can be expanded in a complete basis set consisting of the oscillator wavefunctions Λ_n for the asymmetric mode and the two electronic states $a_i{}^+| \Omega \rangle$ and $a_r{}^+| \Omega \rangle$, The symmetric mode does not effect the localization properties. We write

$$(21) \quad | \psi^{ex} \rangle = \sum_n C_n^r \Lambda_n^r a_r^\dagger | \Omega \rangle + \sum_n C_n^l \Lambda_n^l a_l^\dagger | \Omega \rangle$$

The C_n^r and C_n^l are the expansion coefficients. The polaronic states (5) and (6) correspond to only one term of a displaced harmonic oscillator. The solitary states consist of two terms of differently displaced oscillators. For the exact states it is convenient to choose undisplaced oscillators. The probability of finding the electron on the right site is given by

$$(22) \quad \varrho_r^{ex} = | \psi_r^{ex} |^2 = \sum_n | C_n^r |^2$$

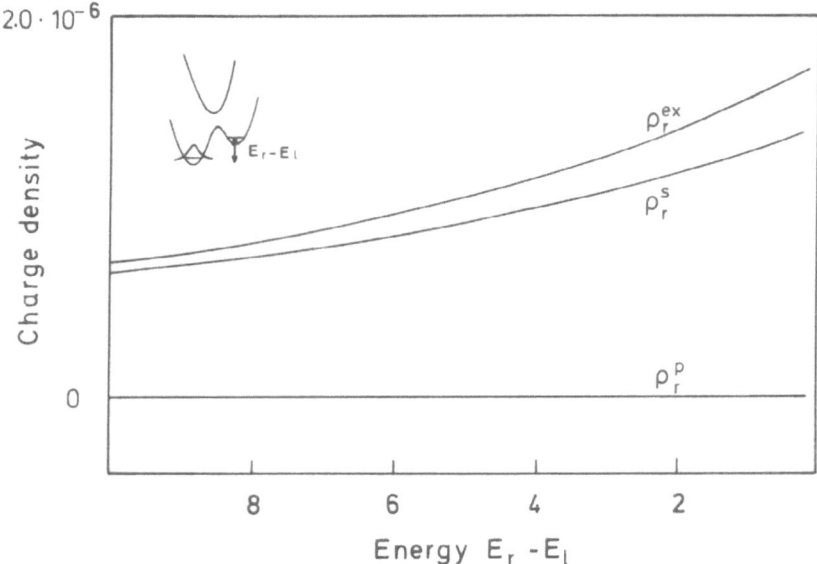

Fig. 1 Charge densities on the right side for the different presented models as a function of the energy separation if the regarded state is mostly localized on the left side. Used parameters $\omega = 1$, $g = 2.9$, $V = 0.02$.

In Fig.1 we show ϱ_r^{ex} as function of $E_r - E_l$ for $E_r - E_l > 10^{-2}$. For comparison also the solitary density $\varrho_r^s = |\varphi_r^s|^2$ is plotted. The delocalized polaronic contribution ϱ_r^p from Eq. (9) is neglegible on that scale. The solitary part ϱ_r^s deviates only little from the exact. It is essentially given by $|C_o^r|^2$. For energy differences comparable to the polaron splitting (8) the exact calculation and the polaron equation agree very well as shown in Fig. 2. The deviation from the exact result (upper part of Fig. 2) is smaller than 1% of the area under the full curve. Due to symmetry the density ϱ_r^p must become 0.5 for $E_l = E_r$. If we define a dynamic solitary charge distribution the density approximates the exact very well over the whole range.

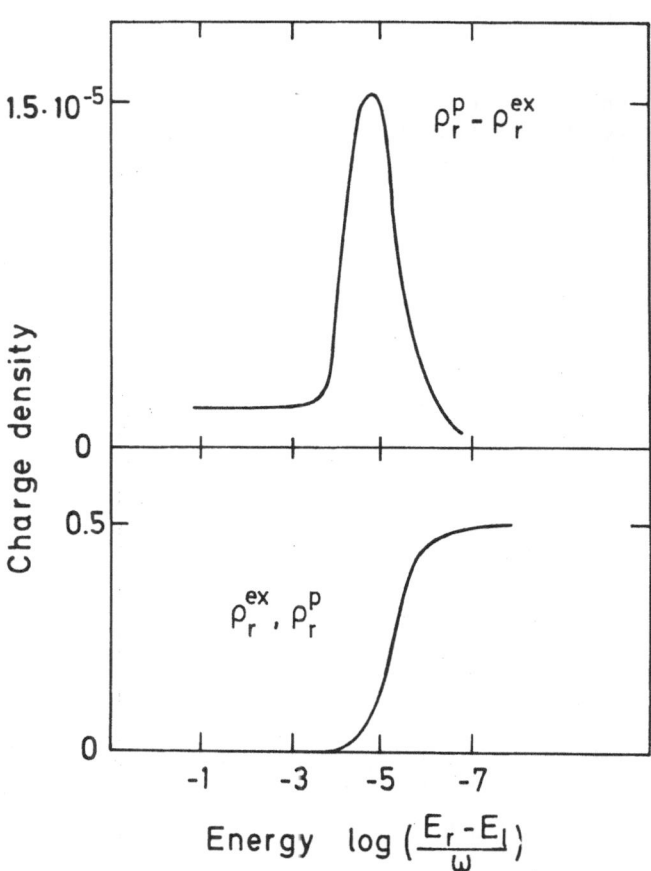

Fig. 2 Comparison of polaron and exact calculation near the position $E_r \approx E_l$. The upper part shows that the difference between polaron and exact charge densities is smaller than 10^{-5}. Same parameters as in Fig. 1.

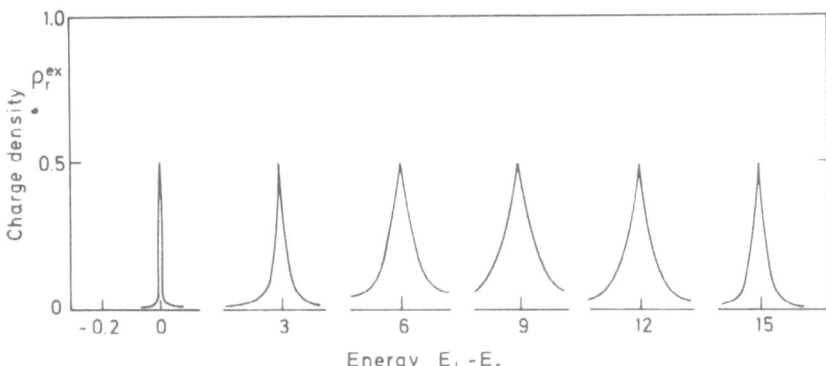

Fig. 3 Charge density on the right side if the energy E_l is greater than E_r. The resonances appear if the interacting vibronic states are nearly degenerated. Used parameters $\omega=1$, $V=0.1$, $g=2.12$

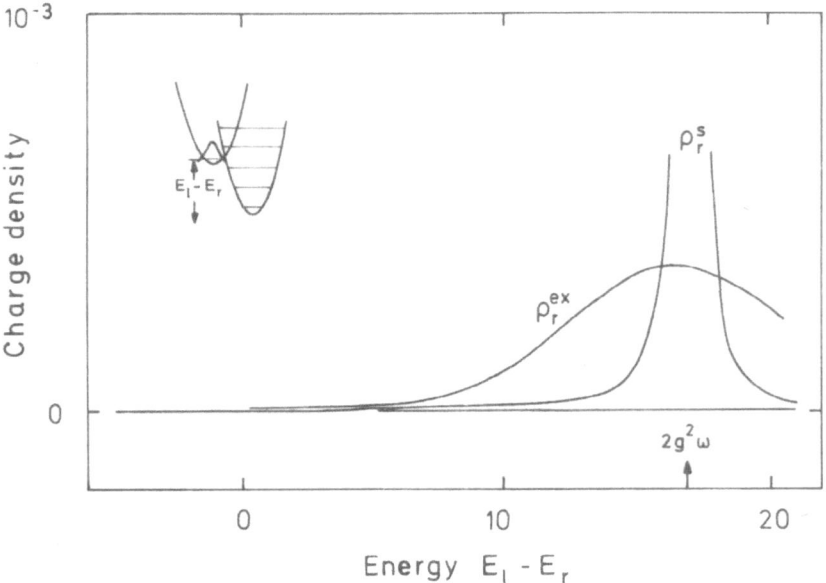

Fig. 4 Comparison of solitary and exact charge distributions. To avoid resonances like in Fig. 3 we used for the exact charge density the values exactly between these resonances. Parameters as in Fig. 1.

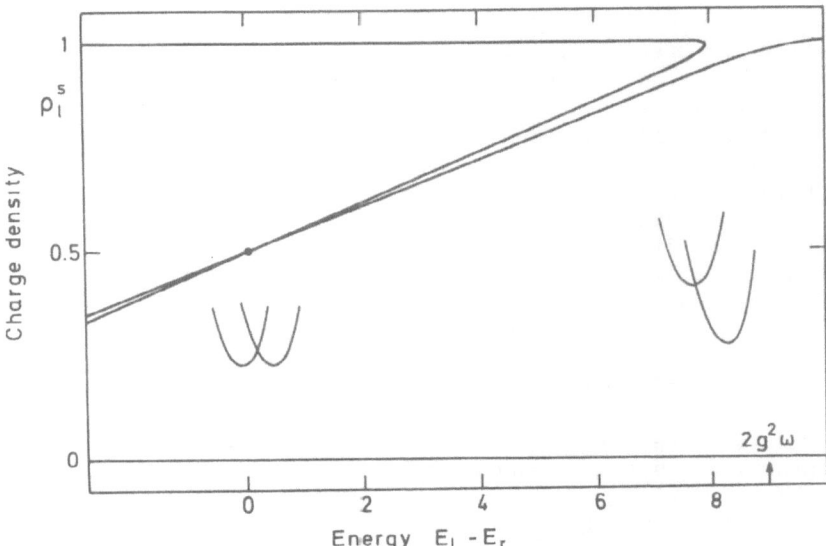

Fig. 5 Exact solution for the charge density of the soliton problem according to Eq. (19). There are two branches which intersect at $\varrho_l{}^s=0.5$. Parameters as in Fig. 3.

For $E_l > E_r$ we expect resonances like delocalization whenever our test state at $E_r + n\omega$ passes through the energy position of E_l. This is shown in Fig. 3. Only every third resonance is shown and the energy spacing between these resonances is not in scale in order to enlarge the width of these resonances. One finds that these widths can be well described by the coupling term (10). That means they have a maximum for $2g^2\omega=E_l-E_r$. In addition we obtain a smooth background which peaks also at this position. In order to compare this background with a corresponding "static" delocalization $\varrho_r{}^{ex}$ of the exact calculation we constructed optimally localized states. They are situated half way between the resonance states. In Fig. 4 we compare $\varrho_r{}^{ex}$ for these optimized resonance states with $\varrho_r{}^s$ using a coupling $2g^2 = 16.8$ and a transfer $V = 0.02$. It is seen that it peaks for $E_l = E_r + 2g^2\omega$ much stronger than $\varrho_r{}^{ex}$. Still the magnitude of the effect is similar. The discrepancy between the exact eigenstate delocalization $\varrho_r{}^{ex}$ and $\varrho_r{}^s$ is largely due to the fact, that the soliton approximation is essentially a B.O.-approximation with the vibronic wavefunction expanded around the minimum of the potential curve. For $E_l - E_r = 2g^2\omega$ this approximation breaks down. The exact states tend to spread the delocalization more strongly over several states. We can interpret the dependence of the electronic delocalization $\varrho_r{}^s$ upon the energy change $E_l - E_r$ as a non-Condon-effect. The solitary states provide a selfconsistent scheme for partial delocalization but they can not account for nonadiabatic corrections. In Fig. 5 we plotted the full solution for $\varrho_l{}^s$ from the self trapping equation (19) as function of $E_l - E_r$. Starting with the electron being localized on the left ($\varrho_r{}^s \ll 1$) one obtains an unstable point at $E_l - E_r \approx 2g^2\omega$. For larger energy differences localization at the left site is only possible if the electronic wavefunction $\psi_i{}^s$ has a node; that means, it must be negative on the right part. The corresponding density $\varrho_l{}^s(E_l - E_r)$ is shown in Fig. 5 as a separate branch intersecting the unstable but continuous branch at $\varrho_l{}^s = 0.5$. There exists a discontinuous transition in $\varrho_l{}^s(E_l - E_r)$ at $E_l - E_r \approx 2g^2\omega$ to the branch which represents an electronically excited charge transfer state. Electron transfer for large free energy changes $E_l - E_r > 2g^2\omega$ has been referred to as the inverted or abnormal electron transfer /3/. We can now identify it as an internal conversion process between an excited and an unexcited (nodeless) charge transfer state.

5. Electron transfer between solitary states

The solitary states (17) incorporate the electronic transfer interaction V. Transitions between such states can only be caused by nuclear fluctuations. We obtain the coupling term from the nondiagonal matrixelement of the Hamiltonian H formed with $\psi_i{}^s$ from (17) and the component of $\psi_f{}^s$, which is perpendicular to $\psi_i{}^s$.

$$(23) \qquad \psi_f{}^{s\,'} = (1 - \langle \psi_i{}^s | \psi_f{}^s \rangle^2)^{-1/2} (\psi_f{}^s - \langle \psi_i{}^s | \psi_f{}^s \rangle \psi_i{}^s)$$

As long as the localization is relatively strong ($\varrho_r{}^i \ll 1$, $\varrho_l{}^f \ll 1$) we obtain for the interaction with $B_a = b_a + b_a{}^+$

$$(24) \qquad \langle \psi_i{}^s | H | \psi_f{}^{s\,'} \rangle = -\sqrt{2} g \omega \varphi_r{}^i [B_a + B_a{}^i]$$

Where $B_a{}^i$ gives the selfconsistently determined equilibrium position of the asymmetric mode B_a :

$$(25) \qquad B_a{}^i = -\frac{g}{\sqrt{2}} (\varphi_l{}^{i\,2} - \varphi_r{}^{i\,2})$$

Expression (24) shows that the electron transfer is promoted by the asymmetric mode and it is proportional to the coupling constant g as well as to the amplitude of the initial solitary state at the acceptor site r.

We can apply again the golden rule formula and evaluate the rate expression to obtain

$$(26) \qquad k^s = \frac{2\pi}{\omega} \; \varphi_r{}^{i\,2} \; (2g^2\omega - n\omega)^2 \; \frac{[\gamma^i - \gamma^f]^{2n}}{\Gamma(n+1)} \; \exp[(-\gamma^i + \gamma^f)^2]$$

with the abbreviations γ^i

$$(27) \qquad \gamma^i = (g_l{}^i - g_r{}^i)/(\sqrt{2})$$

and analogous for γ^f .

In the limit of weak electronic coupling $V \ll g^2\omega$ we get strong localization and can use the approximations $-\gamma^i \approx \gamma^f \approx g/\sqrt{2}$ and substitute for ϱ_r^s the result from (20). To lowest order in $V^2/(E_r-E_l-2g^2\omega)^2$ we obtain the same result as from the polaron approximation (2). For the lowest correction in order of $V^2/(E_r-E_l-2g^2\omega)^2$ the relative rates from (2) and (26) read

$$(28) \qquad \frac{k^s}{k^p} = \frac{(E_r-E_l-2g^2\omega)^2}{(E_r-E_l-2g^2\omega)^2+4V^2}$$

We should keep in mind that the approximations do not hold very well close to the resonance. An exact calculation shows that the deviations from unity in Eq. (28) as a result of the unstability of the solitary wavefunction are small. A resonance shows up if one includes low frequency medium modes and finite temperatures /8/ since then there is no longer such a good cancellation of the promoting mode enhancement and the delocalization.

We like to point out that this approach is not limited to the nonadiabatic weak electronic coupling limit. In the other extreme of large electronic coupling $V^2 \gg (E_l - E_r - 2g^2\omega)^2$ we get the same results as one obtains from the theory of internal conversion. To see this one needs to diagonalize the pure electronic part of H and use the electronic nondiagonal part of the transformed vibrational interaction as perturbation. The change in the equilibrium positions represented by $\gamma^i - \gamma^f$ will be also substancially reduced.

This way one can see that the presented theory applies to all coupling regimes and it agrees in the limiting cases with well known results based on standard perturbation theory.

Acknowledgement: The authors like to thank Prof. M. E. Michel-Beyerle who brought the experiments related to the interrelation between the exchange coupling J and the electronic coupling V to their attention. The work has been supported by the DFG.

References:

/ 1/ For recent review see
      R.A. Marcus, N. Sutin, Biochim. Biophys. Acta, 811, 265 (1985)
      M.D. Newton, N. Sutin, Ann. Rev. Phys. Chem., 35, 437 (1984)
      J. Jortner, J. Am. Chem. Soc., 102, 6676 (1980)
/ 2/ E.W. Schlag, S. Schneider, S.F. Fischer, Ann. Rev. Phys. Chem., 22,
      465 (1971)
/ 3/ R.P. Van Dyne, S.F. Fischer, Chem. Phys., 5, 183 (1974)
      S.F. Fischer, R.P. Van Dyne, Chem. Phys., 26, 9 (1977)
      S. Efrima, M. Bixon, Chem. Phys., 13, 447 (1976)
/ 4/ G. Venzl, S.F. Fischer, Phys. Rev. B, 32, 6437 (1985)
      A.S. Davydov, Physica 3D, 1 (1981)
/ 5/ M.Y. Okamura, R.A. Isaacson, and G. Feher, Biochim. Biophys. Acta,
      546, 394 (1979)
/ 6/ R. Haberkorn, M.E. Michel-Beyerle, Biophys. J., 26, 489 (1979)
/ 7/ R. Haberkorn, M.E. Michel-Beyerle, R.A. Marcus, Proc. Nat. Acad.
      Sci. USA, 76, 4185 (1979)
/ 8/ S.F. Fischer, I. Nußbaum, P.O.J. Scherer in: Antennas and Reaction
      Centers of Photosynthetic Bacteria, 256 (1985), ed. by M.E. Michel-
      Beyerle, Springer, Berlin

ROLE OF STERIC REPULSIONS IN THE ELECTRON TRANSFER BETWEEN FERROUS AND FERRIC IONS IN WATER

Paul Magestro and Neil R. Kestner
Chemistry Department
Louisiana State University
Baton Rouge, LA 70803 USA

ABSTRACT. Using the RWK2 potential to model water-water interactions between two hexahydrated complexes at transition state geometries for the ferrous-ferric reaction, we evaluate the electron transfer rate as a function of separation and angular orientations. Using fixed water geometries and the matrix elements of Newton, the electron transfer occurs primarily around 5.4 A (iron-iron distance). At such distances solvent dynamics are expected to have a minimal influence.

1. INTRODUCTION

The electron transfer process in the precursor state is governed by the magnitude of the matrix element at the most probable distance. That distance is determined by the intermolecular forces involved[1]. In most cases these forces are dominated by the repulsion of the ions and the repulsions of the ligands. In the case of the ferrous-ferric reaction we have the repulsion of plus two and plus three ions and the interactions of the water molecules around the ions. In this first work we will treat the water molecules as fixed at their monomer geometry and unpolarized. These assumptions are not too serious since the interactions of the ligands are primarily repulsive and dominated by the hydrogen hydrogen interactions. Small changes in geometry of the waters should not be too serious.

In many ways this work should be looked upon as an improvement in the interactions used by Tembe, Friedman, and Newton[2]. They used very simple models for the hydration shell but analyzed the entire reaction in much detail. We will concentrate only on the short range interactions and will use all of the matrix elements and other factors from their work. The electronic elements will be taken from that work and other papers of Newton[3], with whom this work originated and will be continued in collaboration.

As part of the early work of one of us, (NRK), along with Mathers and Newton[4], we determined that the best empirical intermolecular potential for water water interactions, especially those in the repulsive regions was the RWK2 potential of Reimer, Watts, and Klein[5]. That multi-parameter potential attempts to treat all regions of the intermolecular separation and yields reasonable dimer as well as fluid properties.

327

J. Jortner and B. Pullman (eds.), Tunneling, 327–332
© *1986 by D. Reidel Publishing Company.*

In this paper we present the thermally averaged interaction energy and the minimum interaction energy for two hexaaquo complexes. This then leads us to further information on the distances of interaction most important in this problem and provides information on the role of steric effects in limiting the available phase space for the reaction.

2. CALCULATIONS

The rate constant of the electron transfer process is dependent on the probability of the donor and acceptor reaching a certain distance and the electronically determined rate based on the matrix elements of the wave functions involved. If we ignore the solvent dynamics and any equilibrium effects leading to the precursor, the rate for a given separation is proportional to

$$k(R) \propto e^{-V(R)/kT} H_{ab}^2 R^2 \qquad (1)$$

where H_{ab} is the appropriate matrix element and we assume that the radial distribution function at short distances can be writtten as a simple exponential. Of course the actual rate constant must be integrated over all separations in order to be compared with experimental data. We shall ignore the variation of the matrix element on the orientiaional angles since it is small[3c] and later will be unimportant in the region of most interest to us, namely close approaches.

Figure 1 illustrates that part of the interaction between two hexaaquo complexes which is most dependent on their angular orientation.The water water interactions are taken as those given by the RWK2 potential. It is composed of three point charges as well as site site repulsions. In this version the bond angles are fixed, although their more recent modifications allow for bond distortion. The long range interactions between the waters include not only the electrostatic terms but dispersion interactions. The irons are assumed to interact via the electrostatic terms. In this paper we do not include the polarization interactions. They will not change the qualitative results. The iron-oxygen distance is fixed at the transition state value estimated by Newton to be 2.06A[6].

The bottom curve is the minimum energy interaction at a given iron iron separation when the complexes were allowed to explore all rotational phase space. It was inferred from a sampling of the results of a direct search optimization that this minimum occurs, at all separations, in a configuration of the pair with their three fold symmmetry axes collinear. This agrees with the prediction of reference 3d based on orbital overlap considerations.

Above this curve is plotted our present state of knowledge concerning the thermal average of the potential at 300K for fixed separations. At separations greater than 6A an estimate of the thermal

average was accessible through sample mean Monte Carlo integration
over the six dimensional angular phase space of the two complexes.
Results were obtained with a 95% confidence limit of less than 5% of
the value based on sample mean variances. From a comparison of the
results obtained with the minimum interaction, it is clear that at
large separations, say 10A or so, the complexes can rotate relatively
freely leading to interactions on the average different from the
lowest one, but with decreasing separation the complexes lose the
ability to change their orientation without drastically altering the
energy. In other words, there are only a few possible orientations
which lead to iron iron separations less than 6A. Just as important,
though, these orientations do exist and the complexes can approach
each other closely.

The determination of the angle averaged potential using Monte
Carlo methods becomes very unstable at separations less than 6A. Two
earlier studies using integration of angle dependent potential
functions chose quadrature methods over Monte Carlo[7],[8], the latter of
these giving the reason that the error analysis was too difficult to
perform. A third[9], presented data with the reservation that the error
estimate given by the Monte Carlo methods had been overly optimistic.
These studies predated the introduction of the two best available
algorithms for speeding convergence of Monte Carlo integrations. The
algorithms of Lepage[10] and Friedman and Wright[11] use respectively,
"importance sampling" and "stratified sampling",two related methods of
finding the optimal distribution of points to reduce the variance.
Recent applications of these methods have involved lower
dimensionality[12], or a more smoothly varying integrand[13],[14],[15] than is
involved here.

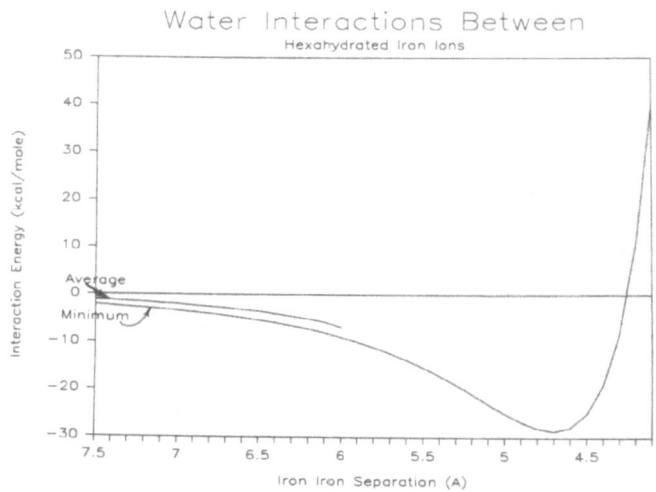

Figure 1. Interaction of two hexahydrated iron complexes
    neglecting the iron iron repulsion and using the RWK2
    potential for the waters

At present, we can not resolve the question of the suitability of
Monte Carlo integration of highly angle dependent, realistic
intermolecular potential functions. Future work will pursue this
point. Qualitative aspects of the model under consideration imply that
near the minimum, the variable over which there is the greatest
orientational freedom is a rotation of the complex about the symmetry
axis directed toward the other ion. We have evaluated the sample mean
average over this variable, with the other angles fixed, and the
results confirm our inference that the thermal average approaches the
minimum energy curve.

In Figure 2 we present the interaction of the irons and of the
waters corrected for the dielectric constant of the water (78.5). The
use of this macroscopic dielectric constant is questionable at these
short distances and needs to be explored further.

Figure 3 compares the relative distance dependent rate constant,
Eq.(1), i.e. that quantity which when integrated over yields the
actual rate constant. It is expressed in relative units, both curves
have their peaks set to unity. This uses the matrix elements from
Logan and Newton (for example, Table I in reference 3b) to arrive at
the relative rates as a function of distances all thermally averaged
for these short distances. It is easily seen that the macroscopic rate
for this step is dominated by iron-iron separations of the order 5.5 A
or higher. It is interesting to compare the rates with and without the
presence of the water interactions. The distances of importance are
much more peaked for the full calculation.

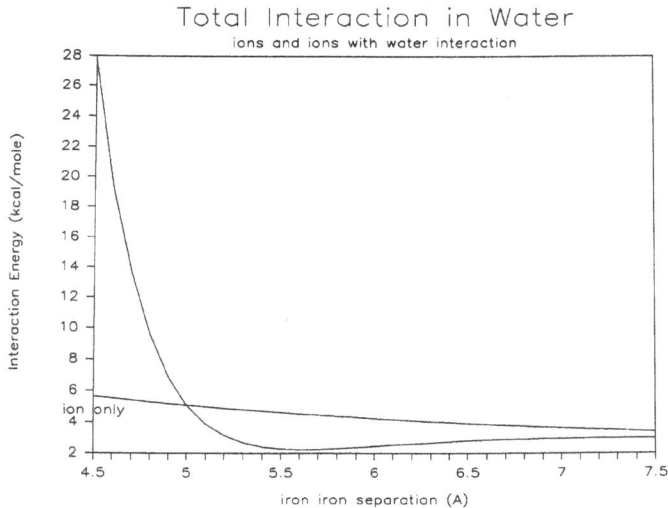

Figure 2. The total interaction compared with the iron
iron interactions in water at 300K, using the
minimum interaction curves for the waters

At shorter distances this is clear since the repulsive interactions cut off the curve. At larger separations we are seeing the more rapid variation of the potential with distance. The curves in Figure 3 are very misleading because of the arbitrary normalization to get them on the same graph. The predicted rates with the water water interactions included are roughly 50 times larger than rates predicted using only iron iron interactions because of the attractive forces or less repulsion between the species. In both cases however, the major effects occur in the range of 5.5A. As Newton and Friedman[2],[3c] have pointed out, for such separations the solvent relaxation rates should not play a major role in determining the overall rate of the reaction.

It is also clear from this work that small iron iron separations can occur as Tembe, Friedman, and Newton[2] inferred from their more elementary model for the water interactions. However it is also clear that the number of configurations of the two species which can lead to such close encounters is also very limited.

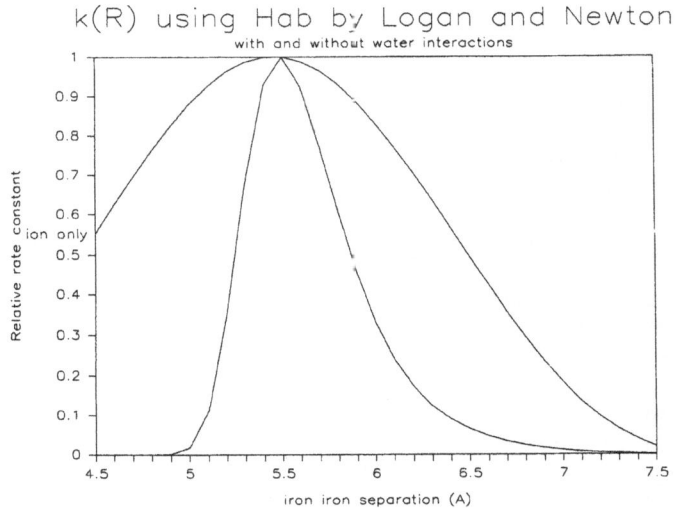

k(R) using Hab by Logan and Newton
with and without water interactions

Figure 3. Relative rates versus separations showing
the effects of the water interactions. The actual
rates when water interactions are included are
increased by a factor of about 50 at the maximum.

To make a direct comparison with the model used by Tembe, et. al.[2] we need to evaluate the entropy term. They assume that corrections for the coordinated water vanish at 6.9 A and follow a switching function down to 4.5 A. From initial comparisons, it is clear that their model includes the general features of this current work but differs in some details.

3. IMPLICATIONS AND FUTURE WORK

This is preliminary work on a system where such steric factors
do not play a major role in determining the course of the reaction and
where solvent dynamics do not play a major role. One reason for this
is the high charge on the ions. We plan to consider the angular
effects and distortions of the water complex further. Some of this has
already been done. We have found that the complex does not distort in
any major way since that would require more energy than one would gain
by increased attractive interactions. More fruitful will be
investigations of less highly charged species where the steric effects
are expected to play a major role in dictating the path and rate of
the reaction, as well as the importance of solvent relaxation.

4. ACKNOWLEDGEMENTS
This work is part of a long term project involving Dr. Marshall
Newton at Brookhaven National Laboratory. The work is supported by DOE
Contract DE-AS05-77ER05399.

REFERENCES.
1. For a general review of electron transfer theory and a survey of
general references see N. Sutin and M.D. Newton, Annual Reviews of
Physical Chemistry,35,437(1984).
2. B.L. Tembe, M.D. Newton, and H.L. Friedman,J. Chem. Phys. 76,1490
(1982).
3a. M. D. Newton,Int. J. Quantum Chem. Symposium 14, 363(1980) as well
as
  b. J. Logan and M.D. Newton, J. Chem. Phys. 78(II),4086(1983) and
  c. H.L. Friedman and M.D.Newton, Faraday Discussion 74,73(1983).
  d. J.A. Jafri,J. Logan, and M.D. Newton, Israel J. Chem.
19,340(1980).
4. N.R. Kestner,M.D. Newton, and T.L. Mathers, Int. J. Quantum
Chem.Symposium 17,431(1983).
5. J.R. Reimer,R.O. Watts, and M.L. Klein, Chem. Phys. 64,95(1982).
6. Discussed in B.S. Brunschwig,J.Logan,M.D. Newton, and N. Sutin,
J.Am.Chem.Soc. 102,5798(1980) as well as earlier Newton papers.
7. D.J. Evans and R.O. Watts, Mol. Phys 28,1233(1974).
8. C.H.J. Johnson and T.H. Spurling, Aust. J. Chem.24,1567(1971).
9. D.J. Evans and R.O. Watts,Mol. Phys.32,995(1976).
10 G. Peter Lepage, J. Comp. Phys. 27,192 (1978).
11. J.H. Friedman and M.H. Wright, ACM Trans. Math. Software
7,76(1981).
12. J.G. Berryman,J.Chem.Phys.82,1459(1985).
13. G.H. Paine and H.A. Scheraga,Biopolymers24,1391(1985).
14. S.C. Faranatos, J.N. Murrell, and J.C.Hajduk, Chem.
Phys.68,109(1982).
15. K.W. Kratky Chem. Phys.57,167(1981).

ROLE OF INTRAMOLECULAR BRIDGES ON ELECTRON TRANSFER
IN DONOR/ACCEPTOR SYSTEMS

H. Heitele and M.E. Michel-Beyerle
Institut für Physikalische und Theoretische Chemie
Technische Universität München
Lichtenbergstr. 4
8046 Garching, FRG

P. Finckh
Institut für Physikalische Biochemie
Ludwig-Maximilians-Universität
Schillerstr. 44
8000 München 2, FRG

W. Rettig
I.N. Stranski-Institut
Technische Universität Berlin
Straße des 17. Juni 112
1000 Berlin 12

ABSTRACT

Intramolecular electron transfer rates for ion formation and recombination in bridged donor/acceptor systems have been determined using
fluorescence lifetime and transient absorption measurements. The data
show a weak dependence of the rate on the donor/acceptor distance as
well as a dependence on the electronic structure of the spacer. These
findings are interpreted as evidence for an electronic coupling between
donor and acceptor through the bridge. The magnitude of the electronic
coupling is estimated on the basis of Extended-Hückel molecular orbital
calculations for model structures simulating the electronic interaction
in the donor/acceptor systems. The calculations can semiquantitatively
explain ratios of rates for different spacers between donor and acceptor. Intramolecular and solvent reorganization energies are derived
from the temperature dependence of the ion formation and the ratio of
the rates for ion formation and recombination.

J. Jortner and B. Pullman (eds.), Tunneling, 333–344

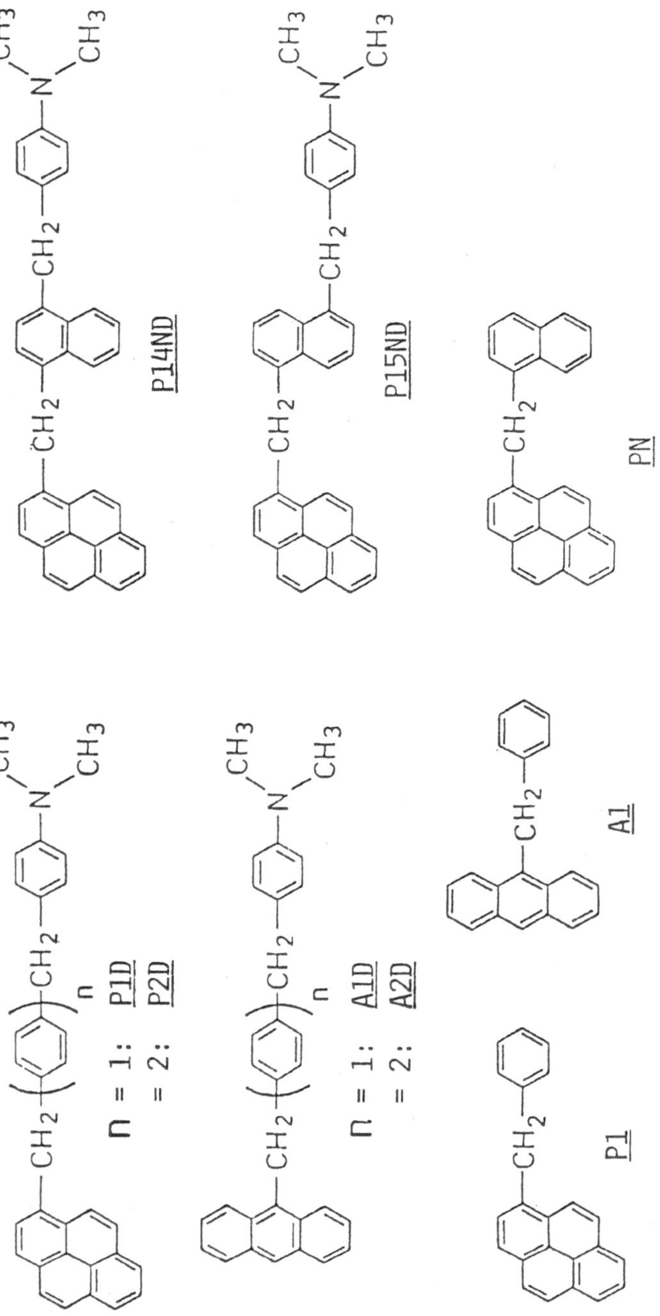

Fig. 1:   List of donor/acceptor systems and reference substances.

1. INTRODUCTION

The X-ray analysis of the photosynthetic reaction center of R.viridis /1/ revealed the arrangement of prosthetic groups which act as electron donors and acceptors in the sequence of primary charge separation reactions. Recent femtosecond, time-resolved studies /2/ showed that the photooxidation of the excited bacteriochlorophyll dimer and the photoreduction of bacteriopheophytin occur simultaneously with a rate of 3 ps. This fast electron transfer between remote donor and acceptor species (center to center distance 17 Å) leads to speculations on the participation of the medium in a superexchange mechanism /3/. Possible candidates supporting superexchange are in principle the proximate accessory bacteriochlorophyll /1/, the phytol chain attached to the bacteriochlorophyll dimer /1/ and the protein environment /4/.

For an experimental test of the predictions of the superexchange approach several bridged donor/acceptor-systems have been synthesized /5/ (Fig.1). In these compounds excited pyrene serves as electron acceptor and dimethylaniline as donor. Donor and acceptor are linked by different aromatic spacers. Al, Pl, and PN are used as reference substances.

Comparing compounds P1D and P2D or A1D and A2D, respectively, the donor acceptor distance is increased by about 4Å. In P1D, P14ND, and P15ND the distances are (almost) the same, but the electronic structures of the spacers are different.

The intramolecular electron transfer rates k in these compounds have been measured and are discussed within nonadiabatic electron transfer theory /6/ using quantum chemical estimates of the electronic matrix element V_{fi} and a simplified calculation of Franck-Condon-factors (FC) which together determine the rate k :

$$k = 2\pi/\hbar \cdot |V_{fi}|^2 \cdot (FC) \tag{1}$$

2. MEASUREMENTS

The kinetic scheme of the photochemical donor/acceptor system is depicted in Fig. 2.

Fig. 2:
Kinetic scheme of the
bridged donor/acceptor
systems (A-S-D).

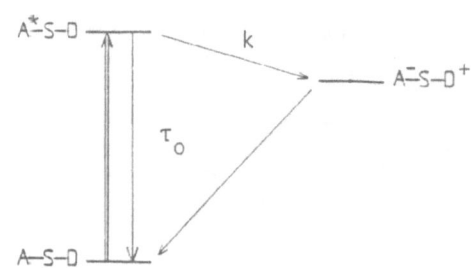

In polar solution the donor transfers an electron to the excited acceptor whereby radical ion pairs are formed. The free energy change of the
reaction in acetonitrile is approximately -0.5 eV /7/. The electron
transfer competes with the fluorescence of the acceptor and reduces its
fluorescence lifetime. By comparison with the fluorescence lifetime
of the reference substance, the electron transfer rate for the ion pair
formation follows from: $k = 1/\tau - 1/\tau_0$. The radical ions recombine
within a few nanoseconds forming the neutral ground states. The free
energy change of this process is about -2.8 eV. The measured fluorescence lifetimes and the corresponding electron transfer rates are
compiled in Table 1:

compd	A1D	A2D	P1D
τ [ns]	0.13 ± 0.03	2.1 ± 0.1	1.2 ± 0.1
k [$10^8 s^{-1}$]	76 ± 18	3.4 ± 0.3	8.3 ± 0.7

compd	P2D	P14ND	P15ND
τ [ns]	28.5 ± 1.5	0.89 ± 0.1	2.1 ± 0.1
k [$10^8 s^{-1}$]	0.30 ± 0.02	11.0 ± 1.3	4.7 ± 0.3

compd	A1	P1	PN
τ_0 [ns]	7.5 ± 0.4	204 ± 7	166 ± 10

Table 1: Fluorescence lifetimes τ, τ_0 in acetonitrile and electron
          transfer rates of ion formation at room temperature.

The recombination rate in compound P1D as determined in transient
absorption experiments is about $5*10^8$ s^{-1} /8/.

For the compounds A1D and P1D the temperature dependence of the fluorescence lifetimes between -83 and +20 °C have been determined. The
measurements were performed in propionitrile because of its lower
freezing point. The results plotted in Fig. 3 show a linear dependence
between $\ln(k*\sqrt{T})$ and $1/T$.

Fig. 3:
Temperature dependence of the
ion formation rate in A1D and
P1D.

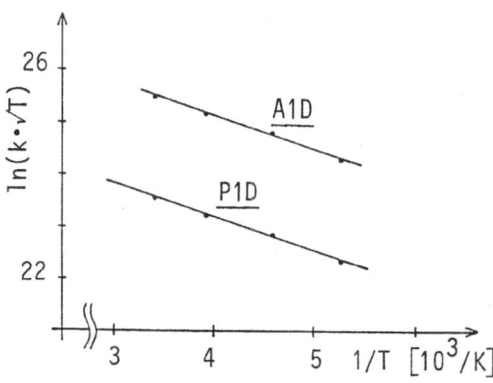

3. CALCULATION OF THE ELECTRONIC MATRIX ELEMENT V_{fi}

For the following discussion two experimental findings are of special interest:

i) There is a surprisingly weak distance dependence of the electron transfer rate. Increasing the distance by 4 Å decreases the rate by a factor of 22-27, which is much less than observed in isotropic media with a statistical distribution of donor and acceptor molecules /9/.

ii) At a fixed distance the rate depends on the electronic structure of the spacer between donor and acceptor.

Both effects indicate that the donor/acceptor interaction responsible for electron transfer is mediated by the spacer.

To evaluate the influence of the spacer on the electronic coupling between donor and acceptor the interaction matrix element V_{fi} was estimated using Extended-Hückel (EH) molecular orbital calculations. To this aim use was made of the relation between the matrix element V_{fi} and the splitting Δ of the adiabatic potential energy surfaces of the system at the intersection of the diabatic potential energy surfaces for reactands and products /10/:

$$\Delta = 2 \cdot |V_{fi}| \tag{2}$$

In a one-electron approximation this corresponds to the splitting of (isoenergetic) donor and acceptor orbitals when their mutual interaction is included in the calculation of the orbital energies. Such a calculation is very complicated for the real donor/acceptor molecules,

as the wavefunctions have to be determined for a configuration corresponding to the intersection region, which is not known. The following
simplified model structures were used instead /11/:

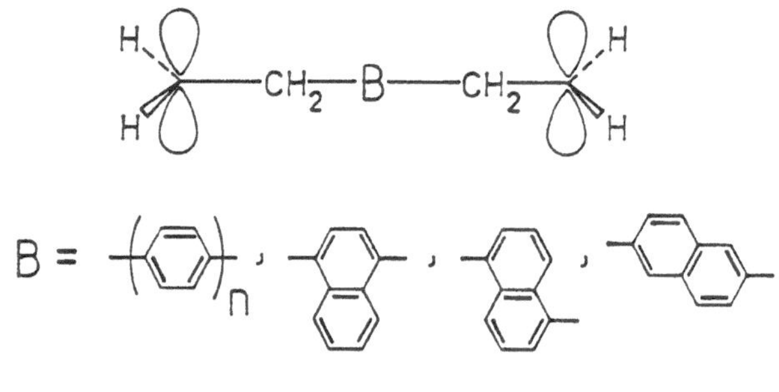

Fig. 4: Model structures for EH calculations.

For the sake of simplicity a fixed twist angle of 44^0 between the two
phenyl rings in biphenyl was assumed. The donor/acceptor-compounds
simulated by the structures 3' and 26N' have not yet been synthesized
but are included for completeness.
The p-orbitals at the terminal CH_2-residues correspond to the atomic
orbitals in the donor and acceptor states next to the bridge. Considering the perturbational expansion of the superexchange interaction /3/

$$V_{fi} = \sum_j \frac{\beta_{dj} \cdot \beta_{ja}}{E - E_j} \qquad (3)$$

the important features of the donor and acceptor states are their
interaction matrix elements β_{dj}, β_{ja} with the electronic states of
the bridge and their energy E at the intersection of the reactands and
products potential surfaces. It is plausible to assume, that the interaction of the states of the spacer with the two p-orbitals in Fig.2
contributes predominantly to the matrix elements β_{dj} and β_{ja}. Therefore, the model structures 1',2',3',14N', 15N', and 26N' should simulate the interaction between donor and acceptor reasonably well, on
condition that the energy of the p-orbitals is adjusted to the correct
value E.

This is done in the following way:
Using spectroscopic and thermodynamic data, the energy difference between E and the highest occupied molecular orbital of the phenyl-spacer,
henceforth called ΔE, has been estimated to lie in the range 1-1.5 eV
/7/. In the EH calculation ΔE is identified with the energy difference
between the average of the molecular orbitals corresponding to the
symmetric and antisymmetric linear combination of the two p-orbitals
and the energy of the next orbital below (see Fig. 5). This energy is
changed by a slight variation of the Coulomb integral for the terminal

p-orbitals in the EH calculation of structure 1'. The Coulomb-integral necessary to obtain a certain ΔE for this bridge is then used for all the other model structures as well. The other parameters in the calculation are standard values.

The interaction with the states of the spacer leads to a splitting Δ' of the energies of the symmetric and antisymmetric linear combinations of the two terminal p-orbitals (and to a weak mixing of those states into these linear combinations) as well as to a shift of the center of the two energies /3c/. Δ' is calculated with the EH method. Multiplication of Δ' with the coefficients C_d and C_a of the two p-orbitals in the donor and acceptor states then gives the splitting Δ in the real molecules:

$$\Delta = C_a \cdot C_d \cdot \Delta' \tag{4}$$

This relation is plausible within the perturbational expansion, but its approximate validity is assumed also in the more exact Extended-Hückel calculation.

Fig. 5:
Energy level scheme of the model structures.

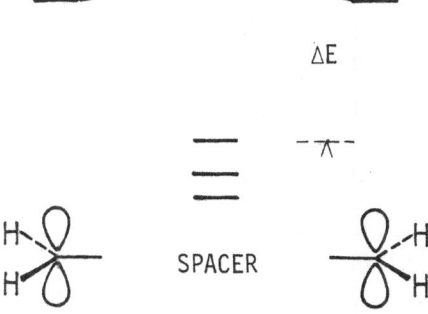

The splittings Δ' have been calculated for several (symmetric) conformations of the model structures. Δ' is strongly dependent on the orientation of the CH_2-groups but in all cases the function $|V_{fi}|^2$ has a pronounced maximum $|V_{fi,max}|^2$ for the conformations:

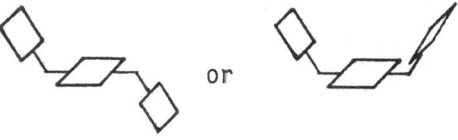

or

It is assumed that averaging $|V_{fi}|^2$ over all conformations leads to common geometric factor f:

$$|V_{fi}| = |V_{fi,max}| \cdot f \qquad\qquad (5)$$

for all spacers. f is probably of the order of 1/10. In the real molecules steric hindrance can reduce the probability of certain conformations. Comparing the phenyl- and biphenyl- or the 1,4-naphthyl- and 1,5-naphthyl-bridged compounds the steric hindrances are the same, whereas the formation of the optimal conformation in the naphthyl-bridged compounds is less probable compared to the phenyl-bridged ones, due to the H-atoms in 5- and 8-position at the naphthyl-residue. Therefore, it is plausible to expect a cancellation of conformational and steric effects in the ratios of the rates for P1D/P2D and A1D/A2D as well as for P14ND/P15ND and a slight decrease of the rates in P14ND and P15ND compared to the calculation.

Table 2 contains the calculated electronic matrix elements $V_{fi,max}$ where ΔE in $\underline{1}'$ has been adjusted to 1.1 eV. The center of the energies of the two relevant orbitals is shifted to higher energies (Fig. 5) by about 0.7 eV.

	$\underline{1}^-$	$\underline{2}^-$	$\underline{3}^-$
V_{fi} [eV]	0.020	0.0053	0.0017
	$\underline{14N}^-$	$\underline{15N}^-$	$\underline{26N}^-$
V_{fi} [eV]	0.026	0.018	0.013

Table 2: Calculated matrix elements for optimal conformations.

When ΔE is varied between 1 and 1.5 eV the calculated electronic factors $|V_{fi}'|^2/|V_{fi}|^2$ for the ratios of the rates are:

	$\underline{1}^-/\underline{2}^-$	$\underline{2}^-/\underline{3}^-$	$\underline{14N}^-/\underline{1}^-$	$\underline{14N}^-/\underline{15N}^-$	$\underline{15N}^-/\underline{26N}^-$
$|V_{fi}^-|^2/|V_{fi}|^2$	21 - 12	13 - 8	1.5 - 1.9	2.6 - 2.2	1.7 - 1.8

Table 3: Calculated electronic contributions to ratios of rates.

The calculation semiquantitatively reproduces the experimental ratios of the rates. The lower experimental value for k(P14ND)/k(P1D) can be explained by the increased sterical hindrance in the naphthyl-bridged compound.

An analysis of the admixture of spacer states to the symmetric and antisymmetric linear combination of the p-orbitals reveals a significant participation of the σ-orbitals of the spacer in the mediation of the interaction between donor and acceptor. This is mainly due to the favorable electronic overlap of the terminal p-orbitals with the p-orbitals in the plane of the aromatic rings of the bridge. This favourable overlap compensates the increased energy mismatch $E-E_j$ for the σ-type orbitals compared to the π-orbitals.

Whereas for the compounds investigated here EH calculations give a sufficient description of the experimental phenomenology in bridged donor/acceptor-systems, the situation with saturated aliphatic bridges is not as clear. Joran et al. /12/ report a ratio of 600 to 1500 for the transfer rates in compounds with one and two aliphatic bicyclooctyl residues as spacers. Warman et al. /13/ investigated the charge recombination in systems where donor and acceptor are separated by a rigid, aliphatic structure with 4, 6, 8, and 10 σ-bonds in between. They find a decrease of the recombination rate by a factor of only 5-10 for every pair of inserted σ-bonds. EH calculations in general give a weak distance dependence, for instance, in the model structures corresponding to the bicyclooctyl-bridged compounds

the electronic contribution $|V_{f i}|^2/|V_{fj}|^2$ to the ratio of the rates is about 5-6 compared to the experimental one /12/ of 500-1600! In poly-methylene-chains in the conformation

the decrease of the electronic factor per $-CH_2-CH_2-$ fragment is approximately 4, which is of the same order of magnitude as in the measurements of Warman et al. /13/. For less ordered conformations the dependence on the length of the bridge is much stronger /14/.

Beratan and Hopfield /15/ developed a different method for the calculation of the distance dependence of the transfer rate in bridged donor/-acceptor-compounds which predicts a stronger distance dependence as well as a stronger dependence on the location of the donor and acceptor orbitals in the energy gap between HOMO and LUMO of the spacer compared to the calculations here. Their method is also based on an EH calculation of electronic wave functions within the bridge but their results were obtained with a different set of parameters and a restricted basis set of orbitals. At the moment it is not yet clear whether the discrepancy between the EH method applied in this paper and in ref. /15/ is due to different parametrization and/or to more profound reasons.

4. ESTIMATE OF REORGANIZATION PARAMETERS

The magnitude of the vibrational reorganization during the electron transfer is given by the reorganization energies of the solvent λ_S and of the intramolecular modes λ_V /16/, which determine the Franck–Condon factor (FC) in the electron transfer rate (1).

a) Solvent reorganization energy λ_S: Information on the reorganization energy λ_S can be obtained from the temperature dependence of the ion formation. λ_S in acetonitrile and propionitrile can be assumed to be almost identical. For moderately exothermic reactions the temperature dependence of the transfer rate is dominated by the first term in the sum for the Franck–Condon factor /17/:

$$(FC) = (4\pi\lambda_S kT)^{-1/2} \cdot \sum_n (e^{-S} \cdot S^n/n!) \cdot \exp\{-(\Delta G^\circ + \lambda_S + nh\omega)^2/4\lambda_S kT\} \quad (6)$$

$$S = \lambda_V/h\omega$$

i.e., in a plot of $\ln(k^*/T)$ vs. $1/T$ the dependence is almost linear and the slope is given by:

$$- \Delta G^*/k = - (\Delta G^\circ + \lambda_S)^2/4\lambda_S k \quad (7)$$

Using $\Delta G^\circ = -0.5$ eV for the ion formation, the measurements yield $\lambda_S = 0.95$ eV for both compounds. According to ref. /16/ λ_S increases with the donor/acceptor distance. Its (estimated) contribution to the ratio of the rates of P1D/P2D or A1D/A2D is a factor of about 2.

b) Internal reorganization energy λ_V: An estimate for λ_V can be obtained from the ratio of the ion formation and recombination rates. The matrix elements V_{fi} for the two processes are of similar magnitude and the ratio is therefore given by the ratio of the corresponding Franck–Condon factors (6). Experimentally it is approximately 2. Theoretically this value can be reproduced with the expression for the Franck–Condon factor and the parameters $\lambda_S = 0.95$ eV, $\omega = 1500$ 1/cm, $\Delta G^\circ = -0.5$, -2.8 eV and $\lambda_V = 0.6$ eV. (The reorganization parameters are assumed to be approximately the same for ion formation and recombination). Due to the approximations in the formula (harmonic oscillators, no frequency change) λ_V is probably overestimated and $\lambda_V = 0.4$-0.5 eV seems more probable.

Using the calculated matrix element V_{fi}, together with the estimated geometric factor $f \sim 0.1$ and $(FC) \cong 3 \cdot 10^{12} (eV)^{-1}$ derived from the values the forward electron transfer rate for P1D at room temperature $k \sim 12 \cdot 10^9$ s^{-1} agrees with the experimental one within an order of magnitude.

As a crude measure for the nonadiabaticity factor of the reaction the relations

$$\kappa = 2 \cdot \frac{|V_{fi}|^2}{\hbar \omega_n} \cdot (\pi^3/(\lambda_s + \lambda_v)kT)^{1/2}$$

$$\omega_n^2 \cong \lambda_v \omega^2/(\lambda_s + \lambda_v)$$

(8)

are taken from ref. /18/. For the same parameters as above the nonadiabaticity criterion $\kappa \cong 0.03 \ll 1$ is fulfilled.

5. CONCLUSIONS

The intramolecular electron transfer rates in the donor/acceptor compounds with aromatic bridges show a remarkably weak distance dependence and a dependence on the structure of the bridging element. Both effects can be explained with an electronic coupling between donor and acceptor through the bridge. The electronic matrix element V_{fi} has been calculated within the Extended-Hückel approximation. These calculations semiquantitatively reproduce experimental ratios of rates for different spacers and give absolute values for electronic matrix elements sufficient for order of magnitude estimates of electron transfer rates. Measurements of both the ion recombination rate and the temperature dependence of the ion formation allow the determination of the reorganization parameters of the reaction.

ACKNOWLEDGEMENTS:

We are highly indebted to Profs. J. Jortner, Tel Aviv, and N. Rösch, Munich, for many stimulating discussions. The financial support of the Deutsche Forschungsgemeinschaft, the Volkswagenstiftung, and the Alfried Krupp von Bohlen und Halbach-Stiftung is gratefully acknowledged.

REFERENCES

/1/ J. Deisenhofer, O. Epp, K. Miki, R. Huber, and H. Michel, J. Mol. Biol. 180, 385 (1984).
/2/ J. Breton, J.L. Martin, A. Migus, A. Antonetti, and A. Olscag, Proc. Nat. Acad. Sci. (1986), in press.

/3/ a) J. Halpern and L.E. Orgel, Discuss. Faraday Soc. 29, 32 (1960).
    b) H.M. McConnell, J. Chem. Phys. 35, 508 (1961).
    c) S. Larsson, J. Am. Chem. Soc. 103, 4034 (1981).
    d) J. Jortner and M. Bixon, Proceedings of the Conference on
       Structure and Functions of Proteins, Philadelphia, March
       1985, in press.
/4/ H. Michel, O. Epp, and J. Deisenhofer (1986), in preparation.
/5/ H. Heitele, M.E. Michel-Beyerle, and G. Wegner, to be published.
/6/ a) V.G. Levich, Adv. Electrochem. Electrochem. Eng. 4, 249 (1966).
    b) N.R. Kestner, J. Logan, and J. Jortner, J. Chem. Phys. 78,
       2148 (1974).
    c) R.P. Van Duyne and S.F. Fischer, Chem. Phys. 5, 183 (1974).
    d) J.Ulstrup, "Charge Transfer Processes in Condensed Media",
       Springer Verlag Berlin, 1979.
/7/ H. Heitele and M.E. Michel-Beyerle, J. Am. Chem. Soc. 107, 8286
    (1985).
/8/ N. Remy-Richter, M. Volk, A. Ogrodnik, H. Heitele, and
    M.E. Michel-Beyerle, in preparation.
/9/ J.R. Miller, J.V. Beitz, and R.K. Huddleston, J. Am. Chem. Soc.
    106, 5057 (1984).
/10/ a) N. Sutin, Prog. Inorg. Chem. 30, 441 (1983);
     b) M.D. Newton and N. Sutin, Ann. Rev. Phys. Chem. 35, 437 (1984);
        and references therein.
/11/ H. Heitele and M.E. Michel-Beyerle in: "Antennas and Reaction
     Centers of Photosynthetic Bacteria", ed. M.E. Michel-Beyerle,
     Springer Series in Chemical Physics 42, p. 250, Springer Verlag,
     Berlin 1985.
/12/ a) A.D. Joran, B.A. Leland, G.G. Geller, J.J. Hopfield, and P.B.
        Dervan, J. Am. Chem. Soc. 106, 6090 (1984).
     b) B.A. Leland, A.D. Joran, P.M. Felker, J.J. Hopfield, A.H.
        Zewail, and P.B. Dervan, J. Phys. Chem. 89, 5571 (1985).
/13/ J.M. Warman, M.P. de Haas, M.N. Paddon-Row, E. Cotsaris, N.S.
     Hush, H. Oevering, and J.W. Verhoeven, Nature 320, 615 (1986).
/14/ S. Larsson, J. Chem. Soc., Faraday Trans. 2, 79, 1375 (1983).
/15/ D.N. Beratan and J.J. Hopfield, J. Am. Chem. Soc. 106, 1584 (1984).
/16/ a) R.A. Marcus, J. Chem. Phys. 24, 966 (1956); J. Chem. Phys. 43,
        679 (1965).
     b) R.A. Marcus and N. Sutin, Biochim. Biophys. Acta 811, 265
        (1985).
/17/ J. Jortner, J. Chem. Phys. 64, 4860 (1976).
/18/ B.S. Brunschwig, J. Logan, M.D. Newton, and N. Sutin, J. Am.
     Chem. Soc. 102, 5798 (1980); and references therein.

ENVIRONMENTAL MODULATION OF THE ELECTRONIC FACTOR IN ELECTRON TRANSFER RATE THEORY, AND IMPLICATIONS FOR THE PRIMARY STEPS IN BACTERIAL PHOTOSYNTHESIS

Alexandr M. Kuznetsov
The A.N. Frumkin Institute of Electrochemistry
of the Academy of Sciences of the USSR
Leninskij Prospect 31, Moscow V-71, USSR

and

Jens Ulstrup
Chemistry Department A, Building 207
The Technical University of Denmark
2800, Lyngby, Denmark

1. INTRODUCTION

The presence of environmental matter in the space between molecular solute reactants in "long-range" electron transfer (ET) affects the electronic exchange integral which induces the process. "Long-range" implies that the ET distance exceeds the geometric extension of the reacting molecules, so that the process is diabatic and the "tails" of the donor and acceptor wave functions exposed to environmental nuclear influence.

Quantum chemical calculations using ab initio and semi-empirical methods lead to approximately exponential distance decay of the electronic factor[1,2], with orbital exponents between 1 and 2 $Å^{-1}$. Such dependence is also obtained by much simpler tunnelling descriptions or approaches which rest on solutions to the stationary Schrödinger equation for localized potential wells of various shapes[3]. Neither of these approaches incorporate directly intervening matter between the donor and acceptor. Exponential distance dependence of the pre-exponential rate factor with orbital exponents in the range 1.0–1.5 $Å^{-1}$ is, however, still observed experimentally for rather different donor-acceptor systems such as trapped electrons[4], solute radical cations and anions[5], and metal–insulator–metal systems.[6a] Lower values, of 0.3–0.5 $Å^{-1}$ have been inferred from photoconduction between excited molecules such as anthracene and chloranil, and a metal cathode[7], and values in the range 0.6–1.0 $Å^{-1}$ have been obtained for electrochemical electron transfer across thin oxide[8] or polymer[9] films.

The electronic effect of molecular groups in addition to the donor and acceptor were investigated early by Halpern and Orgel[10a], and by McConnell[10b]. Incorporation of the nuclear motion as well – for both the initial, final, and intermediate states, and for both high-energy,

J. Jortner and B. Pullman (eds.), Tunneling, 345–360.

"virtual" and low-energy intermediate states – were provided in several reports[11,12]. These still rest on the perturbative approach which is, however, little restrictive when high-energy intermediate states are involved[11g,11h]. The distance dependence for this limit follows an approximate $p^{-(n-1)}$ dependence where n is the number of transmitting orbitals and p the ratio between the exchange integrals and the energy gap between the initial and intermediate orbitals. In the limit of long distances and exponential decay of the exchange integrals "bridge-assisted"[11] mechanisms can therefore also display an approximately exponential distance dependence of the overall process (cf. refs. 13 and 14).

Several reports on the handling of stronger inter-group interactions have appeared recently. Larsson[13] estimated the energy splitting for donor and acceptor orbitals in bridge-separated mixed-valence complexes using an extended Hückel scheme. The numerical estimates reported, however, reduce to those of the perturbative approach. Beratan and Hopfield[14] used a similar approach to a different class of ruthenium complexes, and Ohta, Closs, Morokuma and Green[15] applied ab initio techniques to intramolecular ET between substituent groups on a rigid alicyclic molecular frame (cf. refs. 5c and 5d). A related class of models cover weak-coupling electron transfer between an individual molecular and two strongly interacting substystems. Such models have been applied to molecular centres coupled to a continuum of protein levels in biological environments[16] and to electron transfer via impurity atoms adsorbed on a substrate metal electrode[17].

Environmental effects on the donor and acceptor wave functions have recently been approached from a different angle. Redi and Hopfield[18] calculated the electronic factor for two one-dimensional delta function and for two one-dimensional square-well potentials. The exchange integrals depend on the energies of the levels between which transition occurs and are coupled to Gaussian nuclear level distributions. As a result, widely different values of the electronic factor are obtained, depending on the position of the redox levels relative to the environmental mobility edge.

We have previously approached solvent effects on the electronic factor in a more realistic fashion, namely by considering the interaction of a localized trial wave function with the inertial solvent polarization and the effects on the parameter(s) of the function[19]. Conceptually this is analogous to procedures known from polaron theory[20], but electron transfer wave functions differ in several respects from the polaron description. First, the transferring electron is exposed to two-centre polarization fields. Secondly, the inertial polarization is not the equilibrium value, but the particular non-equilibrium value at the moment of electron transfer, and finally, it is necessary to go beyond simple structureless solvent representations and incorporate vibrational frequency distribution (and possibly spatial dispersion as well), as this is important at low temperatures, and for some solvents perhaps also at high temperatures[21]. In this work we consider the interaction of the electron not only with the polarization and nuclear core charges in the donor region as in our previous results, but in the whole space including field terms at the acceptor site. In certain ways this modifies our previous conclusions, the validity conditions for which are also obtained.

2.   TRIAL WAVE FUNCTIONS AND ELECTRONIC-POLARIZATION FREE ENERGY FUNCTIONAL

In view of the approximately exponential distance dependence of the electronic factor which emerges from both calculations and experimental estimates, we assume that the initial (i) and final (f) state electronic wave functions are suitably represented by single-parameter 1s-like functions of the form

$$\psi_{1s}^i(\rho) = (\lambda_i^3/\pi)^{1\!/2} \exp(-\lambda_i |\vec{\rho}|)$$

$$\psi_{1s}^f(\rho) = (\lambda_f^3/\pi)^{1\!/2} \exp(-\lambda_f |\vec{\rho}-\vec{R}|)$$

(2.1)

where $|\vec{R}|$ is the distance between the spherical donor and acceptor centres. The electron-polarization interaction will thus be reflected in the orbital exponents λ_i and λ_f which are the only parameters involved.

We shall furthermore calculate the orbital exponents within the Born-Oppenheimer and Condon approximations. We thus assume that solvent modulation of the electronic wave functions is solely reflected in the electronic transmission coefficient. Orbital exponents modified by the solvent, however, in turn lead to a modified activation free energy. We have previously given a procedure for such effects,[19c] but presently this and non-Condon effects will not be incorporated in our formalism.

We thus minimize the total electronic energy, represented by the functions in eq. (2.1) in a "preset" field determined by the nuclear core charges and the polarization at the moment of electron transfer. A crucial point is further that electron transfer theory[22,23] prescribes in detail how this polarization value is obtained, when the initial and final state electric inductions are known.

The free energy functional, when the electron is located in the donor state, and for the instantaneous non-equilibrium polarization, $\vec{P}^*(\vec{r})$ - r being the space coordinate - is now

$$F_i(\psi_{1s}^i; \vec{P}^*(\vec{r})) = \frac{\hbar^2}{2m_e} \int |\vec{\nabla}\psi_{1s}^i(\rho)|^2 \, d\vec{\rho} - \int \vec{P}^*(\vec{r})\vec{D}_i^e(\psi_{1s}^i; \vec{r}) \, d\vec{r} -$$

$$- \frac{z_D e^2}{\epsilon_o} \int d\vec{\rho} \, \frac{1}{\rho} |\psi_{1s}^i(\rho)|^2 - \frac{z_A e^2}{\epsilon_o} \int d\vec{\rho} \, \frac{1}{|\vec{\rho}-\vec{R}|} |\psi_{1s}^i(\rho)|^2$$

(2.2)

where m_e is the mass of the electron, e the electronic charge, $z_D e$ and $z_A e$ the core charges of the donor and acceptor, respectively $\hbar$ Planck's constant divided by 2π, and ϵ_o coincides approximately with the optical dielectric constant. $D_i^e(\psi_{1s}^i; \vec{r})$ is

finally the dielectric induction aroused by the electronic charge represented by $\psi_{1s}^{i}(\rho)$, and corresponding to the non-equilibrium orbital
exponent, i.e.

$$\vec{D}_i^e(\psi_{1s}^i; \vec{r}) = -e \int d\vec{\rho} |\psi_{1s}^i(\rho)|^2 \frac{\vec{r}-\vec{\rho}}{|\vec{r}-\vec{\rho}|^3} \tag{2.3}$$

The first term in eq. (2.2) is the kinetic energy of the electron. The
second is its potential energy in the total, non-equilibrium polarization field created by the donor and acceptor ionic centres, the
"excess" electron and the ionic cores, at the moment of electron transfer, and the last two terms are the potential energies of "direct" interaction of the electron with the donor and acceptor ionic cores. From
$F_i(\psi_{1s}^i; \vec{P}^*(\vec{r}))$ is excluded the free energy required to create the inertial polarization $\vec{P}^*(\vec{r})$. This energy contitutes part of the nuclear
potential energy. It is reflected in the activation free energy and independent of the instantaneous values of λ_i and λ_f.

The free energy functional for the final state, when the electron
is located at the acceptor, is determined by an equation analogous to
eq. (2.2). The crucial point now is that $\vec{P}^*(\vec{r})$ can be obtained from
the initial and final state equilibrium polarizations, $\vec{P}_{io}(\vec{r})$ and
$P_{fo}(\vec{r})$, respectively, by prescriptions in electron transfer theory. For
a structureless dielectric in the high-temperature limit this relation
is[19,22]

$$\vec{P}^*(\vec{r}) = \frac{c}{4\pi} \left[(1-\theta^*)\, \vec{D}_{io}^e(\vec{r}) + \theta^* \vec{D}_{fo}^e(\vec{r}-\vec{R}) + \vec{D}_o^D(\vec{r}) + \vec{D}_o^A(\vec{r}-\vec{R}) \right] \tag{2.4}$$

where $\vec{D}_{io}^e(\vec{r})$ and $\vec{D}_{fo}^e(\vec{r})$ are the inductions from the electron only, when
the orbital exponents have their equilibrium values, and $\vec{D}_o^D(\vec{r})$ and
$\vec{D}_o^A(\vec{r})$ the inductions from the donor and acceptor core sites, while
$c = \varepsilon_o^{-1} - \varepsilon_s^{-1}$. θ^* is the "saddle point parameter" of electron transfer
theory in the strong-coupling limit and depends on the free energy of
reaction, the total nuclear reorganization free energy, and the solvent
dispersion features[12,22] (cf. section 5).

At low temperatures, or when solvent vibrational frequency dispersion is otherwise important, $\vec{P}^*(\vec{r})$ takes the more involved form which
reduces to eq. (2.4) in the high-temperature limit

$$\vec{P}^*(\vec{r}) = \frac{2}{\pi} \vec{D}_{io}^e(\vec{r}) \int_0^\infty \frac{d\omega}{\omega} \operatorname{Im}G(\omega) \frac{\operatorname{th}\tfrac{1}{2}\beta\hbar\omega(1-\theta^*)}{\operatorname{th}\tfrac{1}{2}\beta h\omega(1-\theta^*) + \operatorname{th}\tfrac{1}{2}\beta\hbar\omega\theta^*} +$$

$$+ \frac{2}{\pi} \vec{D}_{fo}^e(\vec{r}-\vec{R}) \int_0^\infty \frac{d\omega}{\omega} \operatorname{Im}G(\omega) \frac{\operatorname{th}\tfrac{1}{2}\beta\hbar\omega\theta^*}{\operatorname{th}\tfrac{1}{2}\beta\hbar\omega(1-\theta^*) + \operatorname{th}\tfrac{1}{2}\beta\hbar\omega\theta^*}$$

$$+ \frac{c}{4\pi} \left[\vec{D}_o^D(\vec{r}) + \vec{D}_o^A(\vec{r}-\vec{R}) \right]; \quad \beta = (k_B T)^{-1} \tag{2.5}$$

where k_B is Boltzmann's constant and T the temperature, and $G(\omega)$ is
the Green's function of the solvent polarization at the frequency ω.
$G(\omega)$ satisfies the Kramers=Krönig relation[23]

$$\frac{2}{\pi} \int_0^\infty \frac{d\omega}{\omega} \, ImG(\omega) = \frac{c}{4\pi} \qquad (2.6)$$

The polarization thus consists of two kinds of terms. One kind arises from the electron ($\vec{D}_{io}^{\,e}(\vec{r})$ and $\vec{D}_{fo}^{\,e}(\vec{r}-\vec{R})$). By combining the donor centre contribution with the interaction of the electron with the donor core charge itself (the third term on the right-hand side of eq. (2.2)). We obtain a term which coincides with the latter, except that ε_o is replaced by ε_s. Eq. (2.2) therefore takes the following form at arbitrary temperatures

$$F_i(\psi_{1s}^i; \vec{P}*(\vec{r})) = \frac{\hbar^2}{2m_e} \int |\vec{\nabla}\psi_{1s}^i(\rho)|^2 \, d\vec{\rho} - \frac{z_D e^2}{\varepsilon_s} \int d\vec{\rho} \, \frac{1}{\rho} |\psi_{1s}^i(\rho)|^2 -$$

$$- \frac{2}{\pi} \int d\vec{r} \, \vec{D}_{io}^{\,i}(\vec{r}) \, \vec{D}_i^{\,e}(\psi_{1s}^i; \vec{r}) \int_0^\infty \frac{d\omega}{\omega} \, ImG(\omega) \, \frac{th\frac{1}{2}\beta\,\hbar\omega\,(1-\theta*)}{th\frac{1}{2}\beta\hbar\omega(1-\theta*)+th\frac{1}{2}\beta\hbar\omega\theta*} -$$

$$- \frac{2}{\pi} \int d\vec{r} \, \vec{D}_{fo}^{\,e}(\vec{r}-\vec{R}) \vec{D}_i^{\,e}(\psi_{1s}^i; \vec{r}) \int_0^\infty \frac{d\omega}{\omega} \, ImG(\omega) \, \frac{th\frac{1}{2}\beta\hbar\omega\theta*}{th\frac{1}{2}\beta\hbar\omega(1-\theta*)+th\frac{1}{2}\beta\hbar\omega\theta*} -$$

$$- \frac{c}{4\pi} \int \vec{D}_o^{\,A}(\vec{r}-\vec{R}) \vec{D}_i^{\,e}(\vec{r}) \, d\vec{r} - \frac{z_A e^2}{\varepsilon_o} \int d\vec{\rho} \, \frac{1}{|\vec{\rho}-\vec{R}|} \, |\psi_{1s}^i(\rho)|^2 \qquad (2.7)$$

where the first three terms are "single-centre" interactions referring to the donor region, while the latter three are interactions between the electron, located at the donor, and the inertial polarization prevailing if the electron were located at the acceptor, the inertial polarization from the acceptor core, and the acceptor itself. These terms are two-centre interactions and in general significantly smaller than the former three.

Closer inspection reveals that the final two terms contribute insignificantly for all practical purposes. The fourth term is insignificant for $\theta* \lesssim 0.5$, corresponding to the "normal", "activationless", and strongly exothermic free energy regions, but significant in the $\theta*$-range between 0.5 and unity. This range is important for determination of the final state orbital exponent, λ_f (cf. eq. (2.9)). The equation for the free energy functional therefore becomes

$$F_i(\psi_{1s}^i; \vec{P}*(\vec{r})) = \frac{\hbar^2}{2m_c} \int \|\vec{\nabla}\psi_{1s}^i(\rho)|^2 \, d\vec{\rho} - \frac{z_D e^2}{\varepsilon_s} \int d\vec{\rho} \, \frac{1}{\rho} |\psi_{1s}^i(\rho)|^2 -$$

$$- \frac{2}{\pi} \int d\vec{r} \, \vec{D}_{io}^{\,e}(\vec{r}) \vec{D}_i^{\,e}(\psi_{1s}^i; \vec{r}) \int_0^\infty \frac{d\omega}{\omega} \, ImG(\omega) \, \frac{th\frac{1}{2}\beta\hbar\omega(1-\theta*)}{th\frac{1}{2}\beta\hbar\omega(1-\theta*)+th\frac{1}{2}\beta\hbar\omega\theta*} -$$

$$- \frac{2}{\pi} \int d\vec{r} \, \vec{D}_{fo}^{\,e}(\vec{r}-\vec{R}) \vec{D}_i^{\,e}(\vec{r}) \int_0^\infty \frac{d\omega}{\omega} \, ImG(\omega) \, \frac{th\frac{1}{2}\beta\hbar\omega\theta*}{th\frac{1}{2}\beta\hbar\omega(1-\theta*)+th\frac{1}{2}\beta\hbar\omega\theta*} \qquad (2.8)$$

The final state free energy functional is similarly

$$F_f(\psi_{1s}^f; \vec{P}*(\vec{r})) = \frac{\hbar^2}{2m_e} \int |\vec{\nabla}\psi_{1s}^f(\rho)|^2 \, d\vec{\rho} - \frac{z_A e^2}{\varepsilon_s} \int d\vec{\rho} \, \frac{1}{\rho}|\psi_{1s}^f(\rho)|^2 -$$

$$- \frac{2}{\pi} \int d\vec{r} \, \vec{D}_{fo}^e(\vec{r}-\vec{R})\vec{D}_f^e(\psi_{1s}^f;\vec{r}-\vec{R}) \int_0^\infty \frac{d\omega}{\omega} \, \mathrm{ImG}(\omega) \frac{\mathrm{th}\tfrac{1}{2}\beta\,\hbar\omega\,\theta*}{\mathrm{th}\tfrac{1}{2}\beta\hbar\omega(1-\theta*)+\mathrm{th}\tfrac{1}{2}\beta\hbar\omega\theta*} -$$

$$- \frac{2}{\pi} \int d\vec{r} \, \vec{D}_{io}^e(\vec{r})\vec{D}_f^e(\psi_{1s}^f;\vec{r}-\vec{R}) \int_0^\infty \frac{d\omega}{\omega} \, \mathrm{ImG}(\omega) \frac{\mathrm{th}\tfrac{1}{2}\beta\hbar\omega(1-\theta*)}{\mathrm{th}\tfrac{1}{2}\beta\hbar\omega(1-\theta*)+\mathrm{th}\tfrac{1}{2}\beta\hbar\omega\theta*}$$

$$(2.9)$$

from which the final state orbital exponent is obtained.

3. VARIATIONAL CALCULUS FOR THE ORBITAL EXPONENTS

The free energy functionals can be converted to tractable forms by the following steps:

(A) The wave functions in eq. (2.1) are inserted into eqs. (2.8) and (2.9). With the particular functions chosen, all space integrals become available in analytical form, except the one in the third terms in eqs. (2.8) and (2.9) which is, however, handled numerically in a straight-forward fashion.

(B) The functional is subsequently minimized with respect to the orbital exponents λ_i and λ_f at the fixed, polarization given by eq. (2.5).

We provide details of this scheme elsewhere. The resulting equation for λ_i is

$$\frac{\hbar^2}{m_e} \lambda_o \xi_i - \frac{z_1 e^2}{\varepsilon_s} - 2e^2 \xi_i^{-4} f(1-\theta*) \int_0^\infty \frac{u^2 \, du}{(\xi_i^{-2}+u^2)^2(1+u^2)^3} -$$

$$- \frac{4\pi e}{\lambda_o R} f(\theta*) \frac{1}{(\mu_o^{16}-3\mu_o^{14}+3\mu_o^{12}-1)^2} \{-12\mu_o^{i3}e^{-2\lambda_o R}(1-\mu_o^{i2})^2 +$$

$$+ 4\lambda_o \mathrm{Re}^{-2\lambda_o R}\mu_o^{i3}(\mu_o^{i6}-3\mu_o^{i4}+3\mu_o^{i2}-1)+12\mu_o^{i3}e^{-2M_o^i\lambda_o R}(1-\mu_o^{i2})^2 +$$

$$+2\lambda_o \mathrm{Re}^{-2\mu_o^i\lambda_o R}(3\mu_o^{i8}-2\mu_o^{i7}-10\mu_o^{i6}-6\mu_o^{i5}+12\mu_o^{i4}+6\mu_o^{i3}-6\mu_o^{i2}-2\mu_o^i+1) +$$

$$-2(\lambda_o R)^2 e^{-2\mu_o^i\lambda_o R}(1-\mu_o^{i2})(\mu_o^{i6}-3\mu_o^{i4}+3\mu_o^{i2}-1)\} = 0$$

with a similar equation for λ_f. We have here assumed, for the sake of simplicity that λ_i and λ_f corresponding to equilibrium in the initial and final states, respectively, coincide and take the value $\lambda_{io} = \lambda_{fo} = \lambda_o$. ξ_i in eq. (3.1) is λ_i/λ_o, μ_o^i is λ_i/λ_o and coincides with ξ_i when $\lambda_{io} = \lambda_{fo}$. Finally,

$$f(1-\theta^*) = \frac{2}{\pi} \int_0^\infty \frac{d\omega}{\omega} \, ImG(\omega) \, \frac{th\tfrac{1}{2}\beta\hbar\omega(1-\theta^*)}{th\tfrac{1}{2}\beta\hbar\omega(1-\theta^*)+th\tfrac{1}{2}\beta\hbar\omega\theta^*} \qquad (3.2)$$

Eq. (3.1) and the corresponding equation for λ_f provide the necessary equations for the variation of λ_i and λ_f with θ^*. We return to a determination of this quantity in section 5 and presently provide a set of numerical solutions to eq. (3.1) in the θ^*-range from -0.20 to 1.20. As a representative form of the solvent vibrational dispersion we choose a broad resonance function[24]

$$ImG(\omega) = G_R^o \, \frac{\Omega_R^2+\Gamma_R^2}{2\Omega_R} \, [\frac{1}{[(\Omega_R-\omega)^2+\Gamma_R^2]} - \frac{1}{[(\Omega_R+\omega)^2+\Gamma_R^2]}] \qquad (3.3)$$

where Ω_R and Γ_R are the resonance frequency and width, respectively. The constant G_R^o is determined by noting that

$$f(1-\theta^*) = \frac{2}{\pi} \int_0^\infty \frac{d\omega}{\omega} \, ImG(\omega) = \frac{c}{4\pi} \quad \text{for} \quad \theta^* = 0 \qquad (3.4)$$

Figs. 1-4 show plots of λ_i and λ_f against θ^* calculated numerically from eq. (3.1) and the analogous equation for λ_f. In all cases $\lambda_o = 1 \text{ Å}^{-1}$, R = 10 Å, while $\Omega_R = \Gamma_R = 150 \text{ cm}^{-1}$. Results are shown for both the high-temperature limit (T = 298 °K) and for a low temperature (T = 37 °K) where a substantial part of the frequency distribution is quantum mechanically "frozen". The figures reveal the following features of the orbital exponent variation:

(A) The exponents indeed display variation with the transfer coefficient θ^* (or the free energy of reaction). In the "normal" free energy range, where $0 < \theta^* < 1$, λ_i and λ_f are both smaller than at equilibrium. It is as if the electron cloud has "swollen" due to less favourable inertial polarization. The effect is more pronounced when the donor core charge is negative and helps to "push" the electron into a spatially larger volume, while the electron is more reluctantly given up for positive donor charges. Opposite effects, i.e. a more confined electron density than at equilibrium, is obtained in the strongly exothermic region, where $\theta^* < 0$.

(B) The initial state orbital exponent decreases from its equilibrium value at $\theta^* = 0$, corresponding to activationless electron transfer, to about 40% of this value as $\theta^* \to 1$. The variation is symmetric with respect to $\theta^* = 0.5$, for coinciding equilibrium orbital exponents, in the initial and final states. It is notable and was not inherent in our previous observations[19] that this θ^*-dependence only arises when the polarization from the acceptor (donor) region is included in the equation for $\lambda_i(\lambda_f)$ (the final terms in eqs. (2.8) and (2.9)). This term is important when θ^* exceeds 0.5, and formal instability (negative λ_i) arises as $\theta^* \to 1$ if the term is omitted (fig. 2).

(c) The effects are rather similar in the high-temperature limit and at low temperatures, as long as θ^* remains between zero and unity (figs. 1,

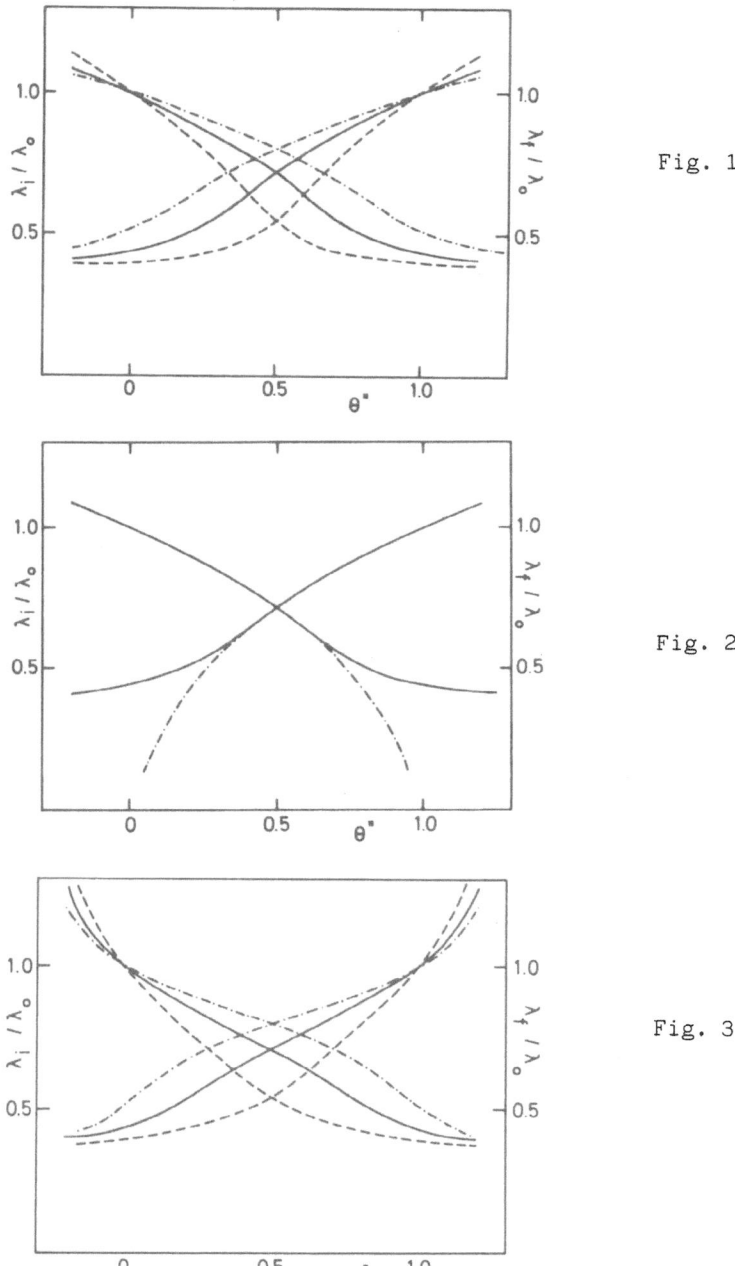

Fig. 1

Fig. 2

Fig. 3

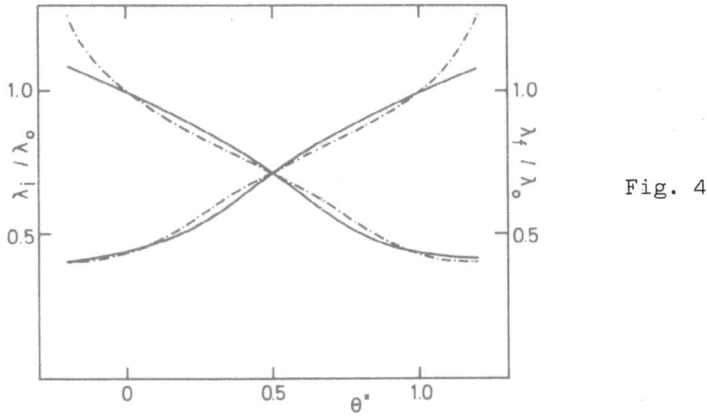

Fig. 4

Fig. 1. Dependence of orbital exponent ratios λ_i/λ_o and λ_f/λ_o on θ^*. System parameters given in the text. (———): $z_D = z_A = 0$. (._._): $z_D/\epsilon_s = z_A/\epsilon_s = +0.1$. (-----): $z_D/\epsilon_s = z_A/\epsilon_s = -0.1$. High-temperature limit.

Fig. 2. Orbital exponents with (———) and without (·—·—) the acceptor (donor) term in eq. (2.8) (eq.(2.9)). High-temperature limit and system parameters as in fig. 1. $z_D = z_A = 0$.

Fig. 3. Orbital exponent variation with θ^*. Same parameters and symbols as in fig. 1. $T = 37°K$.

Fig. 4. Comparison of orbital exponents at 298 °K (———) and 37 °K (·—·—). $z_D = z_A = 0$. Other parameters as in fig. 1.

(D) More pronounced differences between the high- and low-temperature limits appear when θ^* is significantly negative for λ_i, or θ^* larger than unity for λ_f, i.e. for strongly exothermic (λ_i) or endothermic (λ_f) ET processes, or in the high-frequency (λ_i) or low-frequency wing (λ_f) of optical charge transfer absorption bands (figs. 3 and 4). These effects have their origin in the third terms of eqs. (2.8) and (2.9) which rise rapidly when $\theta^* < 0$ and $\theta^* > 1$, respectively.

4.   FREE ENERGY VARIATION OF THE OVERLAP INTEGRAL

The variation of λ_i and λ_f with θ^* obviously affect the lap integral, S_{fi}, of the wave functions in eq. (2.1). However, since λ_i and λ_f vary in opposite directions, the overall effect on S_{fi} is not necessarily large, even if λ_i and λ_f individually vary strongly. The overlap integral between two 1s-like wave functions, with different orbital exponents, is

$$S_{fi} = 32\mu_f^{i\frac{3}{2}} \frac{1}{\lambda_i R} \frac{\mu_f^{i2}}{\mu_f^{i6} - 3\mu_f^{i4} + 3\mu_f^{i2} - 1} \left\{ e^{-\lambda_i R} - e^{-\lambda_f R} - \right.$$

$$\left. - \frac{1}{4}(1 - \mu_f^{i2})\lambda_f R\, e^{-\lambda_f R} - \frac{1}{4\mu_f^{i2}}(1 - \mu_f^{i2})\lambda_i R\, e^{-\lambda_i R} \right\} \qquad (4.1)$$

where $\mu_f^i = \lambda_i / \lambda_f$. Two limiting cases are important. When $\lambda_i = \lambda_f$, eq. (4.1) takes the simpler form

$$S_{fi} = [1 + (\lambda_i R) + (\lambda_i R)^2]\, e^{-\lambda_i R} \qquad (4.2)$$

On the other hand, when $\lambda_i \gg \lambda_f$, for long-range ET in practice when λ_i only slightly exceeds λ_f, then

$$S_{fi} \approx 8\mu_f^{i\frac{3}{2}}\, e^{-\lambda_i R} \qquad (4.3)$$

with a similar equation for $\lambda_i \ll \lambda_f$. Fig. 5 shows the variation of S_{fi}^2 with θ^* for $R = 10$ Å, $\lambda_o = 1$ Å^{-1}, for positive, negative, and vanishing core charges, at high (298 °K) and low (37 °K) temperatures, and for the same solvent vibrational dispersion as in figs. 1–4. The variation of S_{fi}^2 is in fact rather slow, i.e. less than an order of magnitude. This is because we have chosen the equilibrium orbital exponents to be equal, and eq. (4.3) combined with the data in figs. 1–4 shows that much stronger variation emerges if $\lambda_{io} \neq \lambda_{fo}$. S_{fi}^2 can furthermore either rise or fall with increasing θ^* depending on the sign of the core charges.

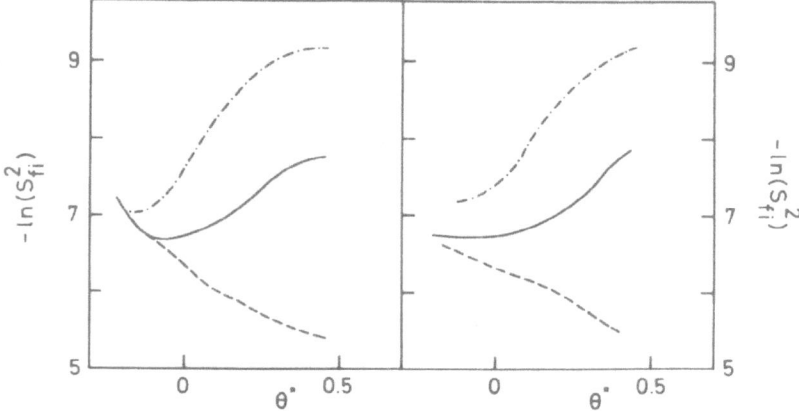

Fig. 5. Overlap integral variation with θ^*. Data and symbols are the same as in figs. 1 and 2. Left: 37 °K. Right: 298 °K.

5. DISTORTION OF FREE ENERGY RELATIONS AND IMPLICATIONS FOR DISTANCE PARAMETERS IN BACTERIAL PHOTOSYNTHESIS

In the limit of strong electronic-vibrational coupling, which is what we presently consider, θ^* is determined by the free energy of reaction, ΔF_o, and the environmental nuclear reorganization free energy, E_r, by the relation[19,22,23]

$$\frac{1}{8} \frac{\Delta F_o}{E_r} = \frac{1}{c} \int_o^\infty \frac{d\omega}{\omega} \, ImG(\omega) \, \frac{sh\frac{1}{2}\beta\hbar\omega(2\theta^*-1)}{sh\frac{1}{2}\beta\hbar\omega} \qquad (5.1)$$

The transition probability per unit time for electron transfer is

$$W_{fi} = \frac{\beta}{\hbar} \{\frac{1}{2\pi} \, \Phi''(\theta^*)|\}^{-\frac{1}{2}} V^2_{fi} \, \exp[-\beta\theta^*\Delta F_o^* - \Phi(\theta^*)] \equiv V^2_{fi} W_{fi} \quad (5.2)$$

V_{fi} being the electron exchange matrix element. The function $\Phi(\theta^*)$ is determined by the reorganization free energy and the environmental frequency dispersion

$$\Phi(\theta^*) = \frac{1}{\hbar c} E_r \int_o^\infty \frac{d\omega}{\omega^2} \, ImG(\omega) \, \frac{sh\frac{1}{2}\beta\hbar\omega(1-\theta^*)sh\frac{1}{2}\beta\hbar\omega\theta^*}{sh\frac{1}{2}\beta\hbar\omega} \qquad (5.3)$$

Within the Condon approximation, and for allowed transitions, optical bandshapes for strong-coupling transitions can be obtained from eqs. (5.1) - (5.3) by replacing V_{fi} by the optical transition dipole matrix element, and ΔF_o by $\Delta F_o - h\nu$, ν being the light frequency of the optical transition.

Free energy relations or bandshapes, represented by eqs. (5.1) - (5.3) are modified when the variation of the electronic factor, described above, is incorporated. More specifically, if [19a]

$$[R|d(\lambda_i + \lambda_f)|/d\theta^*]^2 \gtrless 4\beta E_r \qquad (5.4)$$

and the overlap integral is taken to represent the electron exchange integral, then we can anticipate both distortion and a shift of the maximum of the correlation. The shift is towards lower, and higher energies, when $\lambda_i < \lambda_f$ and $\lambda_i > \lambda_f$, respectively, and more pronounced the lower E_r, since the variation of θ^* with $\hat{\Delta}F_o$ or $h\nu$ is then faster. These expectations are illustrated in figs. 6 and 7. which show free energy relations where the nuclear factor is represented by the same resonance distribution as above, and the electronic factor in the form of the squared overlap integral is included. The variation is "normalized" with respect to the initial state equilibrium polarization, and the following is notable:

(A) When the equilibrium orbital exponents coincide, the solvent effect on the electronic factor is in fact small. There is a 10-15% band broadening for finite ionic core charges, a slight maximum shift towards smaller and larger energies, for positive and negative core charges, respectively, and a slight band narrowing for vanishing core charges.

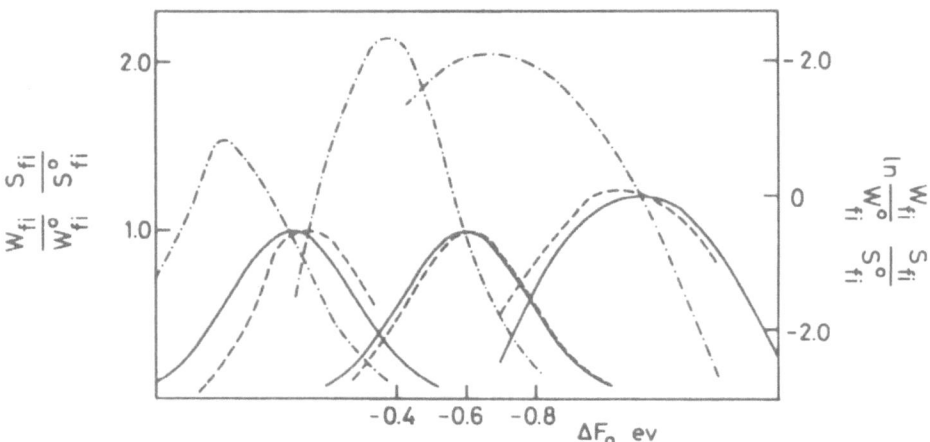

Fig. 6. Free energy plots "normalized" with respect to the transition probability (W^o_{fi}) and overlap integral (S^o_{fi}) at $\theta^* = 0$. Same parameters as in previous figures. $T = 298$ °K and $E_r = 0.6$ ev. (———): constant overlap integral. (----): θ^*-dependence of S_{fi} included, $\lambda_{io} = \lambda_{fo} = 1$ Å^{-1} (eq. (4.1)). (·—·—): θ^*-dependence of S_{fi} included, $\lambda_{io} = 1$ Å$^{-1} < \lambda_{fo}$ (eq. (4.3)). The three families of curves from left to right correspond to $z_D/\varepsilon_S = +0.1$, 0, and -0.1, respectively. The latter groups are shifted to the right by 0.5 and 1.0 ev. The first two groups refer to the scale on the left, the final one ($z_D/\varepsilon_S = -0.1$) to the logarithmic scale on the right.

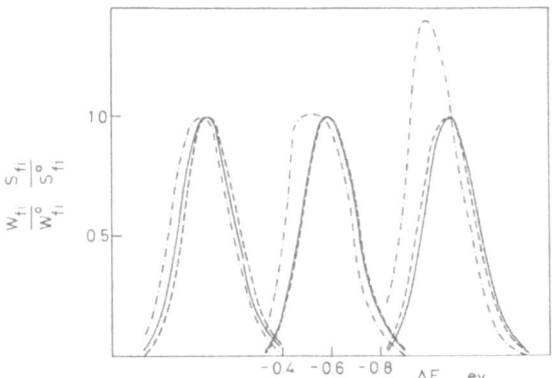

Fig. 7. Same kind of plots as in fig. 6. $T = 37$ °K. Symbols and parameters also as in fig. 6.

(B) Similar, but smaller effects are found at low (37 °K) temperatures.

(C) Rather pronounced distortion and shifts are, however, encountered if one of the orbital exponents significantly exceeds the other one. This corresponds to ET from a "loosely" to a firmly bound electronic state. Examples could be reactions of the solvated electron and in general involving excited electronic states[19,26], ET from cytochrome c to the bacteriochlorophyll cation, or perhaps ET from photoexcited bacteriochlorophyll to its primary electron acceptor. Photoexcitation from a molecular ground state to a charge transfer state may also be represented by notably different exponents. Fig. 7 shows that when $\exp(-\lambda_i R)$ > $\exp(-\lambda_f R)$, so that the overlap integral can be given the form in eq. (4.3), very significant broadening and shifts towards smaller energies appear, and more notably the higher the temperature. Shifts in opposite direction would appear if the inverse inequality is valid.

(D) For very strongly exothermic processes at low temperatures, the transition probability becomes smaller by orders of magnitude than for a constant electronic factor, due to the rapid rise of λ_i with increasingly negative θ^* in these regions (figs. 3 and 4). This effect is hardly detectable for optical bandshapes, but possibly for sufficiently strongly exothermic chemical processes.

Variation of the electronic factor with the free energy of reaction can thus indeed be comparable to the variation of the nuclear factor and should be included for example when distance parameters are extracted from experimental data on long-range ET processes. As an illustration of the possible importance of these effects, we consider, largely in qualitative terms, their implications for two ET reactions in the reaction centre of bacterial photosynthesis.

Depending on the particular representation for the nuclear motion, values of the electron exchange integral, V_{fi}, in the range from $5 \cdot 10^{-3}$ ev to about $5 \cdot 10^{-6}$ ev have been estimated from investigations of the ET reaction between cytochrome c and photo-oxidized bacteriochlorophyll[25] in terms of ET theory. The latter value emerges when the nuclear system is represented by two broad continua centred at 150 and 400 cm^{-1}. A distance dependence of V_{fi} in the form[25b] $V_{fi} = V_{fi}^o \exp(-\alpha R)$, $V_{fi} \approx 1$ ev, $\alpha \approx 1$ Å^{-1} converts these data to an upper limit of 12-14 Å for the ET distance. This is still far below estimates from magnetic interaction between the two centres[26] and from structural data[27] both of which suggest about 25 Å. Inclusion of solvent modulation effects along the lines above lowers the orbital exponent by 15-20 % for $E_r = 1.4$ ev and $\Delta F_o = -0.45$ ev[25], giving an upper limit for the distance of 16-18 Å. The discrepancy is thus smaller when modulation is included, but a notable difference remains.

The ET back reaction from reduced pheophytin to photo-oxidized bacteriochlorophyll is slower by about three orders of magnitude than the forward reactions between photoexcited bacteriochlorophyll and neutral pheophytin, in spite of a much larger standard redox potential separation[28] ("exothermicity"). Effects of this order can be anticipated if the back reaction belongs to the strongly exothermic "inverted" free energy region[25b,c] ($E_r \approx$ 0.3–0.5 ev, $\Delta F_o \approx$ –(0.9–1.0) ev.) A lower rate constant in the "inverted" region is, however, also expected if modulation of the electronic factor is important due to the spatially more confined donor electron cloud at the moment of electron transfer than at equilibrium.

For example, for E_r = 0.5 ev[25], $\theta*$ is –0.41 and –0.34 when ΔF_o = –1.0 and –0.9 ev, respectively, at 298 °K. The corresponding $\theta*$-values at 77 °K are –0.16 and 0.145, and they are –0.09 and –0.08 at 40 °K. At 298 °K the nuclear factor has dropped relative to its maximum value by factors of 80 and 20, respectively, for ΔF_o = –1.0 and –0.9 ev. These drops are factors of $3 \cdot 10^3$ and $3 \cdot 10^2$ at 77 °K and $6 \cdot 10^3$ and $5 \cdot 10^2$ at 40 °K.

ET distances of 6–8 Å were estimated from previous analysis[25 b,c], but 17 Å is now a suitable value emerging from recent crystallographic data for the reaction centre. This value would give the following further rate constant decreases from the modulation effect if the electron exchange integral is given the same form and equilibrium orbital exponents as above. At 298 °K : factors of 10^3 and 300 for ΔF_o = –1.0 and –0.9 ev. At 77 °K : factors of 15 and 12 for the same two ΔF_o-values, and at 40 °K : factors of 5 and 4, respectively. The values at low temperatures are much smaller than at 298 °K due to the smaller $\theta*$-values. All the values furthermore rest on the assumption inherent in eq. (4.3), viz. neutral effective ionic core charges, and temperature independent ΔF_o.

The resulting rate constant decreases amount to $8 \cdot 10^4$, $4,5 \cdot 10^4$, and $3 \cdot 10^4$ at 298, 77, and 40 °K if ΔF_o = –1.0 ev, and to $6 \cdot 10^3$, $3,6 \cdot 10^3$, and $2 \cdot 10^3$ for ΔF_o = –0.9 ev. The estimates are not far from what would reproduce the reported rate pattern. They are also well in line with a reported rate increase of about a factor of four for the back reaction in this temperature range, even though the electronic and nuclear parts contribute rather differently at high and low temperatures.

REFERENCES

1   M.D. Newton, Int. J. Quant. Chem. Symp., 14 (1980) 363.
2   a. M.-H. Wangbo and K. Stewart, Isr.J.Chem., 23 (1983) 133;
     b. W.J. Pietro, D.E. Ellis, T.J. Marks and M.A. Ratner, Mol.cryst.
     liq.cryst., 105 (1984) 273.
3   a. C.B. Duke, in Tunnelling in Biological Systems, B. Chance, D.C.
     DeVault, H. Frauenfelder, R.A. Marcus, J.R. Schrieffer and N. Sutin
     (Eds.), Academic Press, New York, 1979, p. 31. b. B. Brocklehurst,
     J.Phys.Chem., 83 (1979) 536; c. P. Siders, R.J. Cave and R.A. Marcus, J.Chem.Phys., 81 (1984) 5613.
4   a. I.V. Aleksandrov, R.F. Khairutdinov and K.I. Zamaraev, Chem.Phys.,
     32 (1978) 123; b. J.V. Beitz and J.R. Miller, J.Chem.Phys., 71
     (1979) 4579.
5   a. J.R. Miller and J.V. Beitz, J.Chem.Phys., 74 (1981) 6746;
     b. L.I. Calcaterra, G.L. Closs and J.R. Miller, J.Amer.Chem.Soc.,
     105 (1983) 670; c. J.R. Miller, J.V. Beitz and R.H. Huddleston,
     J.Amer.Chem.Soc., 106 (1984) 5057.

6   a. C.B. Duke, Tunnelling in Solids, Solid State Supplement, Vol. 10,
     Academic Press, New York, 1969. b. B. Mann and H. Kuhn, J.Appl.
     Phys., 42 (1971) 4398. c. M. Sugi, T. Fukui and S. Iizima, Appl.
     Phys.Letters, 27 (1975) 559; d. E.E. Polymeropoulos, J.Appl.Phys.,
     48 (1977) 2404.

7   a. H. Killesreiter and H. Baessler, Chem.Phys.Letters, 11 (1971) 411.
     b. Phys.Stat.Sol. b, 51 (1972) 657; c. M. Sugi, K. Nembach, D. Mö-
     bius and H. Kuhn, Sol.St.Comm., 15 (1974) 1867; d. H. Kuhn, J.Photo-
     chem., 10 (1979) 111.

8   a. J.W. Schultze and K.J. Vetter, Electrochim.Acta, 18 (1973) 889;
     b. K.J. Vetter and J.W. Schultze, Ber.Bunsenges.Phys.Chem., 77
     (1973) 945; c. J.W. Schultze and U. Stimming, Z.Phys.Chem.NF, 98
     (1975) 283; d. W. Schmickler and J.W. Schultze, Mod.Asp.Electro-
     chemistry, in press.

9   a. W. Schmickler and J. Ulstrup, Chem. Phys., 19 (1977) 217;
     b. R.R. Dogonadze, A.M. Kuznetsov and J. Ulstrup, Electrochim.Acta,
     22 (1977) 967.

10   a. J. Halpern AND L.E. Orgel, Disc.Faraday Soc., 29 (1960) 32;
     b. H.M. McConnell, J.Chem.Phys., 35 (1961) 508.

11   a. M.V.Vol'kenshtein, R.R. Dogonadze, A.K. Madumarov and Yu.I.
     Kharkats, Dokl.Akad.Nauk SSSR, Ser.Fiz.Khim., 199 (1971) 124;
     b. R.R. Dogonadze, Yu.I. Kharkats and J. Ulstrup, J.Electroanal.
     Chem., 39 (1972) 47; c. J.Theor.Biol., 40 (1973) 259, 279;
     d. J.Chem.Soc.Faraday Trans. II, 70 (1974) 64; e. A.M. Kuznetsov
     and Yu.I. Kharkats, Elektrokhimiya, 12 (1976) 1277; f. Elektrokhi-
     miya, 13 (1977) 1498; g. J. Ulstrup, Surf.Sci., 101 (1980) 564;
     h. A.M. Kuznetsov and J. Ulstrup, J.Chem.Phys., 75 (1981) 2047;
     i. J.Chem.Soc.Faraday Trans. II, 78 (1982) 1497.

12   a. R.R. Dogonadze and A.M. Kuznetsov, Kinetics and Catalysis,
     VINITI, Moscow, 1978, Chapters 9 and 10; b. J. Ulstrup, Charge
     Transfer Processes in Condensed Media, Springer-Verlag, Berlin,
     1979, Chapter 7.

13   a. S. Larsson, J.Amer.Chem.Soc., 103 (1981) 4034; b. J.Phys.Chem.,
     88 (1984) 1321.

14   D.N. Beratan and J.J. Hopfield, J.Amer.Chem.Soc., 106 (1984) 1584.

15   K. Ohta, G.L. Closs, K. Morokuma and N.J. Green, J.Amer.Chem.Soc.,
     in press.

16   a. A.S. Davydov, Phys.Stat.Sol. b, 90 (1978) 457; b. E.G. Petrov,
     The Physics of Charge Transfer in Biosystems, Naukova Dumka, Kiev,
     1984 (in Russian).

17   a. W. Schmickler, J.Electroanal.Chem., 113 (1980) 159; b. A.M. Kuz-
     netsov and J. Ulstrup, J.Electroanal.Chem., 195 (1985) 1.

18   M. Redi and J.J. Hopfield, J.Chem.Phys., 72 (1980) 6651.

19   a. A.M. Kuznetsov, Nouv.J.Chim., 5 (1981) 427. b. A.M. Kuznetsov and
     J. Ulstrup, Faraday Discussions, 74 (1982) 31; c. Chem.Phys.Letters,
     93 (1982) 121; d. Phys.Stat.Sol.b, 114 (1982) 673.

20   S.I. Pekar, Untersuchungen über die Elektronentheorie der Kristalle,
     Akademie Verlag, Berlin, 1954.

21   R.R. Dogonadze, E. Kálmán, A.A. Kornyshev and J. Ulstrup (Eds.),
     The Chemical Physics of Solvation, Elsevier, Amsterdam, 1986, Part B,
     Chapter 6.

22 a. R.R. Dogonadze and A.M. Kuznetsov, Physical Chemistry. Kinetics, VINITI, Moscow, 1973; b. Progr.Surf.Sci., 6 (1975) 1, and references quoted here and in ref. 12.

23 D.N. Zubarev, Irreversible Statistical Thermodynamics, Plenum, New York, 1974.

24 H. Fröhlich, Theory of Dielectrics, 2nd Ed., Clarendon, Oxford,1958.

25 a. J.J. Hopfield, Proc.Nat.Acad.Sci.USA, 71 (1974) 3640;
    b. J. Jortner, Biochim.Biophys.Acta, 594 (1980) 193; c. A.M. Kuznetsov and J. Ulstrup, Biochim.Biophys.Acta, 636 (1981) 50; d. R.R. Dogonadze and M.G. Zaqaraya, Biofizika, 24 (1984) 548. See also literature quoted in refs. 25b and 25c.

26 P.L. Dutton, J.S. Leigh, R.C. Prince and D.M. Tiede, in Tunnelling in Biological Systems, B. Chance, D.C. DeVault, H.Frauenfelder, R.A. Marcus, J.R. Schrieffer and N. Sutin (Eds.), Academic Press, New York, 1979, p. 319.

27 a. J. Deisenhofer, O. Epp, K. Miki, R. Huber and H. Michel, J.Mol. Biol., 180 (1984) 385; b. W. Zinth, E.W. Knapp, S.F. Fischer, W. Kaiser, J. Deisenhofer and H. Michel, Chem.Phys.Letters, 119 (1985) 1.

28 W.W. Parson, R.K. Clayton and R.K. Cogdell, Biochim.Biophys.Acta, 387 (1975) 265.

DOES CYTOCHROME OXIDATION IN BACTERIAL PHOTOSYNTHESIS MANIFEST
TUNNELING EFFECTS?

Mordechai Bixon and Joshua Jortner
Tel Aviv University
School of Chemistry
Tel Aviv 69978, Israel

ABSTRACT. We challenge the traditional interpretation of the temperature dependence of the Chance-DeVault cytochrome oxidation reaction
in Chromatium in terms of a transition from low-temperature nuclear
tunneling to a high-temperature activated electron transfer (ET). We
attribute this unique temperature dependence to two parallel ET processes from two distinct low-potential cytochromes to the bacteriochlorophyl dimer cation. These involve a slow activationless process,
which dominates at low temperatures (T ≤ 120 K) and an activated process, which is practically exclusive at high temperatures. Our analysis implies that low-temperature nuclear tunneling over a nuclear
barrier has not yet been documented in electron transfer reactions.

I.   PROLOGUE

     The temperature dependence of the rates of chemical reactions in a
condensed medium is expected to exhibit serious deviations from the
classical Arrhenius rate expression at low temperatures, where the
effects of zero-point motion are manifested in tunneling effects. A
multitude of chemical reactions, e.g., polymerization,[1] proton transfer
in solution,[2] isomerization[3] and group transfer in biophysical systems,[4] reveal low-temperature, temperature-independent rates, which
originate from nuclear tunneling effects. Similar phenomena were predicted[5] for electron transfer (ET) in condensed phases; however, the
qualitative manifestation of nuclear tunneling on low-temperature ET
rates in physico-chemical systems have not yet been documented. For ET
in biophysical systems Chance and DeVault have discovered[6] a unique
temperature dependence of the rate constant, k, for electron transfer
from the low potential cytochrome c (cytc) to the bacteriochlorophyl
dimer $(BChl)_2$ cation in the photosynthetic bacterium Chromatium
vinosum. A sharp transition is exhibited from a high-temperature
activated region, where k drops by three orders of magnitude (i.e.,
from k ≈ 10^6 sec^{-1} at 300 K to k ≈ 5×10^2 sec^{-1} at 120 K) to a temperature independent k at low-temperatures (T < 120 K). Subsequent
studies[7,8] on the same system revealed the same general behaviour with

J. Jortner and B. Pullman (eds.), Tunneling, 361–371.

an even sharper transition of k between the two regions. The characteristics of this reaction[6-8] were interpreted within the general
framework of the nonadiabatic multiphonon ET theory.[9-13] It was proposed[9,10] that the temperature independent region manifests a nuclear
tunneling process for ET, while the temperature dependence of this
reaction reflects a transition from tunneling at low temperatures to an
activated rate process at high temperatures. The "best fit" of the experimental data in terms of the ET theory is portrayed in Fig. 1. Although the qualitative traditional interpretation[9,10] of the temperature dependence of the Chance-DeVault reaction is very plausible and
appealing, the quantitative physical information, which emerges from
such an analysis[9-13] raises serious conceptual difficulties. The nuclear and electronic parameters, which are deduced from such an analysis,
are unreasonably large. In particular, we note that:

(1) The high transition temperature $T_0 = 120$ K[6-8] implies that
the characteristic vibrational mode, which is effectively coupled to
the electronic process, i.e., undergoes a large configurational change
during ET, has a frequency of $\hbar\omega = 400-500$ cm^{-1}.[9-13] Such a high frequency must correspond to intramolecular vibrational modes of the porphyrin rings in the cytochrome c or/and in the $(BChl)_2$. The effective
coupling of high-frequency intramolecular vibrational modes to this ET
process seems to contradict the results of a recent analysis[14] of the
temperature dependence of two ET reactions in the reaction centers of
bacteria, which are characterized[14] by a low effective frequency $\hbar\omega =$
100 cm^{-1}, and which correspond to coupling with the exterior polar protein medium modes.

(2) A large nuclear reorganization energy $E_c = 18500$ cm^{-1} in conjunction with the high characteristic frequency ($\hbar\omega = 500$ cm^{-1}), which
emerges from the quantitative analysis of the temperature dependence of
the ET rates,[13] manifests huge intramolecular configurational changes
which accompany the ET process. This large value of E_c corresponds to
the normal region of ET. The adaptation of the inverted region of ET
usually results in a lower value of ET. However, it is impossible to
fit the experimental data[6-8] to the inverted region using a high frequency. The large value of E_c implies a change of 0.7 Å in the nuclear
equilibrium configuration of four intramolecular vibrational modes.
This conclusion is in contrast with the available structural and kinetic data. X-ray crystallographic analysis for the two oxidation states
of tuna cytc[15] reveal small configurational changes in the heme groups.
Reorganization energy calculations[16] for the cytc electron exchange
reaction, which are based on the crystallographic data,[15] results in
the low value of $E_c = 350$ cm^{-1} for the intramolecular reorganization
energy. The contribution of the intramolecular configurational changes
within $(BChl)_2$ accompanying ET are also expected to be small, as can be
inferred from the kinetics of ET from electronically excited $(BChl)_2$ to
bacteriopheophytin.[17] The weak temperature dependence of this reaction
between 4-300 K implies that it is activationless.[18] Accordingly, the
reorganization energy is equal to the free-energy ΔE of this reaction.[18] Taking $\Delta E = 1200$ cm^{-1},[17] we infer that for this reaction $E_c \lesssim$
1200 cm^{-1}, which is one order of magnitude lower than E_c deduced[13]
for cytochrome oxidation.

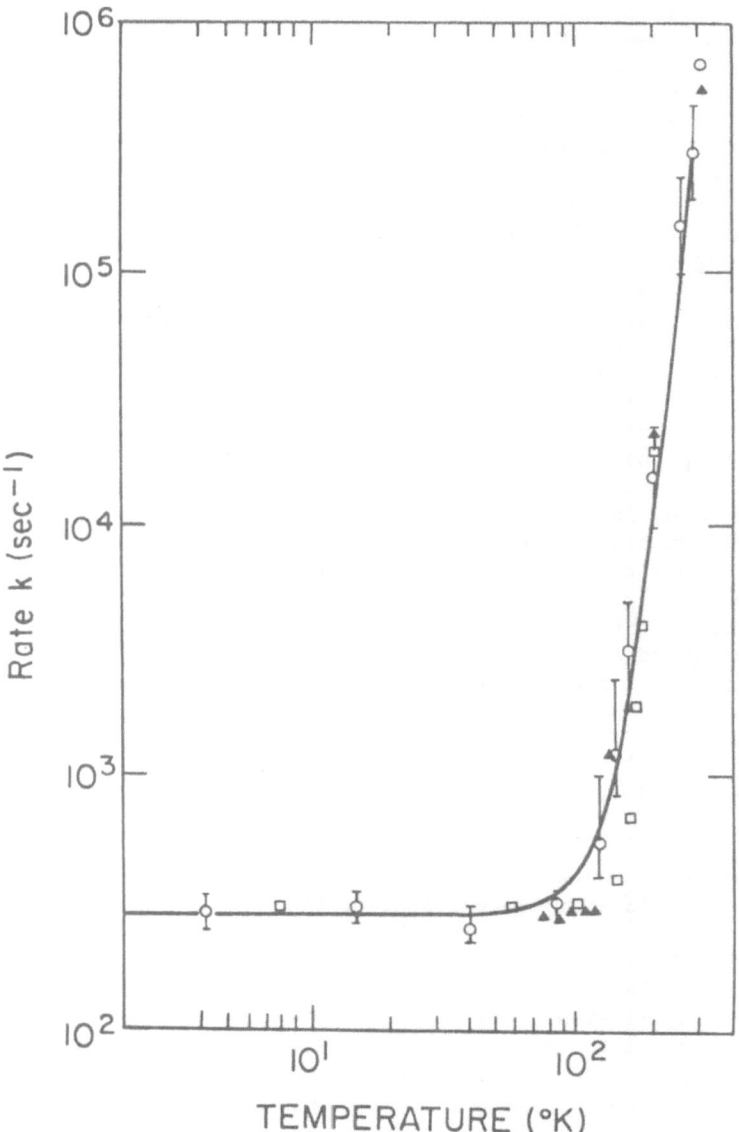

<u>Fig. 1</u>: The traditional fit of the experimental data for the Chance-DeVault reaction[6-8] in terms of a single ET process. The fit was taken from ref. 13. The parameters are $\hbar\omega$ = 500 cm^{-1}, E_r = 18500 cm^{-1} and $|V|$ = 90 cm^{-1}.

(3) A large electronic coupling $V = 90$ cm^{-1} is required for a quantitative fit of the ET data.[13] Such a large interaction can only arise from a close contact between cytc and (BChl)$_2$, which contradicts all the available structural information. ESR data give a center-to-center distance of ~ 25 Å for the separation between these prosthetic groups in Chromatium,[19] which is close to the distance of 23 Å obtained from crystallographic studies for Rps. Viridis.[20] At such distances one expects[9,10,18] that V is lower than 1 cm^{-1}.

(4) A close examination of the fit of the experimental data for Chromatium[6-8] to the multiphonon ET theory[9-13] clearly indicates the inadequacy of the theory in the vicinity of the transition temperature around 120 K (see Fig. 1). The experimental data correspond to a sharp break rather than a smooth change. The only way to achieve a better agreement between theory and experiment is to increase the parameters E_c and V even further, which contradicts both facts (see points (2) and (3)) and intuition.

(5) Some experimental data, which have accumulated regarding analogous ET processes in other photosynthetic bacteria[21-24] reveal that the temperature dependence of the cytc oxidation reaction in Chromatium is by no means universal, as other systems reveal a broad spectrum of temperature dependences of the ET rates.

This information provides compelling evidence that it is highly improbable that a tunneling process prevails at temperatures around 100 K for the ET between cytc and the primary donor. The nature of the Chance-DeVault reaction,[6-8] which served as a touchstone[9,10] for the applicability of the ET theory to biological systems, has to be re-examined.

II. A CONJECTURE ON THE MECHANISM OF CYTOCHROME OXIDATION
                    IN CHROMATIUM

We assert that cytc oxidation in Chromatium does not exhibit a transition from low-temperature nuclear tunneling to a high-temperature activated process, which involves ET from a single cytc to the dimer cation. Rather, we propose that the unique temperature dependence for this reaction originates from the combination of several parallel ET reactions with different activation energies, each of them occurring from a different cytochrome molecule. At least for three photosynthetic bacteria,[19,20,23] it is known that the reaction center includes four cytochrome c molecules involving two high-potential and two low-potential cytcs. Some evidence for parallel ET reactions is available for Chromatium at room temperature, where the ET rate from the low-potential cytochrome and the high-potential cytochrome are (1 μsec)$^{-1}$ and (2 μsec)$^{-1}$, respectively,[17] while at low temperatures oxidation by the high-potential cytochrome is tenfold slower than that by the low-potential cytochrome.[7] Another example for parallel reactions is provided by the high-potential cytochromes in Thiocapsa Pfennigii[25] in which one of the cytochromes is oxidized more rapidly than the second. We propose that both the two low-potential cytochromes in Chromatium can be operative in ET and (at least) two ET processes occur:

(I) A moderately slow activationless process.

(II) An activated process.

At high temperatures reaction (II) is much faster, while at low temperatures reaction (I) predominates. The low-temperature nuclear tunneling contribution from reaction (II) is negligible.

Figure 2 portrays schematic nuclear potential energy curves for the two parallel ET processes. It is important to emphasize that the nuclear coordinate in Fig. 1 represents the configurations of the exterior protein medium as, on the basis of a recent analysis of ET processes,[14] we assert that the dominating nuclear contribution to such ET processes involves the coupling with the low-frequency medium modes. As is apparent from Fig. 2, the activationless reaction (I) involves the crossing of the potential surfaces at the minimum of the initial state, while reaction (II) is characterized by a nuclear barrier. The qualitative differences between the dynamics of reactions (I) and (II), which are reflected in the distinct temperature dependence of their rates, originate from different medium reorganization energies of the protein low-frequency modes. These distinct medium reorganization energies, E_m, for the parallel ET reactions from the cytc molecules are due to: (i) Different local protein configurations of polar groups around the two cytochromes. (ii) Different donor-acceptor distances, which provide different electrostatic contributions to E_m. We shall now proceed to analyze the temperature dependence of reactions (I) and (II) in terms of the nonadiabatic multiphonon ET theory.[9-13,18]

III. APPLICATION OF THE ELECTRON TRANSFER THEORY

The ET rate constant, k, can be expressed in the well-known form of[9-13,18]

$$k = (2\pi/\hbar) \; |V|^2 \; F \; , \tag{1}$$

where V is the two-center, one-electron exchange (or superexchange) integral and F is the thermally averaged nuclear overlap factor. The latter nuclear term involves two contributions,[18] the medium reorganization energy E_m and the intramolecular reorganization energy E_c, with the total nuclear reorganization energy being $E_r = E_m + E_c$. It has been demonstrated[14] that the major contribution to E_r originates from the changes in the equilibrium configurations of the medium modes, i.e., $E_m/E_c \gg 1$. Under these circumstances a single-mode approximation to the ET rate is applicable and Eq. (1) reduces to[18]

$$k = \frac{2\pi |V|^2}{\hbar^2 \omega} \left(\frac{\overline{v} + 1}{\overline{v}} \right)^{p/2} \exp[-S(2v + 1)] \; I_p(2S\sqrt{\overline{v}(\overline{v} + 1)}) \; , \tag{2}$$

where ω is the (average) effectively coupled vibrational frequency, $\overline{v} = [\exp(\hbar\omega/kT) - 1]^{-1}$ is the thermal population of that mode,

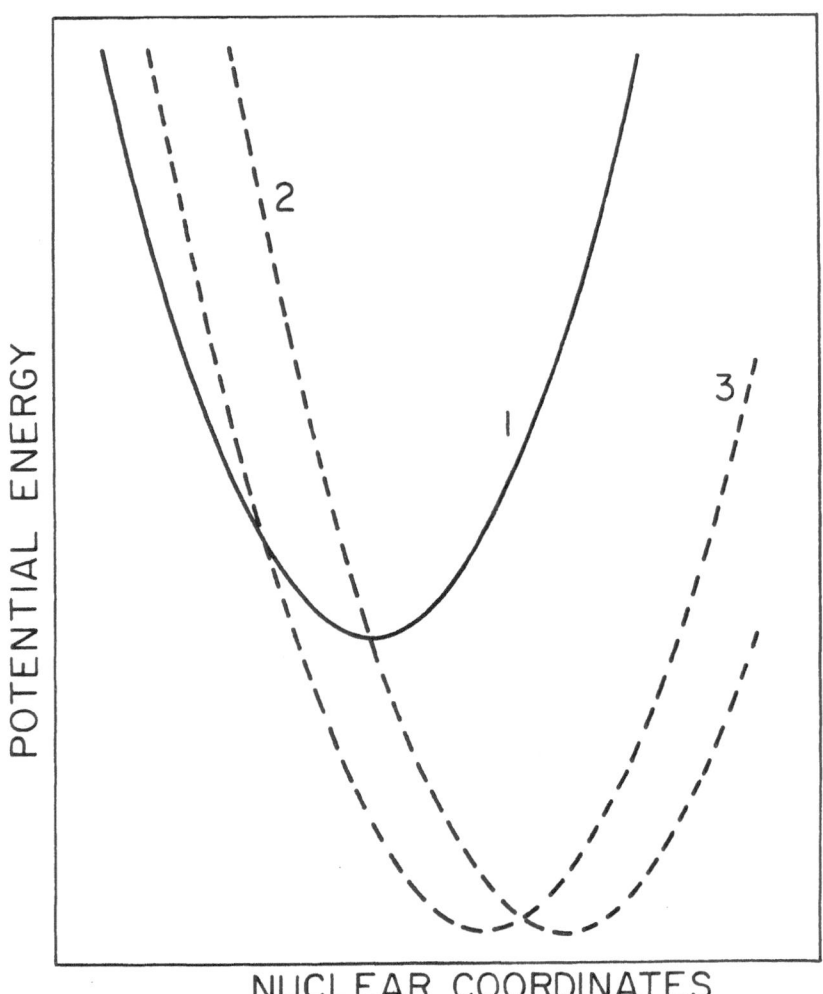

Fig. 2: A schematic representation for the nuclear potential surfaces for the two parallel ET cytochrome oxidation reactions in
Chromatium. Two low-potential cytochromes, labeled cytc$_1$ and
cytc$_2$, can transfer an electron to (BChl)$_2^+$. The activationless
reaction (I) involves the transition from the potential surface
(1) to the potential surface (2). The activated reaction (II)
involves the transition between the potential surfaces (1) and
(3) and corresponds to the inverted region for ET.

$S = E_r/\hbar\omega$ is the nuclear reorganization energy in frequency units and
$p = \Delta E/\hbar\omega$ is the energy gap ΔE, i.e., the free energy of the reaction,
in frequency units. Regarding the parameters required for the quanti-
tative description of reactions (I) and (II), we chose the energy gap

$\Delta E = 3500 \text{ cm}^{-1}$[6] for both reactions. The characteristic average phonon frequency was chosen as $\hbar\omega = 100 \text{ cm}^{-1}$, which constitutes a reasonable average of the frequency spectrum of typical proteins.[26]

Reaction (I) is attributed to an activationless process, whose reorganization energy $E_r^{(I)}$ is equal to the energy gap, i.e., $E_r^{(I)} = \Delta E = 3500 \text{ cm}^{-1}$, so that $p = S = 35$. Equation (2) for an activationless process in the strong coupling limit, i.e., $S \gg 1$, reduces to

$$k = k(T = 0) \left[\frac{\exp(\hbar\omega/kT) - 1}{\exp(\hbar\omega/kT) + 1} \right]^{\frac{1}{2}} \qquad (3)$$

where

$$k(T = 0) = \frac{2\pi |V|^2}{\hbar^2 \omega (2\pi p)^{\frac{1}{2}}} \qquad (4)$$

is the low-temperature ($kT \ll \hbar\omega$) rate. The low-temperature data for the Chance-DeVault reaction, which we attribute to the activationless reaction (I), can well be fit by Eq. (3) with $k_I(T = 0) = 500 \text{ sec}^{-1}$ and $\hbar\omega = 100 \text{ cm}^{-1}$ (Fig. 3). The value of $k_I(T = 0)$ results in the low value of the electronic coupling $|V_I| = 8 \times 10^{-4} \text{ cm}^{-1}$.

Reaction (II) is an activated process, for which Eq. (1) reduces in the high temperature limit ($kT \gg \hbar\omega$) to the well-known form[9-13]

$$k = \frac{2\pi |V|^2}{\hbar (4 E_r kT)^{\frac{1}{2}}} \exp \left[- \frac{(\Delta E - E_r)^2}{4 E_r kT} \right] \qquad (5)$$

The fit of the high-temperature ($T > 120$ K) experimental data (Fig. 3) in terms of the activated process is not unique and different sets of parameters can be utilized. Taking $\Delta E = 3500 \text{ cm}^{-1}$[6] and looking for a reasonable value of E_r, which is not too high, we are forced to choose the nuclear potential surfaces (Fig. 2) as corresponding to the inverted region. In this domain it is possible to obtain a reasonable fit of the data by taking $\omega \leqslant 160 \text{ cm}^{-1}$ and adjusting the values of S and of $|V|$. Some numerical results are presented in Fig. 3, which clearly indicate that it is impossible to achieve a good fit in the inverted region for higher frequencies. A simple fit of the parameters can be misleading, as this research area has previously been fraught with very convincing fits between theory and experiment, which resulted in unreasonable physical parameters. In order to explore the region of reasonable physical parameters, we present in Fig. 4 the frequency dependence of the adjusted parameters E_r and $|V|$, which result in reasonable fits. It is apparent that at frequencies exceeding 140 cm^{-1} the energetic parameters are unacceptable, i.e., the reorganization energy becoming too small ($E_r < 500 \text{ cm}^{-1}$) and the electronic coupling too large ($|V| > 30 \text{ cm}^{-1}$).

The high-temperature ($T > 120$ K) data (Fig. 3), which are attributed to reaction (II), were fit by Eq. (2) using $\hbar\omega = 100 \text{ cm}^{-1}$ and $\Delta E = 3500 \text{ cm}^{-1}$ and adjusting the values of $S(E_r/\hbar\omega)$ and $|V|$. The best

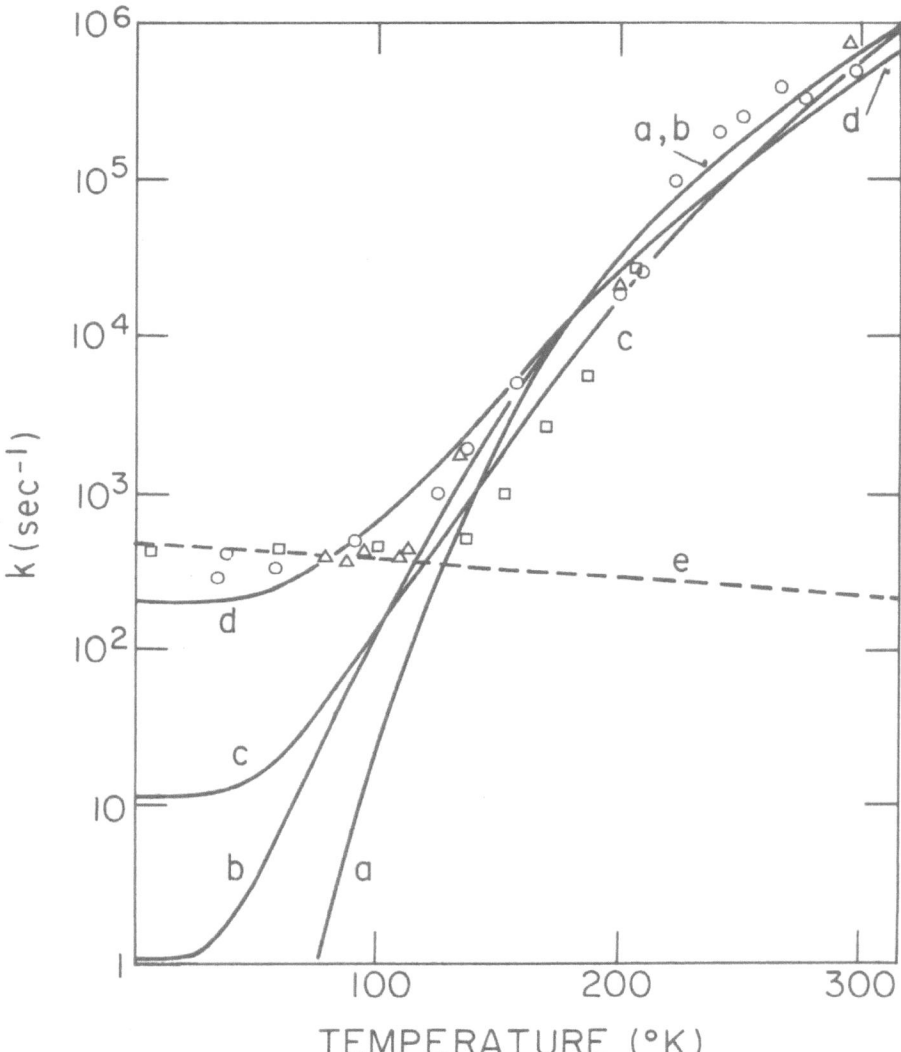

Fig. 3: The temperature dependence of the rates of the Chance-DeVault cytochrome oxidation reactions in Chromatium. The points represent experimental data from the following sources: 0 - reference 6, Δ - reference 7, and □ - reference 8. The dashed curve (e) corresponds to an activationless process, Eq. (3), with $k(T = 0) = 500$ sec^{-1} and $\hbar\omega = 100$ cm^{-1}. The solid curves correspond to an activated process with $\Delta E = 3500$ cm^{-1}, and (a) $\hbar\omega = 60$ cm^{-1}, $E_r = 1000$ cm^{-1}, $|V| = 1$ cm^{-1}; (b) $\hbar\omega = 100$ cm^{-1}, $E_r = 900$ cm^{-1}, $|V| = 1.8$ cm^{-1}; (c) $\hbar\omega = 140$ cm^{-1}, $E_r = 560$ cm^{-1}, $|V| = 32$ cm-1; (d) $\hbar\omega = 180$ cm^{-1}, $E_r = 360$ cm^{-1}, $|V| = 500$ cm^{-1}.

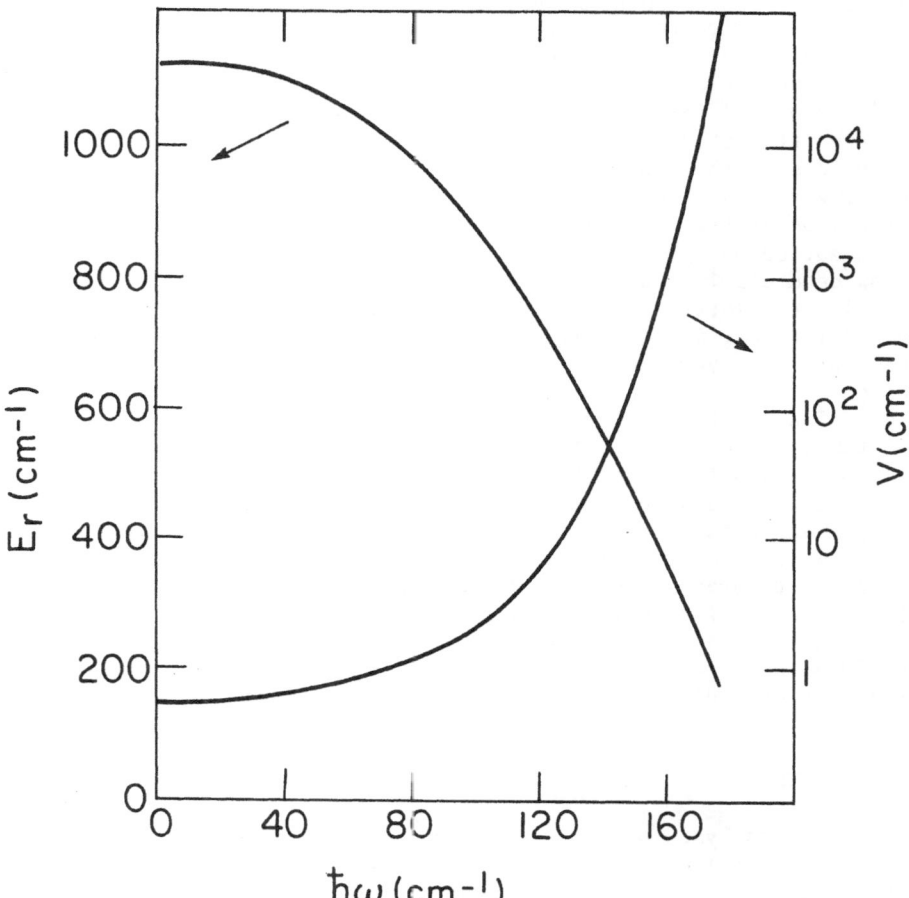

Fig. 4: The values of the reorganization energy E_r and the electronic coupling $|V|$ that give the best fit to the high temperature experimental data, as a function of the assumed phonon frequency.

fit is obtained by using $E_r = 9C0$ cm^{-1} and $|V_{II}| = 1.8$ cm^{-1}. For this low value of ω the low-temperature tunneling limit of reaction (II) should be exhibited below 20 K (Fig. 3). This tunneling process corresponds, according to Eq. (2), to the rate $k_{II}(T = 0) = 1$ sec^{-1}, which is negligible relative to the experimental low-temperature rate (500 sec^{-1}).

The fit of the experimental data in terms of reactions (I) and (II) (Fig. 3) is as good as can be expected. From this analysis, we conclude that:

(1) The low characteristic (mean) nuclear vibrational frequency ($\omega \simeq 100$ cm^{-1}), which is required to fit the data for reactions (I) and (II), implies that the major contribution to the nuclear reorganization involves the protein medium. This nuclear motion involves the disloca-

tion and rotation of polar groups of the protein around the prosthetic groups. This characteristic frequency is close to that extracted recently for other ET processes in the reaction center.[14]

(2) The nuclear reorganization energies $E_r^{(I)}$ = 3500 cm^{-1} and $E_r^{(II)}$ = 900 cm^{-1} for the two parallel ET processes are reasonable. These are attributed to the protein medium reorganization energy. These values of the medium reorganization energy are close to the corresponding values of E_r recently derived[14] for other ET processes in the reaction center. The physically unreasonable huge intramolecular reorganization energy, which has emerged from previous analysis,[13] has now been eliminated.

(3) The electronic coupling terms for reactions (I) and (II) are widely different, i.e., $|V_I/V_{II}| \simeq 0.5 \times 10^{-3}$. This result implies a large separation between the two distinct low-potential cytochromes involved in the low-temperature and in the high-temperature ET processes. The low-temperature activationless ET occurs from a distant cytochrome, while the high-temperature activated ET proceeds from a cytochrome which is closer to the dimer. Invoking the primitive relation for the distance scale (R) of the electronic coupling[9,10,18] $V \alpha \exp(-\alpha R)$ with $\alpha = 0.6$ Å^{-1},[9,10,27] we obtain a rough estimate of $R_I - R_{II} \sim 13$ Å for the differences in the distances of the two low-potential cytochromes from the special pair. This distance is consistent with information derived from ESR data,[19] which implies that the separation between the two low-potential cytochromes in Chromatium exceeds 10 Å, while crystallographic data for Rps. Viridis yield the center-to-center distance of ~ 14 Å between two cytochromes.[20]

IV. EPILOGUE

Our new analysis of the celebrated Chance-DeVault cytochrome oxidation in Chromatium in terms of two parallel reactions resolves some of the mysteries which were prevalent in the theoretical interpretation of ET in bacterial photosynthesis. Low-temperature nuclear tunneling is, of course, a perfectly acceptable physical process. However, we propose that for the cytochrome oxidation this process is masked by a parallel activationless ET reaction, which is efficient at low temperatures. We are aware of one additional report of low-temperature independent ET rate in hybrid hemoglobin,[28] which was attributed to nuclear tunneling. However, the small low-temperature rate in [Zn Fe^{+3}] hybrid hemoglobin[28] was derived on the basis of the assumption that the triplet-ground state intersystem crossing is identical for the [Zn Fe^{+3}] and [Zn Fe^{+2}] systems. Obviously, further work is required to establish the occurrence of low-temperature nuclear tunneling induced ET in [Zn Fe^{+3}] hybrid hemoglobin. From the point of view of general methodology, our iconoclastic proposal regarding the Chance-DeVault reaction carries a disappointing message. It would appear that low-temperature tunneling over a nuclear barrier for ET in biological systems has not yet been documented.

REFERENCES

1. V.G. Goldanskii, Scientific American 254, 38 (1986).
2. W. Siebrand, T.A. Wildman and M.Z. Zgierski, J. Am. Chem. Soc. 106, 4089 (1984).
3. M.L. Applebury, K.S. Peter and P.M. Rentzepis, Biophys. J. 23, 375 (1978).
4. N. Alberding, R. Austin, K. Beeson, S.S. Chan, L. Eisenstein, H. Frauenfelder and T.M. Hordlund, Science 192, 1002 (1976).
5. N.R. Kestner, J. Logan and J. Jortner, J. Phys. Chem. 78, 2148 (1974).
6. D. DeVault and B. Chance, Biophys. J. 6, 825-847 (1966).
7. P.L. Dutton, T. Kihara, J.A. McCray and J.P. Thornber, Biochim. Biophys. Acta 226, 81-87 (1971).
8. B. Hales, Biophys. J. 16, 471-480 (1976).
9. J.J. Hopfield, Proc. Nat. Acad. Sci, USA 71, 3640-3644 (1974).
10. J. Jortner, J. Chem. Phys. 64, 4860-4867 (1976).
11. A.M. Kuznetsov, N.C. Søndergård and J. Ulstrup, Chem. Phys. 29, 383-390 (1978).
12. A. Sarai, Biochim. Biophys. Acta 589, 71-83 (1980).
13. E. Buhks, M. Bixon and J. Jortner, Chem. Phys. 55, 41-48 (1981).
14. (a) J. Jortner and M. Bixon, Comments on Molecular and Cellular Biophysics (in press). (b) M. Bixon and J. Jortner, J. Phys. Chem. (submitted).
15. T. Takano and R.E. Dickerson, J. Mol. Biol. 153, 95-115 (1981).
16. A.K. Churg, R.M. Weiss, A. Warshel and T. Takano, J. Phys. Chem. 87, 1683-1694 (1983).
17. W.W. Parson and B. Ke, in: Photosynthesis: Energy Conversion by Plants and Bacteria, ed. Govindjee, Academic Press, London, New York, Vol. I, pp. 331-385 (1982).
18. J. Jortner, Biochim. Biophys. Acta 594 193-230 (1980).
19. D.M. Tiede, J.S. Leigh and P.L. Dutton, Biochim. Biophys. Acta 503, 524-544 (1978).
20. J. Deisenhofer, O. Epp, K. Miki, R. Huber and R. Michel, J. Mol. Biol. 180, 385-398 (1984).
21. B. Chance, T. Kihara, D. DeVault, W. Hildreth, M. Nishimura and T. Hiyama, in: Photosynthetic Research, ed. H. Metzner, Vol. III, 1321-1346 (1969).
22. D. DeVault, Quart. Rev. Biophys. 13, 387-564 (1980).
23. S.K. Chamorovsky, A.A. Konovenko, S.M. Remennikov and A.B. Rubin, Biochim. Biophys. Acta 589, 151-155 (1980).
24. T. Kihara and J.A. McCray, Biochim. Biophys. Acta 292, 297-309 (1973).
25. R.E.B. Seftor and J.P. Thornber, Biochim. Biophys. Acta 764, 148-159 (1984).
26. N. Gó, T. Noguti and T. Nishikawa, Proc. Nat'l Acad. Sci. USA 80, 3696-3700 (1983).
27. J. Jortner and M. Bixon, Proc. of Conf. on Protein Structure, Philadelphia, 1985 (in press).
28. S.E. Peterson-Kennedy, J.L. McGourty and B. Hoffman, J. Am. Chem. Soc. 106, 5010 5012 (1984).

WHAT CAN BE LEARNED FROM LOW TEMPERATURE REACTIVITY ON ROOM TEMPERATURE REBINDING KINETICS OF HEME PROTEINS?

Noam Agmon
Departement of Physical Chemistry
The Hebrew University of Jerusalem
Jerusalem 91904, Israel.

ABSTRACT. From the low temperature rebinding kinetics of heme proteins we conclude that a single coordinate description of the biophysical process is insufficient. An additional "protein coordinate" is introduced, which is frozen at low temperatures, and relaxes following dissociation at high temperatures. This relaxation causes the barrier to rebinding to increase with time, leading to a biologically significant "self-cooperativity" effect, on the tertiary-structure level, for a single heme subunit. The predicted effect is supported by recent transient Raman scattering experiments. A second conclusion is that the binding rate coefficient must depend on a fractional power of solvent viscosity. The same non-Kramers behavior has recently been observed also in photochemical isomerization. It may therefore be due to the multi-dimensionality of macromolecular dynamics.

1. INTRODUCTION

The kinetics of ligand binding to heme proteins has been followed experimentally from nanoseconds to seconds and for temperatures ranging from 2K to room temperature [1-3]. Below 25K binding seems to occur by quantum mechanical tunneling [1a]. Below 200K the rebinding kinetics is non-exponential, indicating the involvement of a distribution of barrier heights for different protein conformations [1]. For higher temperatures the rebinding is initially slower, becoming exponential at long times. It shows evidence of several sequential processes, where the ligand escapes from the heme pocket to other locations in the protein and ultimately to the solvent.

The question that naturally arises is what is the relevance of the extensive low-temperature experiments to the biological process which, after all, occurs around room temperature? The Illinois group [1] calculates the room-temperature pocket-to-iron binding coefficient as the appropriate average over the low temperature distribution. This tacitly assumes that at the higher temperatures the protein coordinate is completely relaxed at all times, and that during this relaxation process the distribution retains its shape as determined in the frozen-protein limit.

We [4a] have taken an explicit account of protein motion by constructing potential energy surfaces in terms of both ligand-iron and protein coordinates. Dynamics on these surfaces disagrees with the abovementioned assumption, and leads to new physical insights. These include a "self-cooperativity" effect and non-Kramers viscosity dependence. Both follow from the abandonement of a one-dimensional reaction profile in favour of a multi-dimensional surface, as described below.

J. Jortner and B. Pullman (eds.), Tunneling, 373–381.

2. SUBUNIT "SELF COOPERATIVITY"

A schematic description of the binding process is shown in Figure 1: The iron (+2), which is located in the center of the porphyrin moeity, is bound in its fifth coordination site to a nitrogen atom of the imidazole ring of a histidine residue (the "proximal histidine"), which is a constituent of the globin's F-helix. When a ligand, such as O_2 or CO, located in the "heme pocket" between the porphyrin and the "distal histidine", binds to the sixth coordination site of the iron, the iron moves a few tenths of an Ångström from its domed geometry into the porphyrin plane. In this process the iron pulls upon the F-helix, thus perturbing the tertiary structure of the protein. When these perturbations reach the interface between the α and β subunits, a quaternary structural change results. In what follows we shall concentrate on tertiary structural changes in a single subunit.

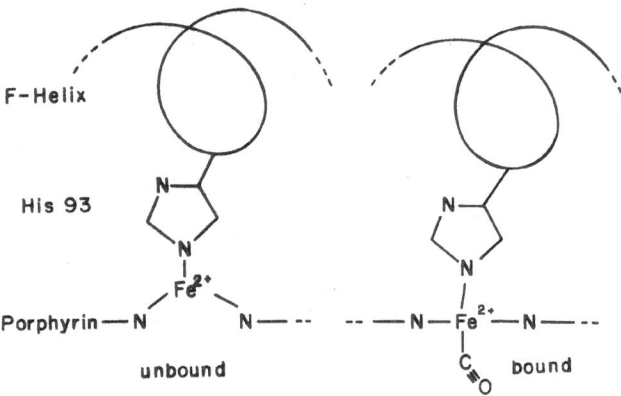

Figure 1. A schematic representation of the heme's binding site in the deoxy and bound states.

The above discussion suggests [4a] that two coordinates are of prime relevance to the binding process. The first is the ligand-iron distance, r, which is the "reaction coordinate" as in a gas-phase dissociation reaction. The second is a "protein coordinate", x. (Such has been recently introduced also by the Illinois group [1e.5]). This may possibly be identified with a stretching or bending mode of the F-helix spring which, by pulling the iron out of the porphyrin plane or increasing the proximal histidine tilt relative to this plane [6], reduces the binding rate coefficient. It is therefore reasonable to assume that the equilibrium position of the protein-spring changes upon ligand-binding, say from $x=0$ in the deligated heme to $x=x_0$ for the bound form. In contrast, the force-constant f for the protein-spring should hardly change by local perturbations at the iron.

An example for such a potential energy surface (for CO binding to the β subunit of adult human hemoglobin) is shown in Figure 2. The details are given in the work with Hopfield [4a]. Let me just mention that the value of x_0 (0.8 a.u.; 0.6 a.u. for the α chain) is indeed in the range of the out/in-plane iron motion. Note that the potential shown is limited to geminate-recombination from the heme pocket: At larger values of r there are additional barriers for leaving the pocket and for entering the solution.

At low temperatures one may assume that the protein motion is completely frozen, while the ballistic ligand motion inside the pocket is not. Before the dissociating laser pulse, one has an equilibrium (Gaussian) distribution over protein coordinate for the ligated heme.

Dissociation, electronic relaxation and iron motion are much faster than rebinding (otherwise our potential energy surface would change with time). Therefore, as in other Franck-Condon processes, the initial distribution is translated horizontally into the deoxy-heme valley. At low temperatures, each protein-conformation x reacts independently of hemes frozen into other conformations. Consequently, the distribution of barrier heights is determined simply from the initial distribution, together with the x-dependence of the barrier height, $V^{\neq}(x)$.

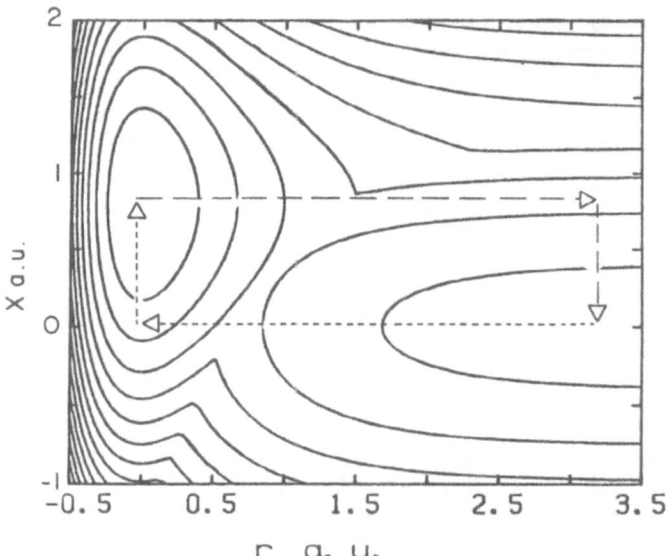

Figure 2. The potential energy surface for CO binding to the β chain of adult human hemoglobin, as fitted [4a] to the low temperature experimental data [1b]. r is the $CO-Fe$ distance, so that the well at $r=0$ represents the bound species. x is a "protein coordinate". The long and short dashed lines depict the dissociation and binding processes, respectively.

For the case considered here, where the two wells are shifted ($x_0 \neq 0$) but the protein force constant f is the same, $V^{\neq}$ increases monotonically (almost linearly) with x. Therefore the distribution of barrier heigths is nearly a Gaussian function of x. Recent work [3] notes the merits of such a distribution in fitting the low-T non-exponential kinetics. Note that if f is independent of ligand binding, x_0 *must* be non-zero, otherwise $V^{\neq}$ becomes x independent and the kinetics exponential. When $x_0=0$ but f varies between the two wells, $V^{\neq}(x)$ has an extremum (maximum or minimum, depending on whether f is larger for the bound or deoxy form) at $x=0$. As a result, the distribution of barrier heights becomes very non-symmetric, with a cutoff energy $V^{\neq}(0)$. Hence by letting f vary (for $x_0 \neq 0$), one may produce less symmetric distributions. For simplicity, we have not utilized this option.

At higher temperatures, protein motion is no longer frozen. The stochastic motion in x can be described [4] by a diffusion (Smoluchowski) equation (or by more complex equations of motion), with an x-dependent rate coefficient, $\kappa(x)$,

$$\partial p(x,t)/\partial t = D L p - \kappa(x) p \tag{1}$$

In Eq. (1), $p(x,t)$ is a probability distribution function, whose temporal evolution is determined by the diffusion operator L, the diffusion constant D, and the sink term $\kappa(x)$. As a result, protein relaxation follows dissociation. It may occur during (for intermediate D) or prior to (for large D) rebinding. The relaxation is towards the minimum in the well of the deoxy state, from which the barrier for recombination is higher, and the rate-coefficient smaller. Similarly, subsequent to rebinding (e.g., from the minimum of the deoxy heme well), there is protein relaxation into the bottom of the ligated heme well, from which dissociation is less probable. This is demonstrated by the dashed lines in Figure 2, which trace schematically the temporal motion of the peak of the distribution function. It is quite in contrast to the assumption made by the Illinois group [1], that the peak in the low T distribution gives the room temperature barrier to the geminate recombination step.

By considering the low-temperature data, we reach the conclusion that there must be a novel "self cooperativity" effect, on the tertiary-structure level, for a single heme subunit. Once ligand binding or unbinding occurs, the motion of the iron triggers a protein relaxation process which decreases the probability of the reverse reaction. This may be of considerable biological significance, since (in addition to quaternary-level cooperativity) it ensures the efficient utilization of heme molecules in oxygen intake and release.

Recent transient Raman scattering experiments [6] provide support to the abovementioned mechanism. In these experiments the iron-proximal histidine bond is monitored subsequent to ligand dissociation. It is found that its Raman frequency ν decreases monotonically during the time regime of tertiary structural change. On the other hand lower ν correlates with smaller oxygen affinity and a slower recombination rate. This has been determined by [6]: (i) Considering quaternary structures (R and T) and species which vary in heme oxygen affinity; (ii) Considering the pH effect: Lower pH (such as under the condition of excess CO_2 in the blood) triggers oxygen release. This decreased affinity correlates nicely with a smaller ν for a given time delay after dissociation, as well as with diminished geminate recombination for lower pH; (iii) Observing, for a variety of conditions, that hemes with smaller iron-histidine frequency ν, have a higher quantum yield for photolysis (averaged over the first 10 ns following dissociation). This again can be ascribed to a smaller well for the ligated species and a larger barrier for recombination.

In the above discussion it is clear that structure-reactivity correlations [7] are operating here: For a given ligand the "intrinsic barrier" is more or less constant, hence a decrease in ligand affinity (decreasing thermodynamic drive for recombination) correlates with an increase in recombination barrier height. For different ligands, such as CO and O_2, the intrinsic barrier varies. Hence the slow binding ligand CO (slow in spite of its larger affinity) has a larger intrinsic barrier. As a result, its barrier is more symmetric in r, and changes in the affinity are more markedly reflected in variations in the kinetics than in the case of fast binding ligands, such as NO and O_2 [7a].

It is tempting to ask to what extent such an auto-catalytic effect may be operating in excited-state biological electron transfer reactions [8], by slowing down the back-transfer in the ground state. For example, the recently deciphered structure of the photosynthetic reaction center of purple bacterium [9], shows that the two subunits (L and M) in which the primary electron donor and acceptor are imbedded, have a "spring-bed" (or "harmonica") structure. with five membrane-spanning helices for each subunit. If, following the initial electron transfer process, this spring structure relaxes, the distance between donor and acceptor increases, thus diminishing the probability for the back electron transfer reaction and increasing the efficiency of the photosynthetic process. Such a mechanism is corroborated by the discovery [10] that the distribution of electron-transfer "distances", in a bacterial reaction center, is peaked at a larger separation for samples which were frozen under illumination (in comparison to samples frozen in the dark).

The importance of conformational changes in protein dynamics [11] is also demonstrated in the field of enzyme kinetics. An example is the "hinge-bending motion" in lysozyme [11]. The cleft between the two globular domains surrounding the binding site, increases prior to binding, and closes down before the enzymatic reaction takes place. Such large amplitude Brownian motion as a prerequisite to enzymatic reaction, results in reaction kinetics which is very much affected by solvent viscosity. This is manifested in the modified form for the Michaelis-Menten equation [12], which should be observed experimentally.

3. NON-KRAMERS VISCOSITY DEPENDENCE

Kramers' rate theory [13] predicts that in a viscous medium the rate coefficient k should be inversely proportional to the viscosity η. Experiments with ligand binding to heme proteins show [1c] that the viscosity dependence of k is better described by

$$k = (A/\eta^a + B)\exp(-\Delta G^{\neq}/k_B T) \tag{2}$$

The preexponential factor is different from the ordinary Kramers behavior, A/η, in two ways: (i) At finite viscosities the rate depends on a fractional power of viscosity, η^a, where $0 < a \leq 1$; (ii) The reaction rate does not vanish at infinite viscosities. This same behavior is qualitatively observed in solving [4b] the reactive-diffusion equation (1).

Let us suppose [4a] that the surrounding solvent affects reactivity mainly by restricting the protein motion from the outside. (The effect of the few, if any, solvent molecules in the pocket, cannot really be described as a "viscosity effect" on ligand motion; their discrete nature should be taken into account). We therefore consider the effect of varying the diffusion constant D (assumed proportional to $1/\eta$), on the solution of Eq. (1).

When $D=0$, the reaction rate is finite (though non-exponential), since each and every conformation can still react with a rate coefficient $\kappa(x)$. For a non-zero D, let us consider two limiting forms for $\kappa(x)$, as shown in Figure 3. In the case of a constant $\kappa(x)$, all protein conformations react with the same rate $k = \kappa$, which is independent of viscosity. This is the limit where the exponent $a = 0$. The other extreme case is $\kappa(x)$ which is a delta function. Then, as in Kramers' model [13], k is inversely proportional to η, and $a = 1$. It is therefore reasonable to assume that intermediate forms of $\kappa(x)$ will give rise to values of a intermediate between 0 and 1, at least in part of the viscosity range.

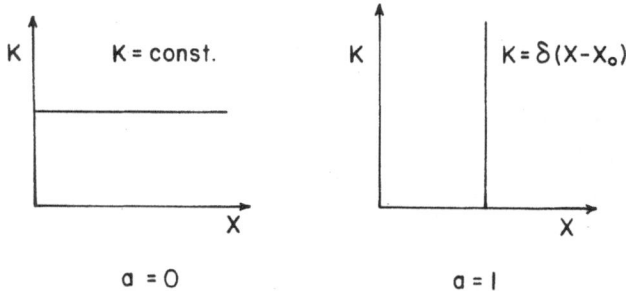

Figure 3. Two limiting forms for the sink term in the reactive-diffusion equation (1).

The one case where the lowest eigenvalue (an estimate to the rate coefficient) for a Smoluchowski equation with a coordinate dependent sink term can be found analytically [14], is diffusion in a harmonic potential with a parabolic $\kappa(x)$. The result is of the form

$$k = C_1[(C_2\eta+1)^{1/2}-1]/\eta \qquad (3)$$

where C_1 and C_2 are constants which depend on the parameters of the abovementioned problem. This η dependence is numerically close to a $\eta^{-1/2}$ behavior, and can explain some of the results [1c] for ligand migration in heme proteins.

The same type of non-Kramers η^{-a} dependence has been observed in photochemical isomerization experiments [15,16]. These deviations from Kramers have only recently been related to the multi-dimensionality of the problem [17]. Although it is possible to fit experiment to Eq. (3), as shown in Figure 4, we seek a physical explanation more suitable to the problem at hand, in which the degrees of freedom along and perpendicular to the reaction coordinate are not of such disparate time scales.

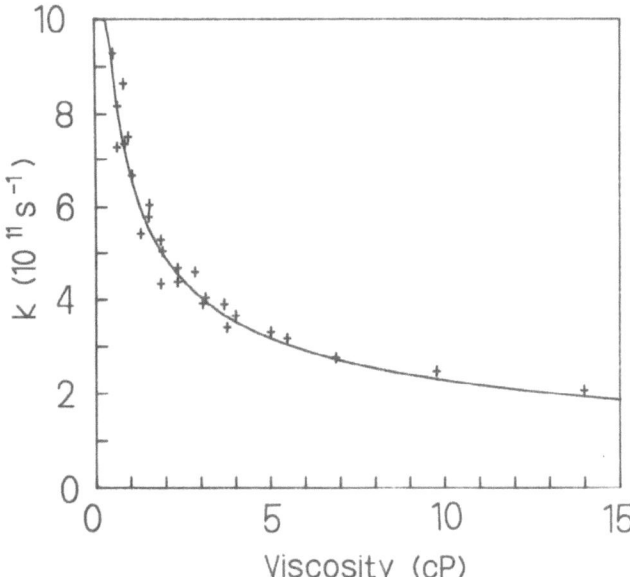

Figure 4. Viscosity dependence of the rate coefficient for photochemical isomerization of the cyanine dye DODCI in its excited electronic state. Experimental data [15a] is denoted by plus signs. Line is a fit to Eq. (3) with $C_1=0.8\,cP$ and $C_2=87\,cP^{-1}$.

Consider the nice example of stilbene vs. "stiff" stilbene [16], as shown in Figure 5. Stiff-stilbene has only one rotational degree of freedom (the angle θ around the double bond), and it was indeed found to obey Kramers' kinetics with $a=1$. In contrast, stilbene has an additional rotational motion possible, around the carbon-phenyl bond (angle ϕ). It was found to deviate substantially from Kramers' model, with a value of a as low as 0.32 in alkanes [15b,16].

One may conjecture that the following effect is operative: At low friction (small viscosity) the angle ϕ assumes a value which minimizes the potential energy. At larger friction, those molecules which rotate their phenyl rings so as to reduce the frictional drag for rotation around the double bond, isomerize faster. Hence the observed positive deviations may be due to a competition between the tendency to minimize the potential energy vs. the need to diminish the friction: At high friction the most probable reaction path may be far removed from the minimum energy path [18].

This effect of friction on the reaction path should decrease with increasing stiffness of the phenyl ring rotation: In "stiff" stilbene it is completely absent; In stilbene, it can be expected to be smaller for larger carbon-phenyl bond-order. This bond-order is larger than unity in the excited-state, on account of the C-C bond whose bond-order has decreased below 2. As a result, the barrier for isomerization around the C-C bond is lower in the excited stated as compared with the ground state. When the solvent is changed from alkanes to the more polar alcohols, the excited-state barrier decreases further. This can possibly be taken as an indication of an increasing carbon-phenyl bond-order. Indeed experiment shows (see Table I in Ref. [15b]) that the exponent a increases from 0.32 in alkanes to 0.6 in alcohols. A similar effect is observed there for the other dye molecules.

Since a decreasing barrier height correlates with a smaller barrier frequency (preexponential factor) [15b], the above trend is also in line with the concept of a frequency dependent friction [19]. My explanation therefore needs support from quantal structure calculations, and isomerization experiments for stilbene with electron donating/withdrawing substituents. We hope [20] to investigate these effects quantitatively, by solving the two-dimensional diffusion equation dynamics for this system.

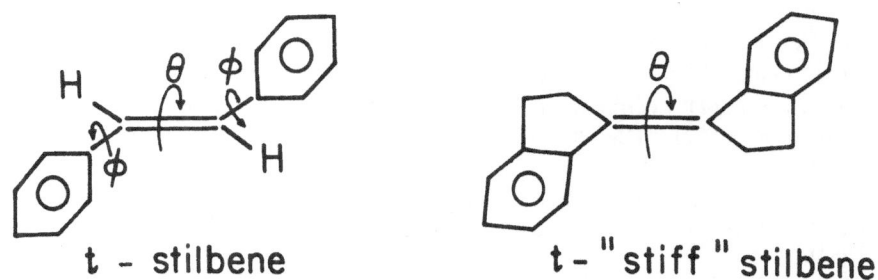

Figure 5. The structure of *trans*- stilbene and *trans*- biindanylidene ("stiff" stilbene).

4. CONCLUSION

In this exposition I have criticized the widespread use of a one dimensional reaction coordinate in biophysical kinetics. A qualitative change is already observed when the number of "active" degrees of freedom is increased from one to two. This has been demonstrated for ligand binding to heme proteins, where the two-coordinate potential energy surface determined from the low temperature data, leads unavoidingly to the description of a new effect. By this "self cooperativity" effect, the protein's tertiary structure relaxes following the binding or unbinding reaction, in a way that makes the reverse reaction less favourable. It has also been suggested, that a multi-dimensional barrier crossing process can lead to deviations from the simple Kramers' theory prediction of rate coefficients which are inversely proportional to solvent viscosity. It is likely that conformational relaxation plays a role in other important processes, such as photochemical isomerization, biological electron transfer, enzyme reactions or even gating of ionic channels in nerve membranes [21].

REFERENCES

[1]   (a) N. Alberding, R.H. Austin, K.W. Beeson, S.S. Chan, L. Eisenstein, H. Frauenfelder and T.M. Nordlund, 'Tunneling in ligand binding to heme proteins', *Science*, **192** , 1002 (1976);
(b) N. Alberding, S.S. Chan, L. Eisenstein, H. Frauenfelder, D. Good, I.C. Gunsalus, T.M. Nordlund, M.F. Perutz, A.H. Reynolds and L.B. Sorensen, 'Binding of carbon monoxide to isolated hemoglobin chains', *Biochem.*, **17** , 43 (1978);
(c) D. Beece, L. Eisenstein, H. Frauenfelder, D. Good, M.C. Marden, L. Reinisch, A.H. Reynolds, L.B. Sorensen and K.T. Yue, 'Solvent viscosity and protein dynamics', *Biochem.*, **19** , 5147 (1980);
(d) H. Frauenfelder, 'Ligand binding and protein dynamics', in E. Clementi, G. Corongiu, M.H. Sarma and R.H. Sarma, *Structure and Motion: Membranes, Nucleic Acids and Proteins*, (Adenine Press, Guilderland NY, 1985) p. 205;
(e) A. Ansari, J. Berendzen, S.F. Bowne, H. Frauenfelder, I.E.T. Iben, T.B. Sauke, E. Shyamsunder and R.D. Young, 'Protein states and proteinquakes', *Proc. Natl. Acad. Sci. USA*, **82** , 5000 (1985);
(f) A. Ansari, E.E. Di Iorio, D.D. Dlott, H. Frauenfelder, I.E.T. Iben, P. Langer, H. Roder, T.B. Sauke and E. Shyamsunder, 'Ligand binding to heme proteins: The relevance of low-temperature data', *Biochem.*, Submitted.

[2]   E.R. Henry, J.H. Sommer, J. Hofrichter and W.A. Eaton, 'Geminate recombination of carbon monoxide to myoglobin', *J. Mol. Biol.*, **166** , 443 (1983).

[3]   W.G. Cobau, J.D. LeGrange and R.H. Austin, 'Kinetic difference at low temperature between R and T state carbonmonoxide-carp hemoglobin', *Biophys. J.*, **47** , 781 (1985).

[4]   (a) N. Agmon and J.J. Hopfield, 'CO binding to heme proteins: A model for barrier height distributions and slow conformational changes', *J. Chem. Phys.*, **79** , 2042 (1983);
(b) N. Agmon and J.J. Hopfield, 'Transient kinetics of chemical reactions with bounded diffusion perpendicular to the reaction coordinate: Intramolecular processes with slow conformational changes', *J. Chem. Phys.*, **78** , 6947 (1983).

[5]   R.D. Young and S.F. Bowne, 'Conformational substates and barrier height distributions in ligand binding to heme proteins', *J. Chem. Phys.*, **81** , 3730 (1984).

[6]   (a) T.W. Scott and J.M. Friedman, 'Tertiary-structure relaxation in hemoglobin: A transient Raman study', *J. Amer. Chem. Soc.*, **106** , 5677 (1984);
(b) M.R. Ondrias, T.W. Scott, J.M. Friedman and V.W. Macdonald, 'A resonance Raman study of the temperature dependence of ligand photolysis and recombination in hemoglobins', *Chem. Phys. Lett.*, **112** , 351 (1984);
(c) J.M. Friedman, 'Structure, dynamics, and reactivity in hemoglobin', *Science*, **228** , 1273 (1985).

[7]   (a) A. Szabo, 'Kinetics of hemoglobin and transition state theory', *Proc. Natl. Acad. Sci. USA*, **75** , 2108 (1978);
(b) N. Agmon, 'From energy profiles to structure-reactivity correlations', *Intern. J. Chem. Kinet.*, **13** , 333 (1981).

[8] (a) J.J. Hopfield, 'Fundamental aspects of electron transfer in biological membranes', in E. Roux, ed., *Electrical phenomena at the biological membrane level* (Elsevier, Amsterdam, 1976) p. 471;
J.J. Hopfield, 'Nonadiabatic electron tunneling: Implications for bacterial photosynthesis and for critical tests of the mechanism', in B. Chance, D. DeVault, H. Frauenfelder, R.A. Marcus, J.R. Schrieffer and N. Sutin, eds., *Tunneling in biological systems* (Academic Press, New York, 1979) p. 417;
(b) J. Jortner, 'Dynamics of electron transfer in bacterial photosynthesis', *Biochim. Biophys. Acta,* **594** , 193 (1980);
(c) D. DeVault, 'Quantum mechanical tunneling in biological systems', *Quart. Rev. Biophys.,* **13** , 387 (1980).

[9] J. Deisenhofer, O. Epp, K. Miki, R. Huber and H. Michel, 'X-ray structure analysis of a membrane protein complex: Electron density map at 3Å resolution and a model of the chromophores of the photosynthetic reaction center from Rhodopseudomonasviridis', *J. Mol. Biol.,* **180** , 385 (1984);
also: 'Structure of the protein subunits in the photosynthetic reaction center of Rhodopseudomonas viridis at 3Å resolution', *Nature,* **318** , 618 (1985).

[10] D. Kleinfeld, M.Y. Okamura and G. Feher, 'Electron-transfer kinetics in photosynthetic reaction centers cooled to cryogenic temperatures in the charge-separated state: Evidence for ligand induced structural changes', *Biochem.,* **23** , 5780 (1984).

[11] J.A. McCammon, 'Protein dynamics'. *Rep. Prog. Phys.,* **47** , 1 (1984).

[12] N. Agmon, 'A diffusion Michaelis-Menten mechanism: Continuous conformational change in enzymatic kinetics', *J. Theor. Biol.,* **113** , 711 (1985).

[13] H.A. Kramers, 'Brownian motion in a field of force and the diffusion model of chemical reactions', *Physica (Utrecht),* **7** , 284 (1940).

[14] G.H. Weiss, 'A perturbation analysis of the Wilemski-Fixman approximation for diffusion-controlled reactions', *J. Chem. Phys.,* **80** , 2880 (1984).

[15] (a) S.P. Velsko, D.H. Waldeck and G.R. Fleming, 'Breakdown of Kramers theory description of photochemical isomerization and the possible involvement of frequency dependent friction', *J. Chem. Phys.,* **78** , 249 (1983);
(b) G.R. Fleming, S.H. Courtney and M.W. Balk, 'Activated barrier crossing: Comparison of experiment and theory', *J. Stat. Phys.,* **42** , 83 (1986).

[16] G. Rothenberger, D.K. Negus and R.M. Hochstrasser, 'Solvent influence on photoisomerization dynamics', *J. Chem. Phys.,* **79** , 5360 (1983).

[17] J. Troe, 'Elementary reactions in compressed gasses and liquids: From collisional energy transfer to diffusion control', *J. Phys. Chem.,* **90** . 357 (1986).

[18] G. van der Zwan and J.T. Hynes, 'Reactive paths in the diffusion limit', *J. Chem. Phys.,* **77** , 1295 (1982);
S.H. Northrup and J.A. McCammon, 'Saddle-point avoidance in diffusional reactions', *J. Chem. Phys.,* **78** . 987 (1983).

[19] R.F. Grote and J.T. Hynes, 'The stable states picture of chemical reactions. II. Rate constants for condensed and gas phase reaction models', *J. Chem. Phys.,* **73** . 2715 (1980).

[20] R. Kosloff and N. Agmon, in preparation.

[21] R.D. Keynes, 'Voltage-gated ion channels in the nerve membrane', *Proc. Roy. Soc. Lond. B,* **220** , 1 (1983).

Model Studies of Protein Dynamics.
The Role of Potential Energy Surface, Friction and Entropy.

E. W. Knapp
Physik Department
Technische Universität München
D8046 Garching, FRG.

ABSTRACT. Myoglobin is considered as a model substance to study protein dynamic. Experimental data of x-ray structure analysis, Mössbauer spectra and ligand rebinding are analysed with the help of a newly developed discretized Brownian oscillator model. This model allows to study the temperature dependence and to establish a relation between friction, fine-structure and entropy of a potential energy surface.

1. Introduction

It is a challenging problem to relate the function of a protein with its molecular structure and dynamics. In the present study it is attempted to make progress towards this goal by correlating data of proteins from different experiments with a common theoretical model. A suitable model substance for such an investigation is myoglobin, where many experimental data are available.
    In particular data from the following experiments are considered: (i) x-ray structure analysis providing the mean square displacement of all non-hydrogen atoms [1-3], (ii) Mössbauer spectra of the iron containing heme group [4-8]. The results of the above experiments are also related with studies on ligand rebinding after photoflash [9-11]. These experiments are very important, since they are so closely connected with the function of myoglobin.
    The scope of this article is as follows. In sections 2 and 3 the experimental information available from x-ray and Mössbauer data is summarized. In section 4 the status of understanding of protein dynamic based on these experiments [12-18] is sketched. In the following sections, which contain new material, protein dynamic is studied with the help of a discretized Brownian oscillator model (see appendix). This allows to establish a relation between friction and fine-structure of the potential energy surface as discussed in section 5. In section 6 it is shown that the calculated intensity loss depends critically on the Mössbauer time window used. Section 7 contains the investigation of the temperature dependence of the discretized Brownian oscillator model and its generalizations. Finally in section 8 the discussed models are related with experiments on ligand rebinding in myoglobin.

383

J. Jortner and B. Pullman (eds.), Tunneling, 383–399.

2.   X-ray Structure Analysis of Proteins.

X-ray structure analysis of proteins provides the mean coordinates $\langle x_i \rangle$ of all non-hydrogen atoms (i), i.e. the average structure. At sufficient high resolution also the mean square displacements from the average structure $\langle (x_i - \langle x_i \rangle)^2 \rangle$ become available. They are henceforth abreviated as $\langle x_i^2 \rangle^x$. The superscript x indicates that they are derived from x-ray structure analysis. Similarly the superscript γ refers to Mössbauer data. The x-ray structure analysis requires a rather long exposure time of single crystals. Therefore the $\langle x_i^2 \rangle^x$ involve not only an ensemble average over the molecules in the crystal, but also a time average. This makes it difficult to discriminate static and dynamic contributions to the $\langle x_i^2 \rangle^x$. Also a method to subdivide static contributions from the crystal disorder and static variations of the molecular structure is not evident.

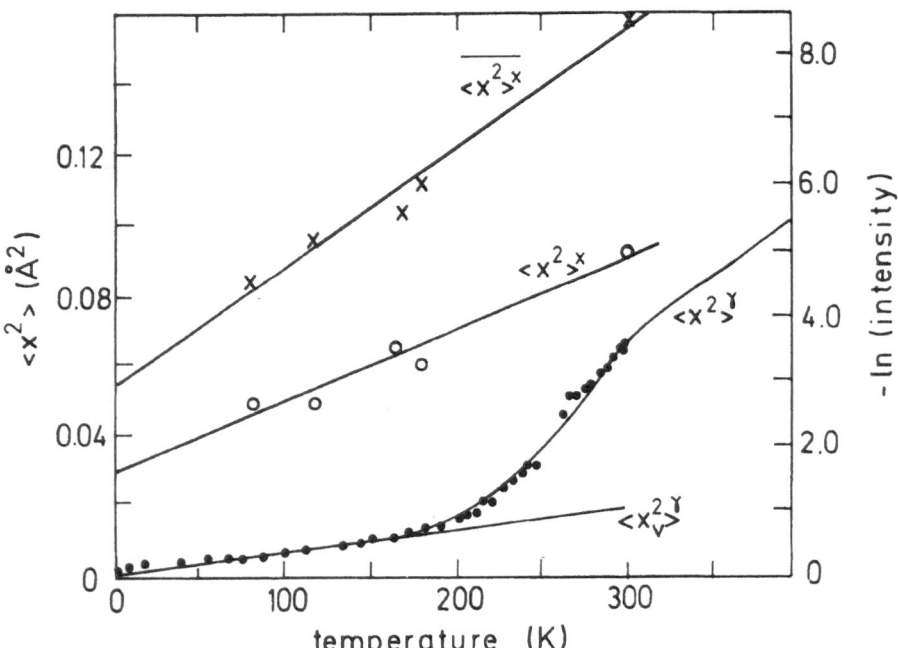

Figure 1.   Temperature dependence of the mean square displacements of myoglobin according to Ref. [3,17]. Mean square displacements derived from x-ray data: the iron atom $\langle x^2 \rangle^x$ and the average over all non-hydrogen atoms $\overline{\langle x^2 \rangle}^x$. Mean square displacements at the iron atom derived from Mössbauer data: total contribution $\langle x^2 \rangle^\gamma$ and vibrational part $\langle x_v^2 \rangle^\gamma$. Experiments are indicated by symbols.

Some information on the dynamic fluctuations of the proteins can be gained by observing the full temperature dependence of the $\langle x_i^2 \rangle^x$ [1-3]. This temperature dependence is depicted in Fig.1 for the average of all non-hydrogen atoms $\overline{\langle x^2 \rangle}^x$ and for the ^{57}Fe atom $\langle x^2 \rangle^x$ [3,17]. Both mean square displacements increase linear with temperature, a behaviour typical for motions in harmonic potentials. However, due to the lack of time resolution one can not discriminate between slow aperiodic motions or almost free oscillations in this potential. This problem can be resolved by Mössbauer spectroscopy.

It is also remarkable that the mean square displacements from x-ray data do not vanish at zero temperature. This can be due to (i) crystal disorder, which in a first approximation is independent of temperature or (ii) a dramatic slowing down of aperiodic motions at low temperatures, such that the protein molecules occupy a static distribution of conformational substates (c-states) [1-3]. With x-ray structure analysis this problem can only be solved by making time resolved measurements of $\langle x_i^2 \rangle^x$ at low temperatures [2-3].

3. Mössbauer Spectra of Proteins

The information, which can be gained from Mössbauer spectra of iron containing proteins, is complementary to the mean square displacements derived from x-ray structure analysis [17]. It provides the mean square displacement of the iron atom only, but allows a time resolution, which is twofold: (1) Dynamic processes slower than $10^{-7}s$ do not contribute to the mean square displacement $\langle x^2 \rangle^\gamma$. It can be derived from the intensity f of the Mössbauer spectrum as follows

$$f = \exp(-k^2 \langle x^2 \rangle^\gamma). \tag{1}$$

For ^{57}Fe the wave number is given by $k^2 = 53$ $Å^{-2}$. At zero temperature, where all contributions to $\langle x^2 \rangle$ are static, the $\langle x^2 \rangle^\gamma$ vanishes. This allows to discard static contributions of protein disorder. (2) Dynamic processes in the time regime $10^{-9} - 10^{-7}s$ (Mössbauer time window) do influence the width and shape of the Mössbauer line. This effect can be used to estimate the rate of the processes, which contribute to $\langle x^2 \rangle^\gamma$. Since the Mössbauer spectra are recorded for equilibrium ensembles the dynamic processes involved are equilibrium fluctuations and not relaxation processes.

Mössbauer spectra of proteins [4-8] and polymers [19,20] exhibit two features, which seem to be typical for such complex systems. For myoglobin they appear above a characteristic temperature of $T_c=200K$: (i) Each line of the Mössbauer spectrum is composed of the usual narrow Mössbauer lines and additional very broad components, which form the so called quasielastic lines. The observed line splitting is due to nuclear interactions not relevant for the present considerations. A typical calculated spectrum exhibiting narrow and broad lines is shown in Fig.2. The width of the broad component increases considerably with rising temperature (Fig.5). (ii) The intensity of the narrow line component f decreases dramatically with increasing temperature as shown in Fig.1. This observation indicates that above T_c additional degrees of freedom are available, which increase the flexibility of the protein.

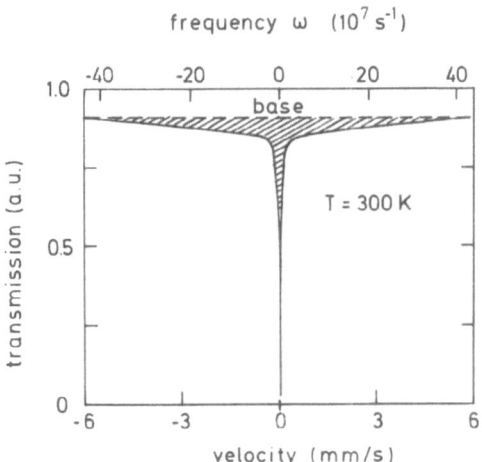

Figure 2. Calculated Mössbauer spectrum for an overdamped Brownian oscillator containing a quasielastic line typical for proteins. The model parameters [$k^2 \langle x^2 \rangle = 3.65$ and $\alpha = 2.6 \ 10^8 s^{-1}$] correspond to T=300K of the calculations in Fig.5. In the experiments the velocity regime (Mössbauer window) is five times larger as indicated in the Fig.2. The intensity, which can be measured, is given by the hatched area. The very broad quasielastic lines contribute to the base line (dashed), which can not be measured.

Below the characteristic temperature T_c the negative logarithm of the line intensity [$-\ln(f)$] i.e. $\langle x_v^2 \rangle^\gamma$ increases linear with T. This is due to harmonic vibrations similar as in other solids denoted by the subscipt v. Since the slope of $\langle x_v^2 \rangle^\gamma$ is rather small [14], the iron must be bound very strong to the heme group such that the increase of $\langle x^2 \rangle^\gamma$ at higher temperatures involves larger parts of the protein. Therefore one can conclude that the quantity

$$\langle x_t^2 \rangle^\gamma = \langle x^2 \rangle^\gamma - \langle x_v^2 \rangle^\gamma \tag{2}$$

is a probe of protein specific motions rather than of the Fe atom alone.

4.   Protein Dynamic Derived from X-ray and Mössbauer Data.

The linear temperature dependence of the mean square displacements from x-ray data suggests a motion in a harmonic potential. The quasielastic lines in Mössbauer spectra correspond to processes with a rate constant of $10^8 - 10^9 s^{-1}$ much smaller than typical vibrational frequencies in such systems. This points to an overdamped aperiodic motion, which is subject

to large friction. It is in fact possible to describe the appearence of
the quasielastic line in a Mössbauer spectrum by a Brownian oscillator
[5,7,8,12-14], whose equation of motion is given by

$$\ddot{x}^2(t) + 2\beta \, \dot{x}(t) + \omega^2 \, x(t) = F(t).$$ (3)

F(t) is the random force, which accounts for the coupling of the oscillator to the degrees of freedom of the protein. For overdamped motions
the friction constant β is much larger than the frequency ω giving rise
to a quasielastic line, whose width is in the order of [7,12-16]

$$\alpha_o = \frac{\omega^2}{2\beta}.$$ (4)

The reciprocal of this rate is the time required to explore all parts of
the harmonic potential surface. With a reasonable estimate of the corresponding frequency [$\omega = 30cm^{-1}$] the measured α values yield a friction
constant $\beta = 10^{16}s^{-1}$ much larger than expected from a collision mechanism [14]. We will see that this large friction constant is due to fine-
structure in the potential energy surface (see section 5).

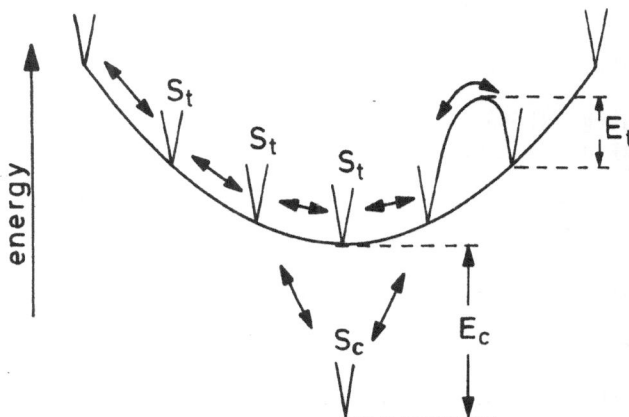

Figure 3.   Scheme for the discrete Brownian oscillator. The t-states
forming the transition state are arranged along a harmonic potential
envelope. At an energy E_c below the potential minimum a conformational
substate (c-state) has been included. For the transfer from one state
(c or t) to another state a barrier energy E_t must be surmounted. The
state entropies S_c and S_t determine the preexponentials of the transfer
rates. The scheme allows to by-pass the c-state.

It has turned out that the temperature dependence of the Mössbauer
spectra can not be understood by the Brownian oscillator without making
further assumptions [7,13,14] (see also section 7). This has motivated a
generalization of the Brownian oscillator model [13]. The essence of this
model is to introduce deep traps in the harmonic potential surface, which

interrupt the motion of the oscillator at random time and position. For
overdamped motions the oscillator can be represented by shallow traps
arranged regularly along the envelope of the potential energy surface.
The shallow traps are intermediate states in the sense discussed recently
[21]. Since these states mediate the transition between nearby deep traps
the ensemble of shallow traps has been named transition state (t) [7,13,
14]. A special case with a single deep trap in the center of the harmonic
potential well is visualized in Fig.3 and discussed in detail later on.
The deep traps correspond to conformational substates (c) of the protein,
which have been introduced first by Frauenfelder to understand the time
dependence of ligand rebinding [9].

Since Mössbauer spectra probe equilibrium properties of the sample
one can characterize the dynamics of this generalized Brownian oscillator
by an equilibrium quantity namely the probability p_t to meet the oscillator in the transition state i.e. in one of the shallow traps. This probability can be expressed in terms of the rates R_{ct} and R_{tc} describing
the transfer from a conformational substate (c) to the transition state
(t) and vice versa as follows

$$p_t = [1 + R_{tc}/R_{ct}]^{-1}. \tag{5}$$

With an Arrhenius rate law it follows

$$p_t(T) = [1 + \exp(\Delta G/k_B T)]^{-1}, \tag{6}$$

where the free energy difference between the t and c states is given as
energy and entropy difference

$$\Delta G = \Delta E - T\Delta S. \tag{7}$$

The analysis of the generalized Brownian oscillator model provides the
following renormalized quantities [13]. They are

$$\alpha_t = \alpha_0 \, p_t \tag{8}$$

for the width of the quasielastic line and

$$\langle x_t^2 \rangle = \langle x_0^2 \rangle \, p_t \tag{9}$$

for the mean square displacement of the protein specific motions, Eq.(2),
occuring in the transition state.

To reproduce the temperature dependence of the mean square displace-
ment $\langle x^2 \rangle^\gamma$ it was necessary to assume not only a large change of energy
$\Delta E = k_B 2300K$ but also of entropy $\Delta S = 9.2 k_B$. This entropy change corresponds to a Brownian oscillator containing 10^4 times more shallow traps
than deep traps. The resulting T dependence of $\langle x^2 \rangle^\gamma$ is shown in Fig.1.
Above 300K the increase of the calculated $\langle x^2 \rangle^\gamma$ slows down and reaches a
flat plateau at 350K with a constant slope. This type of plateau appears
at high temperatures, where the oscillator is most of the time in the
transition state, such that $\langle x_t^2 \rangle = \langle x_0^2 \rangle$. This plateau can not be observed because myoglobin denaturates before the appropriate temperature is

reached. However, similar plateaus have been observed in Mössbauer spectra of polymers [19,20]. The plateau at high temperatures takes care that the mean square displacement derived from Mössbauer data do not exceed the corresponding values taken from x-ray data. By varrying the model parameters it turns out that the steep rise of $\langle x^2 \rangle'$ and the plateau behaviour is only obtained by introducing a large change of entropy between the conformational substates and the transition state.

5. Friction, Trap Density and Entropy.

In this section a relation between friction and trap density of a potential energy surface will be established. For this purpose a discretized Brownian oscillator model is considered, which can be solved exactly (see appendix). This oscillator consists of a regular arrangement of narrow traps (t-states) of equal depth at positions

$$x_n = (n - N/2) \, \Delta x, \quad n=0,1,2...N. \tag{10}$$

They follow the envelope of a harmonic potential with given $\langle x^2 \rangle$ (Fig.3). The Mössbauer nucleus is allowed to jump between neighbour traps with Arrhenius rates

$$R_{n,n+1} = R \exp \left[\frac{\min(0, x_n^2 - x_{n+1}^2)}{2\langle x^2 \rangle} \right], \tag{11}$$

accounting for the shape of the potential envelope.

The calculated Mössbauer spectrum consists of a series of Lorentzians with increasing width (HWHH) (see appendix)

$$\alpha_n = \gamma + n\alpha , \quad n=0,1,2..., \tag{12}$$

where the exploration rate α is given by

$$\alpha = R\pi/N_t^2 . \tag{13}$$

In the last expression the number N_t of thermally accessible trap states [see (A22) of appendix] has been introduced

$$N_t = \frac{(2\pi \langle x^2 \rangle)^{1/2}}{\Delta x} \tag{14}$$

This trap number is also proportional to the density of traps ϱ. Hence the width of the quasielastic Mössbauer line increases with ϱ^{-2}. For the same reason the friction constant β in Eq.(4) is proportional to ϱ^2. This may provide an explanation for the large values $(10^{16}s^{-1})$ of the friction constant obtained from the exploration rate α. The linear dependence of α_n on N_t^{-2} is depicted in Fig.4. Typically several hundred traps are involved and the distance between neighbour traps is about $\Delta x = 0.005\text{Å}$. This distance is very small, but will become larger for a three-dimensional model. Nevertheless it suggests that not all of the traps are real dips in the potential energy surface. Some of them may also account for a dynamical slowing down due to invisible degrees of freedom. Finally the number of t-states N_t can directly be converted into entropy of the transition state, which is made up of the ensemble of t-states

$$S_t = k_B \ln(N_t). \tag{15}$$

Hence there is an intimate relationship between friction, fine-structure (dips) of the potential energy surface and entropy of the transition state.

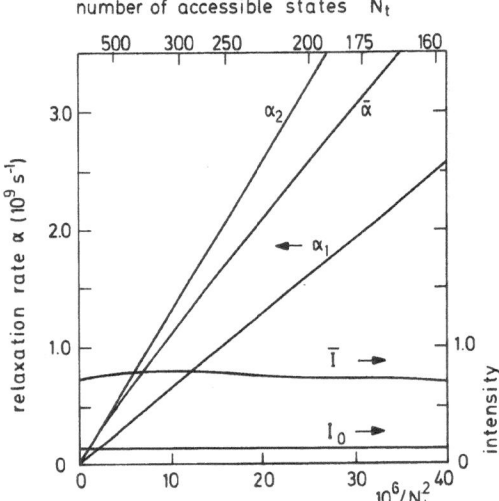

number of accessible states N_t

Figure 4. The discrete Brownian oscillator model as a function of the number of accessible t-states N_t. α_1 and α_2 are the line widths (HWHH) of the first two broad Lorentzians and $\bar{\alpha}$ is the width of the resulting quasielastic line. I_0 and $\bar{I}$ are the intensities of the elastic and quasielastic line respectively. The model parameters are: $k^2\langle x^2 \rangle = 2$ and $R = 2 \; 10^{13} \; s^{-1}$.

6.   Intensity Loss and Mössbauer Time Window.

Normally the Mössbauer spectra are recorded in the mm/s velocity regime corresponding to a window of frequencies smaller than $10^8 s^{-1}$. To measure the broad quasielastic line the window has been enlarged by one order of magnitude [5-8]. The intensity of the narrow and broad line components are obtained by evaluating the area of the lines within the Mössbauer window. Since part of the quasielastic line may fall outside of the window the intensites depend on the size of the window [22]. In the present application the calculated intensities refer to a Mössbauer spectrum with two Lorentzian profiles fitted in the window in exactly the same way as done for the experiments. Discrepancies can arise if other methods are used [14-16,18].

    The intensity lost in a Mössbauer spectrum is intensity, which has moved out of the window. Two different mechanisms contribute to this intensity loss: (1) Inelastic processes give rise to lines, which are completely outside of the Mössbauer window. (2) Quasielastic processes can give rise to broad lines. These lines may become so broad that they appear completely flat within the Mössbauer window. Then they contribute

only to the Mössbauer spectrum by changing the base line (dashed line in
Fig.2), which can not be measured. The protein specific changes in a
Mössbauer spectrum seem to be due to the latter mechanism, which gives
rise to quasielastic lines.

The discrete Brownian oscillator model introduced in the last section exhibits a quasielastic line of width $\bar{\alpha}$ (Fig.4). The width of this
line is between the widths α_1 and α_2, Eq. (12), of the first two broad
lines. By reducing the number of accessible trap states N_t the wings of
this line move only gradually out of the Mössbauer window. Therefore the
intensity of this line $\bar{I}$ decreases only very slowly.

7.   The Temperature Dependence of the Discrete Brownian Oscillator.

It has been stated several times that without further assumptions a
Brownian oscillator can not account for the temperature dependence of
Mössbauer spectra from proteins. Nevertheless the continuous as well as
the discrete Brownian oscillator can account at least for the T dependence of the width $\bar{\alpha}$ of the quasielastic line as depicted in Fig.5. There
it has been assumed that the transfer between nearest neighbour traps
(t-states) is a thermally activated process governed by an Arrhenius rate
law, which in the absence of the harmonic potential is given by

$$R_{tt} = A_t \exp(-E_t/k_B T). \tag{16}$$

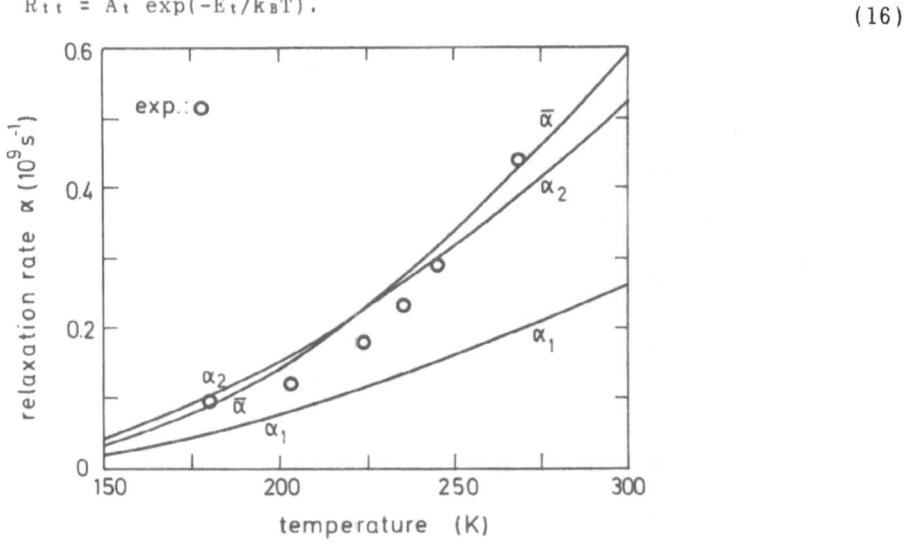

Figure 5.   Temperature dependence of the line widths for the discrete
Brownian oscillator. α_1, α_2 and $\bar{\alpha}$ are the line widths (HWHH) of the first
two broad Lorentzian profiles and the quasielastic line respectively. The
preexponential factor has been fixed to the value $A_t = 10^{13} s^{-1}$. The other
model parameters are adjusted to the Mössbauer data: $k^2\langle x^2 \rangle = 0.012$ T/K,
$\Delta x = 0.007$Å, $E_t = k_B 1000K$. Note that the T dependence of $\langle x^2 \rangle$ includes
the vibrational part $\langle x_v^2 \rangle$ as well.

With the preceding expression the exploration rate α, Eq. (13), can be written as

$$\alpha = A_t \; \Delta x^2/2\langle x^2\rangle \; \exp(-E_t/k_B T). \tag{17}$$

Based on the transition state theory [23] we can assume that $A_t=10^{13}s^{-1}$. Then the nearest neighbour trap distance must be $\Delta x = 0.007\text{Å}$ to account for the width of the quasielastic line (Fig.5). This distance is somewhat larger than the value 0.002Å, which will be used in the other figures. In the absence of conformational substates (c-states) (see Fig.3) the T dependence of the elastic line can not be reproduced. The negative logarithm of its intensity is simply given as $k^2\langle x^2\rangle$, which depends linearly on temperature (Fig.6).

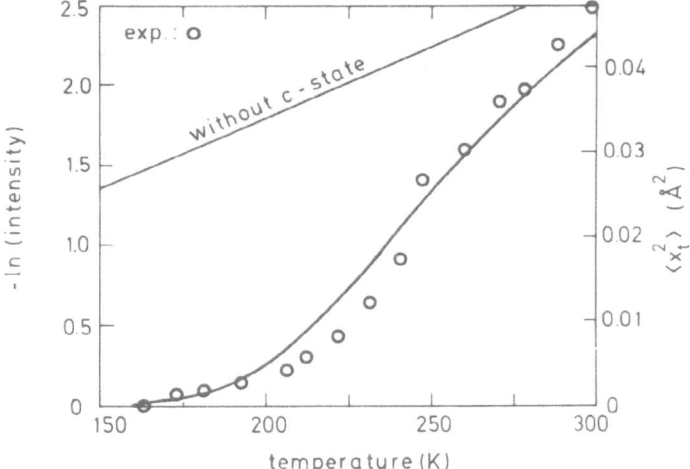

Figure 6.   Intensity of the elastic line in a Mössbauer spectrum as a function of temperature. The discrete Brownian oscillator without and with conformational substate (c-state) in the center of the harmonic potential is considered. The parameters are the same as in Figs. 5 and 7 respectively.

To remove this deficiency a single deep trap (conformational substate) is introduced in the center of the harmonic potential. Note that the restriction to one c-state only does not allow to understand the x-ray data [17]. The c-state is connected with the t-states, such that the oscillator can by-pass the c-state via another t-state (Fig.3). It turns out that without this by-pass the oscillator behaves similar as without any c-state [22]. Similar to the transfer rates between t-states, Eq. (16), one has

$$R_{tc} = R_{tt} \quad \text{and} \quad R_{ct} = A_c \exp[-(E_c+E_t)/k_B T]. \tag{18}$$

The preexponential factor A_c is given by the entropy change between c-

and t-state as follows

$$A_c = A_t \exp[(S_c-S_t)/k_B].$$ (19)

The calculations are performed along the lines sketched in the appendix by diagonalizing the resulting rate matrix numerically. The parameters A_t, E_t and A_c, E_c are adjusted to account for the temperature dependence of the intensity of the elastic line (Fig.6) and the width of the quasi-elastic line (Fig.7).

The full entropy difference between the ensemble of t-states and the c-state is given by

$$\Delta S = k_B \ln(A_t/A_c) + k_B \ln(N_t).$$ (20)

Due to the second term this entropy change involves a weak temperature dependence. For the parameters used (see Fig.7) we have $\Delta S(200K) = k_B\ 9.6$ and $\Delta S(300K) = k_B\ 9.9$. Hence energy and entropy change of the present model are practically the same as in the earlier studies [13].

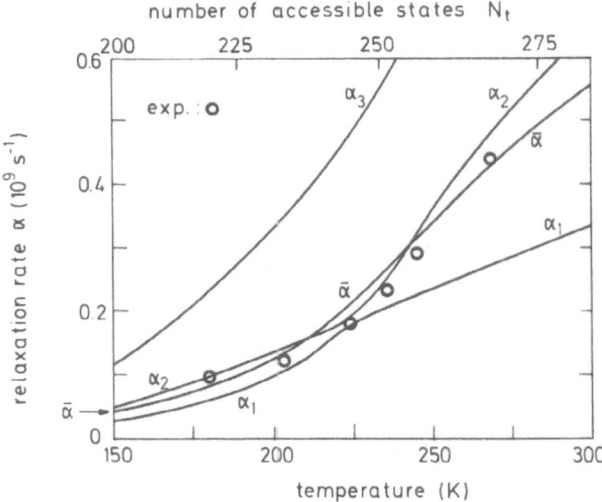

Figure 7. Temperature dependence of the line widths of the discrete Brownian oscillator including a c-state. Notations correspond to Fig.5. In contrast to Fig.5 the mean square displacements used correspond only to the protein specific part, Eq. (2), $k^2\langle x^2\rangle = 0.009\ T/K$. The nearest neighbour trap distance is fixed to $\Delta x = 0.002\text{Å}$. The other model parameters are adjusted: $A_t = 6.\ 10^{13}s^{-1}$, $E_t = k_B\ 800K$ and $A_c = 9.\ 10^{11}s^{-1}$, $E_c = k_B\ 2200K$.

Unfortunately the calculated intensity $\overline{I}$ of the quasielastic line does not exhibit the right temperature dependence. This is evident from Fig.8, which demonstrates that a large fraction of intensity, lost in the narrow elastic line, appears in the broad quasielastic line of the

Mössbauer spectrum. This is not a problem of the present model alone. It is due to the fact that for the first time the intensities of the model calculation are evaluated in the same way as the experimental data. A least square fit with two Lorentzians in the Mössbauer window has been performed (see section 6). Due to the large frequency window the intensity of the quasielastic line can escape only gradually from the window. Other fit methods may lead to wrong conclusions [13,15,16,18,22].

This problem demonstrates that the quasielastic line and the intensity loss of the Mössbauer spectrum can not be due to a single mode of motion. There are other modes of motion, which are also strongly damped. They may have a quasielastic line, which is so broad that it can not be observed in the Mössbauer window. These modes contribute only to the intensity loss of the Mössbauer spectrum. About 50% of the protein specific $\langle x_t^2 \rangle$ must be due to such modes. Since the temperature dependence of the different modes is so similar, they probably are from the same mechanism.

An interesting feature shown in Fig.8 is the cross-over of the intensity of the first two broad Lorentzians, which contribute to the quasielastic line. One of the lines is due to relaxation processes between the c- and t-states the other between t-states only. At low temperatures the c-t relaxation is slower than the t-t relaxation. Above T = 225K the two lines interchange.

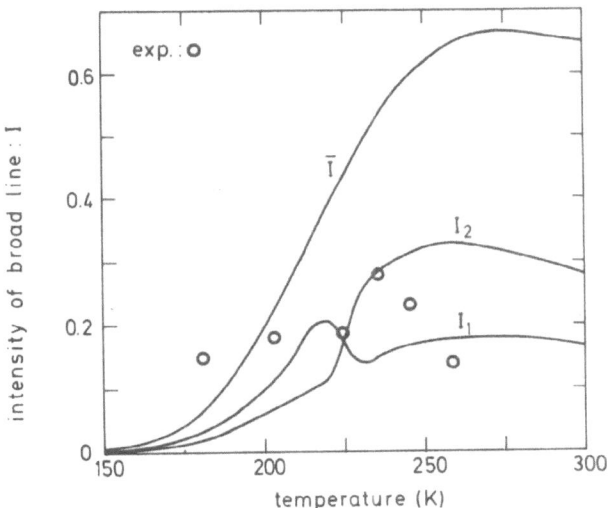

Figure 8. Temperature dependence of the intensities of the first two broad Lorentzians (I_1, I_2) and the quasielastic line ($\bar{I}$) of the discrete Brownian oscillator including a c-state. Only the protein specific part of $\langle x^2 \rangle$ is considered. The model parameters are the same as in Fig.7.

8.   Relation to Ligand Rebinding in Myoglobin.

Rebinding of ligands like CO and O_2 after photoflash exhibits many different dynamical processes, which range from picoseconds to seconds

[9-11]. These processes can be visualized by dynamics in a hierarchical arrangement of conformational substates in myoglobin (Mb) [11] similar as in a glass [24,25]. One can discriminate relaxation processes between different tiers called functional important motions (FIM) and equilibrium fluctuations (EF) within each tier. There is evidence for at least four different tiers [11].

The present study is focused on one specific relaxation process between two tiers of substates, which is most closely related to the EF probed by Mössbauer spectra. In the bound state (MbCO) the heme group is planar and the iron lies close to the heme plane. In the deoxy state (Mb) the heme plane is domed and the iron is displaced 0.5Å from the plane [26]. This amplitude of motion is twice as large as $[\langle x_t^2 \rangle]^{1/2}$ at room temperature.

At low temperatures the protein does not immediately relax to the equilibrium deoxy state but, reaches an excited state Mb* [27]. In this state the porphyrin iron complex is essentially in its deoxy configuration, whereas the surrounding protein matrix is not yet relaxed completely [10,11]. A suitable probe for this protein relaxation (FIM2) is a weak charge transfer absorption band (band III) in Mb, which after photodissociation is shifted 10nm to the red [27]. The major findings are [11]: (1) The shift is non-exponential in time. (2) The band does not exhibit isosbestic points. The latter excludes a simple two state dynamic involving initial and final state only. Instead both features point towards a complicated relaxation process, where intermediate states are involved as well.

Frauenfelder et al. have succeeded to interpret time and temperature dependence of this band shift by using a simple model [10,11]. They assume that the relaxation is a thermally activated process, which can be described by a superposition of Arrhenius rate laws

$$R(T) = \int_{E_{min}}^{E_{max}} \varrho(E_A) \; A \; \exp\left[\frac{-E_A}{k_B T}\right] \; dE_A , \tag{21}$$

with a constant preexponential factor $A = 10^{13} s^{-1}$ and a box distribution $\varrho(E_A) = (E_{max} - E_{min})^{-1}$ for the activation energies. By using activation energies from $E_{min} = k_B$ 1500K to $E_{max} = k_B$ 5000K the rate constant falls into the Mössbauer window. The value of the preexponential factor A suggests that no entropy change is involved in this relaxation process. This seems to contradict the behaviour found from Mössbauer spectra [17]. However, some precaution is necessary for the interpretation of complicated relaxation processes involving intermediate states. We will see in the following how this problem can be resolved.

The most obvious comparison is to relate the relaxation rates of FIM2 from ligand rebinding with the exploration rate α derived from the width of the quasielastic line. As we have seen in section 7 the temperature dependence of the line width can also be understood without the involvement of entropy changes (Fig.5). In contrast to the quasielastic line ligand rebinding requires a distribution of activation energies with a minimum energy slightly larger than the one needed for the quasielastic line. However, the FIM2 process is only recorded below the characteristic temperature, where no quasielastic line has been observed. It is likely that at higher temperatures the width of this distribution collapses, leaving a single activation energy at about the minimum energy of the distribution. This motional narrowing process can be connected with the increased protein flexibility at these temperatures. Experimental investigations of FIM2 at higher temperatures are needed to clarify this point.

9. Conclusions.

The present model studies allow to draw the following conclusions.
(1) Intensity loss and quasielastic line of a Mössbauer spectrum of proteins are due to at least two distinct modes of motion. Both are of the same type of mechanism (overdamped, aperiodic). In order to come to this conclusion it is important to evaluate the line intensities of the calculated Mössbauer spectra by considering explicitly the Mössbauer time window used for the experiments.
(2) The large friction constant derived from the width of the quasielastic line is due to many shallow traps in the potential energy surface. This fine-structure is also responsible for a large entropy difference between the conformational substates (deep traps) and the transition state (ensemble of shallow traps).
(3) The decrease of the line intensities with rising temperature can only be understood by assuming a large entropy change between the conformational substates and the transition state.
(4) The FIM2 relaxation rates observed by ligand rebinding may be related to the rate derived from the quasielastic line. The temperature dependence of the two rates do not require large entropy changes.

Acknowledgement
I like to express my gratitude to Prof. S.F.Fischer for continuous support and valuable discussions. A fruitful cooperation with Prof. F.Parak has contributed also considerably. This work was supported by the Deutsche Forschungsgemeinschaft with a Heisenberg fellowship.

Appendix: the Discrete Brownian Oscillator.

In this appendix an analytical model is presented, whose Mössbauer spectra contain elastic and quasielastic lines. In this model a regular arrangement of N+1 narrow traps at positions

$$x_n = (n - N/2) \Delta x , \quad n=0,1,2...N, \tag{A1}$$

is assumed, which are situated in a harmonic potential well

$$V(x) = m\omega^2/2 . \tag{A2}$$

Neighbour traps at x_n and x_{n+1} are connected by rates $R_{n,n+1}$ and $R_{n+1,n}$, which fulfill

$$Q_n = \frac{R_{n,n+1}}{R_{n+1,n}} = \exp\left[\frac{x_n^2-(x_n+\Delta x)^2}{2\langle x^2\rangle}\right] . \tag{A3}$$

At temperatures $k_B T > \hbar\omega$ the corresponding mean square displacement can be written as

$$\langle x^2\rangle = \frac{k_B T}{m\omega^2} . \tag{A4}$$

For small $\Delta x^2/\langle x^2\rangle$ the exponential function in (A3) can be expanded yielding

$$Q_n = 1 - \frac{\Delta x^2}{2\langle x^2\rangle}(2n+1-N). \tag{A5}$$

The same expression is obtained from an analytically solvable model, where the rate matrix is given by

$$(R)_{m,n} = 2R\left[(1-\frac{m}{N})\,\delta_{m+1,n}+\frac{m}{N}\,\delta_{m-1,n}-\delta_{m,n}\right] . \tag{A6}$$

A more general case is treated in Ref. [28]. The corresponding rate quotient can be expanded around $n=N/2$ and yields

$$Q'_n = 1 - 2/N\,(2n+1-N). \tag{A7}$$

Given the values of $\langle x^2\rangle$ and Δx one is still free to choose the number N of traps involved. For

$$N = \frac{4\langle x^2\rangle}{\Delta x^2} \tag{A8}$$

the expressions (A5) and (A7) do match.

The rate matrix R, (A6) can be diagonalized by a similarity transform P

$$R_d = PoRoP^{-1} \tag{A9}$$

yielding the eigenvalues

$$\lambda_n = (R_d)_{n,m} = -\delta_{n,m}\,n\,2R/N, \quad n=0,1,2...N. \tag{A10}$$

The matrix elements $p_n(m) = (P)_{m,m}$ are obtained by expanding a suitable generating function

$$G_n(z) = 2^{-N}\,(1-z)^n\,(1+z)^{N-n} = \sum_{\alpha=0}^{N} z^\alpha\,p_n(\alpha) . \tag{A11}$$

The eigenvalue zero corresponds to the equilibrium state. Its eigenvector components including a position dependent phase factor $\exp(ikx_n)$ are

$$(P_i)_n = 2^{-N}\binom{N}{n}\,\exp(ink\Delta x) \quad \text{and} \quad (P_f)_n = \exp(-ink\Delta x) \tag{A12}$$

for left and right side eigenvectors respectively. The phase factors account for effects of spacial coherence relevant for the short wavelength $2\pi/k$ of the Mössbauer radiation. After these preparations the relaxation function of a Mössbauer nucleus moving in the discrete harmonic potential can be written as follows

$$\bar\Phi(t) = P_i \circ \exp(Rt) \circ P_f = (P_i \circ P^{-1}) \circ \exp(R_d t) \circ (P \circ P_f). \quad (A13)$$

In the above expression an average over initial P_i and final P_f state distributions has been performed. With the help of the generating function (A11) we have

$$(P \circ P_f)_n = G_n(e^{-ik\Delta x}) \quad \text{and} \quad (P_i \circ P^{-1})_n = \binom{N}{n} G_n(e^{ik\Delta x}). \quad (A14)$$

Hence the relaxation function can be written as

$$\bar\Phi(t) = \sum_{n=0}^{N} \binom{N}{n} \mid G_n(e^{ik\Delta x}) \mid^2 e^{-\lambda_n t} = \left[f(k\Delta x) + (1-f(k\Delta x)) e^{-\alpha t} \right]^N, \quad (A15)$$

where

$$f(k\Delta x) = [1 + \cos(k\Delta x)]/2, \quad (A16)$$

and

$$\alpha = 2R/N. \quad (A17)$$

For small $k\Delta x$ we can expand $f(k\Delta x)$ yielding

$$f(k\Delta x) = 1 - k^2 \langle x^2 \rangle / N. \quad (A18)$$

The Fourier transform of the relaxation function $\exp(-\gamma t)\bar\Phi(t)$ provides the Mössbauer spectrum. The first two narrow Lorentzian lines have intensities of

$$I_0 = f = \left[1 - \frac{k^2 \langle x^2 \rangle}{N} \right]^N \xrightarrow[N \to \infty]{} e^{-k^2 \langle x^2 \rangle} \quad (A19)$$

and

$$I_1 = k^2 \langle x^2 \rangle \left[1 - \frac{k^2 \langle x^2 \rangle}{N} \right]^{(N-1)} \xrightarrow[N \to \infty]{} k^2 \langle x^2 \rangle \, e^{-k^2 \langle x^2 \rangle} \quad (A20)$$

respectively. The line widths, half width at half height (HWHH), are γ and $\gamma+2R/N$ respectively. These expressions are identical to the results derived from the normal (continuous) Brownian oscillator [14], where $2R/N$ is given by $\omega^2/2\beta$, Eq. (4). They correspond to the elastic and smallest quasielastic line in a Mössbauer spectrum.

An important parameter absent in the normal Brownian oscillator is the effective number of accessible states N_t given as temperature dependent partition function

$$N_t = \sum_{n=0}^{N} \exp \left[-\frac{x_n^2}{2\langle x^2 \rangle} \right]. \quad (A21)$$

For large N the sum (A21) can be written as an integral yielding

$$N_t = \frac{1}{\Delta x} \int_{-\infty}^{+\infty} \exp \left[-\frac{x^2}{2\langle x^2 \rangle} \right] dx = \left[\frac{2\pi \langle x^2 \rangle}{\Delta x^2} \right]^{1/2}, \quad (A22)$$

or by using (A8)

$$N_t = (N\pi/2)^{1/2}. \quad (A23)$$

References

[1] Frauenfelder,H., Petsko,G.A., and Tsernoglou,D.: 1979,Nature (London)280,p.558.

[2] Hartmann,H., Parak,F., Steigemann,W., Petsko,G.A., Ringe Ponzi,D., and Frauenfelder,H.: 1982,Proc.Natl.Acad.Sci.(USA)79,p.4967.

[3] Parak,F., Hartmann,H., and Nienhaus,G.U.: 1985,in "Proceedings of the Conference on Protein Structure: Electronic and Molecular Reactivity", Chance,B. and Austin,B., Eds.

[4] Keller,H. and Debrunner,P.G.: 1980,Phys.Rev.Lett.45,p.68.

[5] Cohen,S.G., Bauminger,E.R., Nowik,I., and Ofer,S.: 1981,Phys.Rev. Lett.46,p.1244.

[6] Parak,F., Frolov,E.N., Mössbauer,R.L., and Goldanskii,V.I.: 1981, J.Mol.Biol.145,p.825.

[7] Parak,F., Knapp,E.W., and Kucheida,D.: 1982,J.Mol.Biol.161,p.177.

[8] Bauminger,E.R., Cohen,S,G., Nowik,I., Ofer,S., and Yariv,J.: 1983, Proc.Natl.Acad.Sci.(USA)80,p.736.

[9] Austin,R.H., Beeson,K.W., Eisenstein,L., Frauenfelder,H., and Gunsalus,I.C.: 1975,Biochemistry 14,p.5355.

[10] Frauenfelder,H.: 1985 in "Structure and Motion: Membranes, Nucleic Acids and Proteins", Clementi,E., Corongiu,G., Sarma,M.H., and Sarma,R.H., Eds., Adenine Press.

[11] Ansari,A:, Berendzen,J., Bowne,S.F., Frauenfelder,H., Iben,I.E.T., Sauke,T.B., Shyamsunder,E., and Young,R.D.: 1985,Proc.Natl.Acad. Sci.(USA)82,p.5000.

[12] Shaitan,K.V. and Rubin,A.R.: 1980,Biofizika 25,p.796; 1981, Biophysics 25,p.809.

[13] Knapp,E.W., Fischer,S.F., and Parak,F.: 1982,J.Phys.Chem.86,p.5042.

[14] Knapp,E.W., Fischer,S,F., and Parak,F.: 1983,J.Chem.Phys.78,p.4701.

[15] Nadler,W. and Schulten,K.: 1983,Phys.Rev.Lett.57,p.1712.

[16] Nadler,W. and Schulten,K.: 1984,Proc.Natl.Acad.Sci.(USA)81,p.5719.

[17] Parak,F. and Knapp,E.W.: 1984,Proc.Natl.Acad.Sci.(USA)81,p.7088.

[18] Nowik,I., Bauminger,E.R., Cohen,S.G., and Ofer,S.: 1985,Phys.Rev. A31,p.2291.

[19] Litterst,F.J., Lerf,A., Nuyken,O., and Alcala,H.: 1982,Hyperfine Interactions 12,p.317.

[20] Litterst,F.J.: 1982,Nuclear Instr.Meth.199,p.87.

[21] Frauenfelder,H. and Wolynes,P.G.: 1985,Science 229,p.4711.

[22] Knapp,E.W. and Parak,F.: 1986,in preparation.

[23] Glasstone,S., Laidler,K.J., Eyring,H.: 1941, in "The Theory of Rate Processes", McGraw-Hill, New York.

[24] Palmer,R.G., Stein,D.L., Abrahams,A., and Anderson,P.W.: 1984,Phys. Rev.Lett.53,p.958.

[25] Stein,D.L.: 1985,Proc.Natl.Acad.Sci.(USA)82,p.3670.

[26] Dickerson,R.E. and Geiss,I.: 1983, in "Hemoglobin: Structure, Function Evolution and Pathology", Benjamin-Cummings, Menlo Park.

[27] Iizuka,T.,Yamamoto,H., Kotani,M., and Yonetani,T.: 1974,Biochim. Biophys.Acta.371,p.126.

[28] Knapp,E.W. and Fischer,S.F.: 1981,J.Chem.Phys.74,p.89.

STRUCTURAL DISORDER IN MYOGLOBIN AT LOW TEMPERATURES

Fritz Parak and Hermann Hartmann
Institut für Physikalische Chemie der Universität
Schloßplatz 4-7
4400 Münster
Bundesrepublik Deutschland

ABSTRACT. This contribution discusses mainly low temperature aspects of protein dynamics. X-ray structure investigations of myoglobin between 300 K and 80 K give mean square displacements, $\langle x^2 \rangle$, of the residues in the molecule. The linear temperature dependence of the $\langle x^2 \rangle$ -values can be extrapolated to T = 0. The results show that the molecule cannot be frozen into one well defined structure but in a large number of conformational substates. This result is supported by Mössbauer spectroscopy and specific heat measurements.

Since several years the investigation of protein dynamics is in the center of interest of a number of research groups. Stuctural disorder and structural fluctuations have been analyzed by a large number of experimental techniques. Measurements are reported in a temperature range from 300 K to very low temperatures like 0.15 K. In spite of the fact that there exists a practically infinite number of proteins great efforts have been concentrated on the investigation of only a very limited number of substances. As an example we refer to myoglobin. This protein is rather stable. The relatively low molecular mass (M = 17816) allows a very precise X-ray structure determination. The iron atom in the center renders possible the application of several spectroscopical methods. X-ray structure analysis combined with Mössbauer spectroscopy have given a consistent picture /1/. Protein dynamics can be understood as a relatively slow (characteristic time 10^{-9} s and slower) overdamped Brownian type motion of segments of the molecule in a restricted space. In this contribution we will give preliminary specific heat measurements supporting the general picture given in /1/. We will focus our attention to the low temperature limit where protein specific motions are already frozen in. The discussion is mainly based on the extrapolation of the results of the X-ray structure analysis of myoglobin performed between 300 K and 80 K. Mössbauer spectroscopy and the measurement of the specific heat are used as complementary methods.

Fig 1 shows the mean square displacement obtained at the position of the iron in myoglobin. Results were obtained from X-ray structure

J. Jortner and B. Pullman (eds.), Tunneling, 401–405.
© *1986 by D. Reidel Publishing Company.*

analysis and Mössbauer spectroscopy on crystals. X -ray data represent
the disorder in the molecule which may come from static or dynamic
displacements while the $\langle x^2 \rangle$ -values obtained by Mössbauer spectroscopy
stem from dynamic displacements only occuring in a characteristic time
faster than 10^{-7} s /2/. This explains the different temperature dependence of the $\langle x^2 \rangle$ -values. Recently we started to measure the specific
heat of myoglobin /3/. In Fig. 1 a preliminary result is shown. Even
with the present accuracy a peak in the c_p -values is seen around 180 K
indicating a diffuse phase transiton which is clearly correlated with
the beginning of protein specific dynamics as labeled by Mössbauer
spectroscopy. Detailed investigations are in progress.

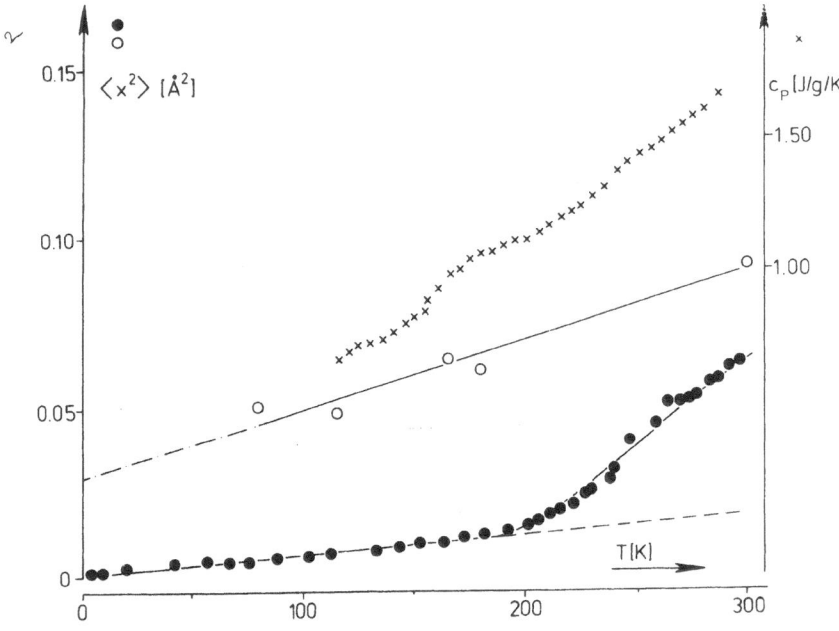

Fig. 1 Mean square displacements of the iron in myoglobin crystals as a
function of temperature. Full circles: Mössbauer spectroscopy;
open circles: X-ray structure analysis. The crosses give specific heat measurements on myoglobin powder. The hydration shell
of the myoglobin was about 150 H_2O molecules per myoglobin
molecule.

High resolution structure analysis of proteins allows the determination of the individual mean square displacements, $\langle x^2 \rangle$, of all different atoms of the myoglobin molecule /4,5/. However, the error bars
are rather large. Therefore , average values are discussed normally. It
is common to give $\langle x^2 \rangle$ -values for the different residues in the protein averaged over the backbone atoms N-C-C and the atoms of the side-
chains, respectively. Individual $\langle x^2 \rangle$ -values measure the structural
disorder at different parts of the molecule. As already mentioned they

allow no decision if the disorder is static or if it stems from dynamic displacements. Since motions can be frozen in at low temperatures the temperature dependence of the $\langle x^2 \rangle$ -values is important for the understanding of the physical nature causing the displacements. Fig. 2 gives as an example $\langle x^2 \rangle$ -values for the residues E41 (Glu) and H113 (His) of myoglobin at 5 different temperatures. Straight lines represent a linear regression. Linear extrapolation to $T = 0$ K gives the lower limit of $\langle x^2 \rangle$ -values at this temperature since it neglects zero point vibrations and the typical Debye behaviour of a solid. The linear temperature dependence of the $\langle x^2 \rangle$ -values is representative for most of the residues. The two examples represent, however, a different behaviour of the extrapolation to $T = 0$ K. In the case of E41 the linear extrapolation gives practically $\langle x^2 \rangle = 0$ A^2 at $T = 0$ K. This is expected. It indicates that the mean square displacement comes from motions which can be frozen out at low temperatures. If the molecule as the whole would have one well defined structure determined by a clear minimum in energy all residues should behave similar. The residue H113 proves the contrary. Extrapolation to $T = 0$ K yields $\langle x^2 \rangle = 0.07$ A^2 which is clearly larger than the zero point vibration. Moreover, it represents the lower limit because the Debye behaviour as indicated by dashed lines is not taken into account. The residue has even at $T = 0$ K not one well defined position but can be frozen in differently in the molecules of the sample. The residue H113 is representative for most residues in myoglobin. In the average for all residues $\langle x^2 \rangle$ becomes 0.055 A^2 for a linear extrapolation to $T = 0$ K. A detailed analysis of the X- ray data is in progress.

Reserving the word conformation for the structure obtained by X-ray methods as the average over about 10^{15} molecules one can say that each individual molecule can be frozen into a conformational substate at $T = 0$ K. Molecules in different substates have a rather similar but not identical structure. This result strongly confirms a picture infered from photodissotiation of CO-myoglobin /6/.

The extrapolation of mean square displacements to $T = 0$ K suggests that a protein crystal used in X-ray structure analysis shows some features of a glass. A crystal of anorganic or small organic compounds is a periodic arrangement of identical molecules. At one instance, variations of the positions of equivalent atoms in different unit cells occur only because of thermal vibrations. Besides of the zero point vibrations these motions can completely be frozen out. In contrast, a glass shows no long range periodicity but local order. In myoglobin crystals the long range periodicity of unit cells is as good as in other crystals. On the scale of the distances of atoms, however, the molecules show structural differences even at very low temperatures which remember on a glass like state.

The existence of different conformational substates is supported by Mössbauer experiments on myoglobin crystals. Mössbauer spectroscopy is not sensitive to static disorder. Consequently, the mean square displacement of the iron obtained from the Lamb Mössbauer factor approaches zero point vibrations if T goes to zero (compare Fig. 1). Differences in the structure of the molecules in an ensemble under inveatigation reveal themselves in an inhomogenity broadening of the

width of the Mössbauer absorption line. For myoglobin crystals a broadening of 0.074 mm/s (natural linewidth : 0.097 mm/s)was found independent of temperature between 4.2 K and 200 K /7/.

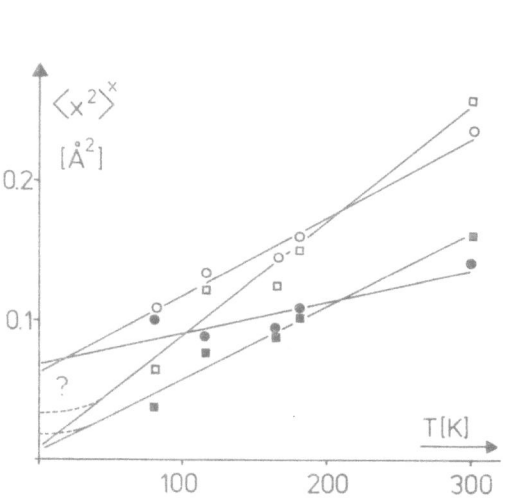

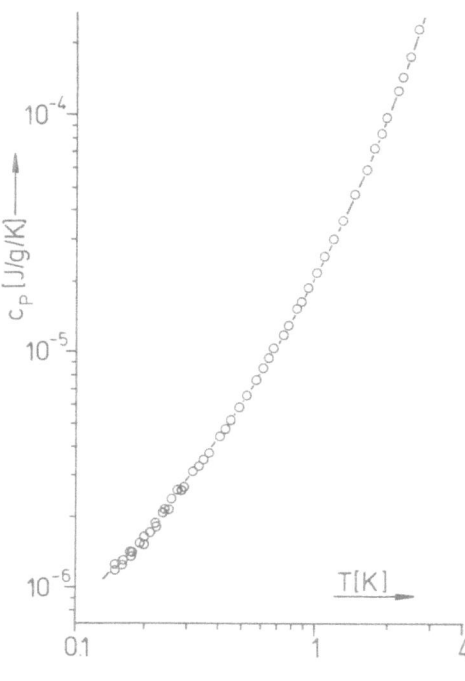

Fig. 2 Mean square displacements, $\langle x^2 \rangle$,
determined by X-ray structure ana-
lysis of myoglobin at 5 temperatures.
Squares: E41 (Glu); circles: H113
(His); full symbols: backbone
averages; open circles: side chain
averages.

Fig. 3 Specific heat
measurements on myoglobin
crystals.

At very low temperatures the irregular network of a glass can be
understood as a large number of different configurations of small
volume elements. Amorphous solids exhibit characteristic anomalies   in
their thermal properties like the specific heat. The X-ray data on
myoglobin crystals extrapolated to T = 0 K let us expect similar anomalies. Fig. 3 gives a specific heat measurement on myoglobin crystals
performed between 0.15 K and 3 K /8/. Below 0.4 K the specific heat
approaches a $T^{1.3}$ -dependence. A similar temperature dependence is
obtained in pure vitreous silica at very low temperatures. The temperature dependence of the specific heat between 0.15 K and 3 K can be
approached by: c = $C_1 T + bT^3$. A number of macromolecules show indications of such a temperature dependence of the specific heat /9/. In
amorphous systems c_p- anomalies are generally explained with the model
of a two level tunneling system (TLS) /10/. From this model C_1 can be
calculated depending on the densitiy of states of the TLS. Assuming
that the density extends to energies corresponding to 100 K the c_p -
data give only 3 TLS per volume of a myoglobin molecule. In comparison
with the large number of conformational substates indicated by X-ray

structure analysis this number looks surprisingly small. However, one has to have in mind that tunneling of large masses is highly improbable. Only conformational substates which differ only slightly by the position of one or two atoms can give rise to a two level tunneling system. It has to be emphasized that the TLS scheme is not the only way to understand the $T^{1.3}$ dependence of the specific heat . Recently it was shown that a fractal model can also give this temperature dependence /11/. In this picture c_p is proportional to T^{d^*} where d^* is a fractional dimensionality. Calculations which are based on the measurements of EPR relaxations rates have supported this picture /12/.

/1/ Parak F. and Knapp E.W. (1984) Proc. Natl. Acad. Sci. USA 81
    7088-7092
/2/ a discussion of this problem has recently been given:
    Mössbauer R.L. (1986) Hyperfine interactions, in press
/3/ Parak F. and Gmelin E. to be published
/4/ Frauenfelder H., Petsko G.A. and Tsernoglou D. (1979) Nature
    (London) 280 558-563
/5/ Hartmann H., Parak F., Steigemann W., Petsko G.A., Ringe Ponzi D.
    and Frauenfelder H. (1982) Proc. Natl. Acad. Sci. USA 79 4967 -
    4971
/6/ Austin R.H., Beeson K., Eisenstein L., Frauenfelder H., Gunsalus
    I.C. and Marshall V.P. (1974) Phys. Rev. Lett. 32 403-405
/7/ Parak F., Knapp E.W. and Kucheida D. (1982) J. Mol. Biol. 161
    177-194
/8/ Singh G.P., Schink H.J., Löhneysen H.V., Parak F. and
    Hunklinger S. (1984) Z. Phys. 55 B 23-26
/9/ Goldanskii V.I., Krupjanskii Yu.F., Flerov V.N. (1983) Doklady
    Biophysics 272 209-212
/10/ for a review compare: Topics in Currant Physics, Amorphous Solids,
    ed. : Phillips W.A., Springer Verlag (1981)
/11/ Jortner J.
/12/ Elber R. and Karplus M. (1986) Phys. Rev. Letters 56 394-397

TWO-LEVEL SYSTEMS IN PROTEINS: THERMODYNAMIC AND DYNAMIC EFFECTS

D.L. Stein
Princeton University
Dept. of Physics
Jadwin Hall/P.O. Box 708
Princeton, N.J. 08544

ABSTRACT. The evidence for two level systems (TLS) in compact proteins such as myoglobin is reviewed, and a model of conformational substates based on strongly-interacting TLS is suggested. Thermal and dynamical properties of this model are explored.

There exists at present a growing number of experiments that indicate
the presence of conformational substates (CS) in certain proteins (Mb,
Hb, CaMod) at sufficiently low temperature (T<200°K). These include
X-ray scattering[1], Mössbauer studies[2], and dynamical experiments such
as ligand recombination[3] and bound ion luminescence decay[4]. The general picture that has emerged from these studies is that many inequivalent nearly degenerate states exist which are essentially geometric
fluctuations about a roughly fixed tertiary structure. The barriers
between the substates vary considerably but characteristically induce
freezing on experimental time scales at approximately 200°K. For a review, see Frauenfelder[5].
An approach[6] toward understanding this behavior has applied concepts
from condensed matter physics, in particular the theory of disordered
systems. The initial observation is that the protein may be considered
as a three-dimensional disordered solid of approximately 10^4 atoms,
with a very close-packed structure; the compressibility of Mb of order 10^{-10} m^2/N[7] indicates that many of the interior atoms are pushed
right to their neighbors at their van der Waals radii. The disorder
implies that the protein should contain many local, internal degrees of
freedom corresponding to the famous two-level systems (TLS) of
Anderson, Halperin and Varma (AHV)[8] and Phillips[9]. The close-packed
nature of the interior atoms implies that adjacent TLS may in fact be
strongly interacting, so that local rearrangement or motion of one TLS
may disturb a neighboring TLS.
Let us examine these points in greater detail. TLS were first proposed by AHV-Phillips to account for the ubiquitous linear (in temperature T) specific heat in glasses at low T and a thermal conductivity
that went as T^2. These, in fact, remain the signature of TLS and are a
necessary prediction of any theory that posits them. AHV-Phillips ar-

J. Jortner and B. Pullman (eds.), Tunneling, 407–411.

gued that in a glassy system with no unique, well-defined lattice
structure there must exist some groups of atoms, or clusters of atoms,
that possessed two, or possibly more, locally stable positions within a
roughly defined unit cell. If these low-energy, local excitations possess a roughly constant density of states within the relevant energy
range, then the entropy and specific heat will be linear in T. The exact microscopic nature of TLS has remained a puzzle for most systems,
but their existence has been verified by a range of thermal and dynamic
experiments, such as ultrasonic attenuation[10]. In a protein, it seems
reasonable to think of TLS as atoms or possibly clusters of atoms in
amino acid side chains, having two (or more) locally stable positions.
The three-dimensional covalent topology of the protein remains largely
intact upon such local shifts in atomic position, in that each atom's
nearest neighbor distribution remains invariant. We shall denote a TLS
at site i by $S_i = \pm 1$, where +1 represents a state with energy lower by 2h
than the state -1; that is, with no interactions $E_i = -hS_i$.

Of course, we expect interactions to be present, especially in
light of the close-packed nature of atoms in the interior. Suppose one
TLS shifts from one locally metastable position to another. It may not
be able to do so without disturbing neighboring atoms, some of which
may also correspond to TLS. Any rearrangement of states of TLS will
cause some regions to increase their local strain energy, while others
will lower theirs. The inability to lower the energy of all TLS simultaneously (and thereby arrive at a unique lowest energy conformational
substate) is reminiscent of frustration in spin glasses or glasses.

The interactions between neighboring TLS are presumably complex
enough so that they can be effectively modeled by a probability distribution of random couplings, such as bimodal or Gaussian. Since $S_i^{2n} = 1$
and $S_i^{2n+1} = S_i$ for any integer n we may expand the energy of a given configuration $S^{(\alpha)}$ of N TLS as a sum of multilinear terms:

$$E\{S^{\alpha}\} - E_0 = -h \sum_{i=1}^{N} S_i - \sum_{\langle ij \rangle} J_{ij} S_i S_j + 0(S^3) \qquad (1)$$

where E_0 is the average protein conformational energy and

$$J_{ij} = \frac{\partial^2 E\{S^{(\alpha)}\}}{\partial S_i \partial S_j}$$ is taken to be a random variable. Eq. (1) then rep-

resents small fluctuations of the conformational strain energy about

the average due to conformational substates. We expect $\sqrt{\langle J_{ij}^2 \rangle} \gg h$ and

throw out higher-order couplings for simplicity, and also because in-

cluding them will not affect our final results. We thus arrive at a
formula for the conformational strain energy which is reminiscent of
the spin glass Hamiltonian.

An immediate consequence of such a picture is the existence of a
gradual freezing transition centered at a temperature of the order of
the width of the J_{ij} distribution. The "freezing temperature" is timescale dependent, but varies only as the logarithm of the observation
time. The number of metastable states scales as $\exp(N)$, where N is the
number of TLS. This picture agrees well with experimental
observations based on x-ray scattering[1] and Mossbauer studies[2], which
also allow one to arrive at a rough estimate of the number of TLS in a
given protein; for Mb, with amino acids, the number of side-chain atoms
with mean square fluctuations greater than .025 $Å^2$ is 95.

The effect on dynamic properties, such as recombination and
luminescence lifetime of bound ions, is pronounced and has been discussed in detail elsewhere[4][6]. It was a dynamical experiment[3] that
first led to the conformational substates conjecture: the number of un-
recombined Mb molecules following dissociation with an O_2 or CO ligand
via flash photolysis falls off exponentially at high temperature
(>200°K) and non-exponentially at low temperatures. The conformational
substates picture suggests that at high temperature, the protein flips
rapidly among its many substates during a typical recombination time
and only a single, averaged structure (and hence barrier) is therefore
seen. At low temperatures, each protein is frozen into a separate
substate and hence a distribution of barriers is observed. The
luminescence lifetime experiment[4] for Ca^{++} in CaMod finds similar
results in a more direct manner, by directly tying observations into
the protein structure itself, without evoking poorly understood
reaction barriers.

Another important consequence of a glass or spin glass picture is
the manner of protein relaxation following ligand binding, as one
example. As glassy relaxation itself is still poorly understood,
little progress has been made in this area to date, but it stands out
as a particularly important area of investigation should the glass
picture of proteins prove applicable.

TLS were originally invoked to explain the low temperature thermal
properties of glasses, in particular linear specific heat, and we
expect that similar predictions can be made for proteins. It is therefore interesting to note that Gol'danskii and collaborators[11] have
examined low-temperature specific heat of a variety of polymers and
have found clear evidence of linear specific heat, one of the
signatures of TLS. (Gol'danskii et al.[11] have proposed a glass model
of proteins which is quite similar in spirit to the one discussed
here.) The biopolymers included melanin, melanosoma, poly-L-alanine,
polyglycine, DNA and collagen. The coefficient of the linear T term is,
characteristically a factor of 3-4 larger than that typical for
polymeric systems, except for DNA and collagen whose coefficients are
approximately 10^3 times as large. Such large coefficients seem somewhat
suspicious and are poorly understood. The primary reason for a larger
linear coefficient is related to the fact that, unlike inorganic

polymeric glasses, each <u>individual</u> protein molecule has glasslike properties[11].

In our model many of the TLS are strongly coupled, which leads us
to ask whether deviations from pure linear temperature dependence of
the specific heat are to be expected. This point has been investigated
for TLS in glasses in a paper by Pietronero[12], who observes that in a
system where neighboring TLS tend to constrain one another the
available phase space for TLS relaxation becomes temperature-dependent.
The essential argument can be summarized by noting that if TLS 1 has
gap E_1 and is not allowed to relax unless TLS 2 has, then TLS 1 will
contribute to thermal properties during an experiment of duration
τ if $\varepsilon_1 \ll T \ln[\tau_e / \tau_R^{(2)}]$, where $\tau_R^{(2)}$ is the characteristic relaxation time
of TLS 2. This leads to an additional temperature-dependent term in the
TLS density of states which is proportioanl to $T \ln (\tau_R / \tau_R)$, where τ
is some characteristic relaxation time for the system. Hence there is
a small T^2 term added to the specific heat which may explain why the
"linear" term in ordinary glasses is not really linear, but has an
exponent somewhat larger than one[13]. Of course, other models such as
the fracton model[14] may also explain the discrepancy.

There are several useful experiments yet to be done which can
further clarify the question of whether TLS exist in proteins, and if
so, what their nature might be. Of course, low-temperature calori-
metric experiments (specific heat, thermal conductivity, thermal expansion) on myoglobin, hemoglobin, and calmodulin will be highly
valuable. Transmission of ultrasound is another important experiment;
the existence of TLS implies saturation at sufficiently high amplitude[8,9].
These low temperature experiments may ultimately provide information on
the behavior of proteins at physiological temperatures.

REFERENCES

1. H. Frauenfelder, G.A. Petsko, and D. Tsernoglou, Nature (London) 280, 558 (1979)

2. F. Parak, E.N. Frolov, R.L. Mössbauer, and V.I. Gol'danskii, J. Mol. Biol. 145, 824 (1981)

3. R.H. Austin, K.W. Beeson, L. Eisenstein, H. Frauenfelder, and I.C. Gunsalus, Biochemistry 14, 5355 (1975)

4. R.H. Austin, D.L. Stein, and J. Wong, preprint

5. H. Frauenfelder, in Structure and Dynamics: Nucleic Acids and Proteins, eds. E. Clementi and R.H. Sarma, (Adenine Press, N.Y. 1983), p. 369

6. D.L. Stein, Proc. Nat'l Acad. Sci. USA, 82, 3670 (1985)

7. B. Garish, E. Gratton, and C.J. Hardy, Proc. Nat'l. Acad. Sci. U.S.A. 80, 750 (1983)

8. P.W. Anderson, B.I. Halperin, and C. Varma, Phil. Mag. 25, 1 (1972

9. W.A. Phillips, J. Low. Temp. Phys. 7, 351 (1972)

10. S. Hunklinger and W. Arnold, in Physical Acoustics, vol. XII, W.P. Mason and R.N. Thurston, eds. (Academic Press, N.Y. 1976), p. 155

11. V.I. Gol'danskii, Yu. F. Krupyanskii, and V.N. Flerov, Doklady Akademii Nauk SSSR 272, 978 (1983)

12. L. Pietronero, preprint.

13. J.C. Lasjaunias, A. Ravex, M. Vandorpe, and S. Hunklinger, Sol. St. Comm. 17, 1045 (1975)

14. S. Alexander, C. Laermans, R. Orbach, and H.M. Rosenberg, Phys. Rev. B 28, 4615 (1983)

RESONANCE TUNNELING AND ITS APPLICATIONS

M. Azbel
School of Physics and Astronomy
Tel Aviv University
Tel Aviv 69978
Israel

ABSTRACT. Resonance tunneling via localized states dominates in mesoscopic systems. Then transport coefficients are determined by a single space fluctuation, are highly specific for each system, may change by orders of magnitude from system to system and be sensitive to a shift in the position of a single impurity. Non-linearity and instability may develop in very weak fields.

In one- and two-dimensional (1D and 2D) systems any randomness localizes quantum states. So, transport is related to tunneling and activated transitions only. At sufficiently low temperatures this leads to a behaviour of a micron-size system which under ordinary conditions is characteristic only of a microscopic system.

First demonstrate it in the case of tunneling at zero temperature. Start with the 1D little transparent potential barrier. Then quasiclassical calculation provides the transmission coefficient t,

$$t \sim exp\left(- \int_{Barrier} |k| dx\right),$$ (1)

where k is the (imaginary) wave vector inside the classically unavailable barrier.

Now consider two little transparent barriers. The potential well between them yeilds quasi-eigenstates. Their quasi-eigenenergies $\mathcal{E}_s$ are determined by the Bohr quantization

$$\oint_{well} k dx = 2\pi\left(n + \frac{1}{2}\right),$$ (2)

where n is an integer, and the integral refers to the classically available region inside the well. The quasi-eigenenergy half-width

413

J. Jortner and B. Pullman (eds.), Tunneling, 413–418.
© *1986 by D. Reidel Publishing Company.*

$\delta \mathcal{E}_s$ is related to the tunneling via the barriers. If their transmission coefficients are t_1 and t_2, then

$$\delta \mathcal{E}_s / \mathcal{E}_s \sim t_1 + t_2 \ll 1 . \tag{3}$$

A quasiclassical calculation proves that a particle with the energy $\mathcal{E}$, which is different from $\mathcal{E}_s$ (i.e., $|\mathcal{E} - \mathcal{E}_s| \gg \delta \mathcal{E}_s$), tunnels through these two barriers with the transmission coefficient

$$t_{12} \sim t_1 t_2 . \tag{4}$$

Thus,

$$\ln t_{12} \sim \ln t_1 + \ln t_2 . \tag{5}$$

A particle density decreases by $1/t_1$ in the first barrier and by $1/t_2$ in the second. The situation drastically changes when $\mathcal{E} \approx \mathcal{E}_s$ (i.e., $|\mathcal{E}_s - \mathcal{E}| \lesssim \delta \mathcal{E}_s$). Then particles, tunneling through the first barrier, accummulate in the potential well eigenstate, where the probability density reaches a maximum. (This is known as "localization". Note that the exact spectrum is continuous, and a potential well can accommodate any number of particles). If, e.g., $t_2 < t_1$ (for simplicity, we assume it thereafter), then an incident particle density <u>increases</u> by t_1^{-1} in the first barrier and <u>decreases</u> by t_2^{-1} in the second barrier. The resulting transmission coefficient t_{12} is

$$t_{12} \sim t_2 / t_1 \tag{6}$$

and

$$\ln t_{12} \sim -|\ln t_1 - \ln t_2| . \tag{7}$$

It may change between $t_{12} \sim 1$ for symmetric barriers and $t_{12} \sim t_2$ for $t_1 \sim 1$. In a general case,

$$\frac{1}{t_{12}} \sim \frac{1}{t_1 t_2} \cos^2 \left(\int_{well} k \, dx \right) + \left(\frac{t_1}{t_2} + \frac{t_2}{t_1} \right) . \tag{8}$$

This yields Eqs.(4,6,2,3).

Now consider any set of n potential barriers with quasi-eigenstates in each of their potential wells. If an incoming particle energy is different from any of the quasi-eigenenergies, then, successivley applying Eq.(5), one obtains:

$$\ln t_{12\ldots n} \sim \sum_{m=1}^{n} \ln t_m = n(\ln t)_{average} \equiv -L/L_o ,$$
(9)

where L is the total length of a system and L_o is the so called "localization" length,

$$L_o = \lambda / |\ln t|_{average} .$$
(10)

Here λ is the system length per barrier. Equations (9,10) remain valid in the case of any 1D barriers if their number is sufficiently large to combine a certain number of highly transparent barriers into one little transparent barrier.

The exponential decrease of the transmission with the length in Eq.(9) is typical for the Anderson localization. This determines the nature of any transport phenomena, and in particular, according to Landauer[1], a conductivity σ :

$$\sigma = (e^2/h) t/(1-t) ,$$
(11)

and

$$\sigma \approx e^2 t/h , \quad if \quad t \ll 1 .$$
(11a)

If an incoming particle energy coincides with the energy of the quasi-eigenstate in the q-th potential well, then one may apply Eq.(1) to the sets of barriers to the left and to the right of the well. Denoting the q-th well and eigenenergy by the superscript, one obtains:

$$\ln t_{12\ldots n}^{(q)} \sim -\left| \sum_{m=1}^{q} \ln t_m - \sum_{m=q+1}^{n} \ln t_m \right| \sim$$
(12)

$$\sim -\left| (2q-n)(\ln t)_{average} \right| .$$

The transmission changes between ~ 1 and $\exp(-n |\ln t| \text{average})$ depending on the location of the eigenstate. In a random system, eigenenergies are randomly located. Thus, the transmission coefficiently randomly fluctuates between the exponentially different values. These "super-fluctuations" are perfectly reproducible, as long as the barriers do not change. However, the shift of even one "impurity" barrier may change the eigenenergy location, and thus the transmission coefficient. The knowledge of the transmission as a function of energy allows, in principle, for the determination of the eigenstate positions.[2,3] Computer simulations verify these conclusions.[2,3]

The situation in higher dimensionalities is similar. The resonance tunneling chooses the shortest quasi-1D cut. It may be considered according to the previous reasoning. One should only account for the "effective area" k^{-2} of the shortest cut. A non-resonant tunneling has a "conventional" small transmission coefficient, but it occurs through the whole system surface. When a system is very large, resonance tunneling is negligible.

In a conductor finite temperatures T provide a natural thermal width of the Fermi energy, which may exceed the eigenenergy difference. This leads to unusual fluctuations in the resistance dependence on temperature[4] and on the Fermi energy.[5]

Finite temperature leads to the Mott activated hopping between specific eigen-states. P. Lee demonstrated[6] that in mesoscopic systems this also implies giant fluctuations in conductance with the Fermi energy. For simplicity, start again with 1D case. Present eigenenergies and their space location in a plane ($\mathcal{E}/T$, x/L_0). The localization length L_0 determines the overlap $\exp(-\Delta x/L_0)$ of probability densities of eigenstates separated by Δx. The temperature determines the probability $\exp(-\Delta\mathcal{E}/T)$ of activated hopping between eigenenergies separated by $\Delta\mathcal{E}$. Thus, the total probability of hopping between the eigenstates separated by $\Delta\mathcal{E}$, ΔX is

$$w \sim \exp\left(-\frac{\Delta\mathcal{E}}{T} - \frac{\Delta x}{L_0}\right).$$

(13)

In the plane ($\mathcal{E}/T$, x/L_0) this probability is related to the air distance $\Delta\ell$ between eigenstates. The total area, occupied by all states below the Fermi energy $\mathcal{E}_F$, and thus the area per state, is $\propto T^{-1}$. Hence the average $\Delta\ell_{av}$ is $\propto T^{-1/2}$. (For more details see ref.7). When temperature is low, then $\Delta\ell$ is large, w's are exponentially small and exponentially different for different hops. Naturally the system chooses mainly the smallest available, i.e., the nearest-neighbour, $\Delta\ell$'s, and thus a quasi-1D Mott path. Each hop yields, by Eq.(11a), the conductance $\propto w$, and thus resistance $\propto w^{-1}$. A series of hops leads to the total resistance R,

$$R \propto \sum_m w_m^{-1} \sim \sum_m exp(\Delta \ell_m).$$ (14)

When the number of hops is very large compared to $exp(\Delta \ell_{av})$, then Eq.(14) leads to R linear in the number of hops (i.e., R proportional to the length L of the system) and proportional to $[exp(\Delta \ell)]_{average}$, i.e., to the conventional Mott result.[8] When the number of hops $L/\Delta \ell_{av}$ is not exponentially large, then R in Eq.(14) is dominated by a single hop:

$$R \sim exp(\max_m \times \Delta \ell_m)$$ (15)

In this case the resistance yields exponentially large Lee oscillations with the Fermi energy. A higher dimensionality is considered in exactly the same way, but eignestates should be presented in the (D+1)-dimensional space of $\mathcal{E}/T$ and $\vec{r}/L_o$. Then the total volume below the Fermi energy is $\propto T^{-1}$, the average $\Delta \ell_{av}$ is $\propto T^{-1/(D+1)}$; the Mott $\ell nR \propto \Delta \ell_{av}$, and the total number of states in Eq.(14), which competes with $exp(\max_m \Delta \ell_m)$, is proportional to the total space volume of the system.

Giant resistance fluctuations were indeed observed experimentally. For instance in the Si inversion layer 10 μm x 300Å x 100Å the conductance changed by two orders of magnitude[9] when few electrons were added to the system. In agreement with Eq.(9), the envelope of ℓnR rather than of R (as in the Ohm's law) was proportional to the electron density. In agreement with Eq.(15), the slope of the log R dependence on $T^{-1/2}$ was different for the conductance peaks and valleys.

Since each truly localized state can accommodate only one electron, even a rather low current may "overcrowd" the "best" states. When the best states are occupied, an electron must switch to the "second best" choice of a hop with higher resistance. This may lead to current instabilities, which were observed experimentally.[10]

Such "fluctuation kinetics" may be expected in all cases of localization in mesoscopic systems. They include, in particular, those of acoustic and electromagnetic waves, thermal conductivity, electric breakdown, etc. Their main feature is the individual behaviour of a system, which may be of importance both in fundamental and applied physics.

REFERENCES

1. R. Landauer, Phil. Mag. 21, 863 (1970).
2. M. Ya. Azbel, P. Soven, Phys. Rev. B27, 831 (1983).
3. M. Ya. Azbel, Phys. Rev. 328, 4106 (1983).

4.   D. P. DiVincenzo and M. Ya. Azbel, Phys. Rev. Lett. 50, 2102
     (1983).
5.   M. Ya. Azbel, A. Harstein, and D. P. DiVincenzo, Phys. Rev. Lett.
     52, 1641 (1984),   and R. F. Kwasnick, M. A. Kostner, J.
     Melngailis, P. A. Lee, Phys. Rev. Lett. 52, 224 (1984).
6.   P. A. Lee, Phys. Rev. Lett. 53, 2042 (1984).
7.   M. Ya. Azbel, in : Localization and Metal-Insulator Transitions,
     p. 451-458, Plenum Co., N.Y., 1985, and refs. therein.
8.   N. F. Mott, E. A. Davis, Electronic Processes in Non-Crystalline
     Materials, Oxford University Press, Oxford, 1971. In a 1D case
     the Mott formula was refined by J. Kurkijarvi, Phys. Rev. B8, 922
     (1973); see also W. Brenig, G. H. Döhler and H. Heyszenan, Phil.
     Mag. 27, 1093 (1973).
9.   A. B. Fowler, A. Harstein, R. A. Webb, Phys. Rev. Lett. 48, 196
     (1982), and Physica 117B-118B, 661 (1983), and A. Harstein, R. A.
     Webb, A. B. Fowler, J. J. Wainer, Surf. Science, 142, 1 (1984).
10. K. S. Ralls et al., Phys. Rev. Lett. 52, 228 (1984), and J. H.
     Claassen et al., in Proc. Int. Conf. on Localization,
     Interaction and Transport Phenomena, Braunschweig, 1984, p.11.

INDEX OF SUBJECTS